AF598025

Methods in Molecular Biology™

Series Editor
John M. Walker
School of Life Sciences
University of Hertfordshire
Hatfield, Hertfordshire, AL10 9AB, UK

For further volumes:
http://www.springer.com/series/7651

JAK-STAT Signalling

Methods and Protocols

Edited by

Sandra E. Nicholson and Nicos A. Nicola

Walter and Eliza Hall Institute, Parkville, VIC, Australia

Editors
Sandra E. Nicholson
Walter and Eliza Hall Institute
Parkville, VIC, Australia

Nicos A. Nicola
Walter and Eliza Hall Institute
Parkville, VIC, Australia

ISSN 1064-3745 ISSN 1940-6029 (electronic)
ISBN 978-1-62703-241-4 ISBN 978-1-62703-242-1 (eBook)
DOI 10.1007/978-1-62703-242-1
Springer New York Heidelberg Dordrecht London

Library of Congress Control Number: 2012952103

Printed on acid-free paper

Humana Press is a brand of Springer
Springer is part of Springer Science+Business Media (www.springer.com)

Preface

A series of discoveries in the late 1980s and early 1990s described how the receptor-associated JAK protein kinases and their substrates, the STAT transcriptional activators, transmitted signals following cytokine and growth factor binding to receptors expressed on the cell surface, to result in specific transcriptional and cellular responses. In the intervening two decades, the JAK/STAT cascade of tyrosine phosphorylation, protein–protein interactions, and transcriptional activation has been studied extensively. The field has progressed from identification of the individual components through to an understanding of the activation and deactivation mechanisms and the complex structural detail of the proteins involved. We now know that these pathways are important in many biological processes, including growth and development, hematopoiesis, and the innate and adaptive immune response. We have gained a greater understanding of the role these proteins may have in human diseases such as cancer, with the exciting prospect of pathway inhibitors being available for clinical use in the near future.

Protocols in JAK/STAT Signaling provide detailed methodology for examining many aspects of the pathway, which we anticipate will be of use not only to those working in the area but also to new investigators who are led to delve into the complexities of JAK and STAT responses. For that reason, we have included methods for the simple analysis of activation status, as well as more complex protocols. It is divided into four sections: two sections distinguish JAK- and STAT-specific approaches, a third section is dedicated to the negative regulators of the pathway, the Suppressor of Cytokine Signaling (SOCS) proteins, while the final section deals with the production and crystallization of JAK and STAT proteins. This last, apart from underpinning our mechanistic understanding, has been critical to the development of specific inhibitors for therapeutic use. The field continues to evolve and there is still much to be learnt. We hope that this volume can in some way serve those involved in such endeavors.

We would like to sincerely thank our many contributors who have generously shared their time and expertise to provide such comprehensive protocols.

Parkville, VIC, Australia

Sandra E. Nicholson
Nicos A. Nicola

Contents

Contributors

JEFFREY J. BABON • *Cancer and Haematology/Structural Biology Divisions, Walter and Eliza Hall Institute of Medical Research, Parkville, VIC, Australia; University of Melbourne, Parkville, VIC, Australia*

REBECCA BAMERT • *Department of Biochemistry and Molecular Biology, School of Biomedical Sciences, Monash University, Clayton, VIC, Australia*

FLORENCE BAUDIN • *European Molecular Biology Laboratory, Heidelberg, Germany; Unit of Virus Host-Cell Interactions, UJF-EMBL-CNRS UMI 3265, Grenoble, France*

E. JOANNA BAXTER • *Cambridge Institute for Medical Research, Cambridge, UK; Department of Hematology, University of Cambridge, Cambridge, UK*

ANTHONY J. BENCH • *Department of Hematology, Addenbrooke's Hospital, Cambridge, UK*

STEPHEN BROWN • *Sheffield RNAi Screening Facility (SRSF), Department of Biomedical Science, University of Sheffield, Western BankSheffield, UK*

CHRISTOPHER J. BURNS • *Chemical Biology Division, Walter and Eliza Hall Institute of Medical Research, Parkville, VIC, Australia; University of Melbourne, Parkville, VIC, Australia*

VELASCO CIMICA • *Department of Molecular Genetics and Microbiology, Stony Brook University, Stony Brook, NY, USA*

ANNA DITTRICH • *Department of Systems Biology, Institute of Biology, Otto-von-Guericke-University, Magdeburg, Germany*

MICHAEL A. FARRAR • *Center for Immunology, Masonic Cancer Center, University of Minnesota, Minneapolis, MN, USA; Department of Laboratory Medicine and Pathology, University of Minnesota, Minneapolis, MN, USA*

KATHERINE H. FISHER • *MRC Centre of Developmental Biology and Genetics, Department of Biomedical Science, University of Sheffield, Western BankSheffield, UK*

GEORGIALINA RODRIGUEZ • *Department of Biological Sciences, The University of Texas at El Paso, El Paso, TX, USA; Border Biomedical Research Center, The University of Texas at El Paso, El Paso, TX, USA*

TORSTEN GINTER • *Center for Molecular Biomedicine (CMB), Department of Biochemistry, University of Jena, Jena, Germany*

GABRIELLE L. GOLDBERG • *Reid Rheumatology Laboratory, Inflammation Division, Walter and Eliza Hall Institute, Parkville, VIC, Australia; University of Melbourne, Parkville, VIC, Australia*

ANTHONY R. GREEN • *Department of Hematology, Addenbrooke's Hospital, Hills RoadCambridge, UK; Cambridge Institute for Medical Research, University of Cambridge, Cambridge, UK; Department of Hematology, University of Cambridge, Cambridge, UK*

CLAUDE HAAN • *Life Sciences Research Unit, University of Luxembourg, Luxembourg, UK*

SERGE HAAN • *Life Sciences Research Unit, University of Luxembourg, Luxembourg, UK*

THORSTEN HEINZEL • *Center for Molecular Biomedicine (CMB), Department of Biochemistry, University of Jena, Jena, Germany*

LYNN M. HELTEMES-HARRIS • *Center for Immunology, Masonic Cancer Center, University of Minnesota, Minneapolis, MN, USA; Department of Laboratory Medicine and Pathology, University of Minnesota, Minneapolis, MN, USA*
CURT M. HORVATH • *Department of Molecular Biosciences, Northwestern University, Evanston, IL, USA*
NADIA J. KERSHAW • *Cancer and Haematology/Structural Biology Divisions, Walter and Eliza Hall Institute of Medical Research, Parkville, VIC, Australia; University of Melbourne, Parkville, VIC, Australia*
REZA KHOROOSHI • *Department of Neurobiology Research, Institute of Molecular Medicine, University of Southern Denmark, Odense, Denmark*
ROBERT A. KIRKEN • *Department of Biological Sciences, The University of Texas at El Paso, El Paso, TX, USA; Border Biomedical Research Center, The University of Texas at El Paso, El Paso, TX, USA*
TATIANA B. KOLESNIK • *Inflammation Division, Walter and Eliza Hall Institute of Medical Research, Parkville, VIC, Australia*
OLIVER H. KRÄMER • *Center for Molecular Biomedicine (CMB), Department of Biochemistry, University of Jena, Jena, Germany*
ARTEM LAKTYUSHIN • *Cancer and Haematology/Structural Biology Divisions, Walter and Eliza Hall Institute of Medical Research, Parkville, VIC, Australia*
ISABELLE S. LUCET • *Department of Biochemistry and Molecular Biology, School of Biomedical Sciences, Monash University, Clayton, VIC, Australia*
CHRISTOPH W. MÜLLER • *European Molecular Biology Laboratory, Heidelberg, Germany*
JAMES M. MURPHY • *Cancer and Haematology/Structural Biology Divisions, Walter and Eliza Hall Institute of Medical Research, Parkville, VIC, Australia*
JANE MURPHY • *Reid Rheumatology Laboratory, Inflammation Division, Walter and Eliza Hall Institute of Medical Research, Parkville, VIC, Australia*
SANDRA E. NICHOLSON • *Inflammation Division, Walter and Eliza Hall Institute of Medical Research, Parkville, VIC, Australia; University of Melbourne, Parkville, VIC, Australia*
TREVOR OWENS • *Department of Neurobiology Research, Institute of Molecular Medicine, University of Southern Denmark, Odense, Denmark*
JEAN-PATRICK PARISIEN • *Department of Molecular Biosciences, Northwestern University, Evanston, IL, USA*
NANCY C. REICH • *Department of Molecular Genetics and Microbiology, Stony Brook University, Stony Brook, NY, USA*
JEREMY A. ROSS • *Department of Biological Sciences, The University of Texas at El Paso, El Paso, TX, USA; Border Biomedical Research Center, The University of Texas at El Paso, El Paso, TX, USA*
FRED SCHAPER • *Department of Systems Biology, Institute of Biology, Otto-von-Guericke-University, Magdeburg, Germany*
DAVID SEGAL • *Chemical Biology Division, Walter and Eliza Hall Institute of Medical Research, Parkville, VIC, Australia; University of Melbourne, Parkville, VIC, Australia*
ELMAR SIEWERT • *Department of Systems Biology, Institute of Biology, Otto-von-Guericke-University, Magdeburg, Germany*
UWE VINKEMEIER • *School of Biomedical Sciences, University of Nottingham Medical School, Nottingham, UK*

NIKOLA WENTA • *School of Biomedical Sciences, University of Nottingham Medical School, Nottingham, UK*
ANDREW F. WILKS • *SYNthesis medchem, Parkville, Victoria, Australia*
MARTIN P. ZEIDLER • *MRC Centre of Developmental Biology and Genetics, Department of Biomedical Science, University of Sheffield, Western BankSheffield, UK*
JIAN-GUO ZHANG • *Cancer and Haematology Division, Walter and Eliza Hall Institute of Medical Research, Parkville, VIC, Australia; University of Melbourne, Parkville, VIC, Australia*

Part I

JAK Protein Tyrosine Kinases

Chapter 1

Analysis of Janus Tyrosine Kinase Phosphorylation and Activation

Jeremy A. Ross, Georgialina Rodriguez, and Robert A. Kirken

Abstract

Activation of Janus kinases (Jaks) occurs through autophosphorylation of key tyrosine residues located primarily within their catalytic domain. Phosphorylation of these tyrosine residues facilitates access of substrates to the active site and serves as an intrinsic indicator of Jak activation. Here, we describe the methods and strategies used for analyzing Jak phosphorylation and activation. Tyrosine-phosphorylated (active) Jaks are primarily detected from cell extracts using anti-phosphotyrosine-directed Western blot analysis of Jak-specific immunoprecipitates. Additionally, receptor pull-down and in vitro kinase assays can also be utilized to measure cellular Jak catalytic activity. In addition to tyrosine phosphorylation, recent evidence indicates Jaks can be serine phosphorylated upon cytokine stimulation, however the lack of commercially available antibodies to detect these sites has hindered their analysis by Western blot. However, phosphoamino acid analysis (PAA) has been employed to monitor Jak serine and threonine phosphorylation. Over the past decade, remarkable advances have been made in our understanding of Jak function and dysfunction, however much remains to be learned about their complex regulatory mechanisms.

Key words: Janus kinase (Jak), Tyrosine phosphorylation, Immunoprecipitation, SDS-PAGE, Immunoblot, Kinase assay, Phosphoamino acid analysis (PAA)

1. Introduction

Reversible protein phosphorylation is a fundamental aspect of cell signaling and is a major means by which information is transmitted from outside the cell and between components within the cell. Protein kinases are enzymes that transfer the γ-phosphate of adenosine triphosphate (ATP) to the hydroxyl groups of their select amino acid substrates and therefore are designated serine/threonine, tyrosine, or dual-specific kinases. Janus kinases (Jaks) are a family of cytoplasmic tyrosine kinases that associate with a variety of cell surface receptors to perform essential roles for transducing intracellular signals (1, 2). Following receptor engagement by

Sandra E. Nicholson and Nicos A. Nicola (eds.), *JAK-STAT Signalling: Methods and Protocols*, Methods in Molecular Biology, vol. 967, DOI 10.1007/978-1-62703-242-1_1, © Springer Science+Business Media New York 2013

cytokines or growth factors, the activation of Jaks occurs through autophosphorylation of key tyrosine residues located within their activation loop. Activated Jaks then phosphorylate tyrosine residues on the associated receptors to produce docking sites for SH2 or PTB containing proteins, including Signal transducers and activators of transcription (Stats) (3), leading to their tyrosine phosphorylation and subsequent activation. There are four Jak family members in vertebrates: Jakl, Jak2, Jak3, and Tyk2. While Jakl, Jak2, and Tyk2 are ubiquitously expressed, Jak3 is predominantly expressed in hematopoietic cells (4–6). As hematopoietic cytokines and growth factors primarily utilize Jaks for signal transduction, Jaks are critical for cell growth, survival, development, and differentiation of immune cells (7). Effective immune responses require functional Jak signaling and mutations leading to their loss of function make up some of the most common inherited immune deficiencies (8). Conversely, Jak-activating mutations have been shown to cause malignant transformation in a variety of cell types. Indeed, activated Jaks have been identified in hematopoietic, breast, prostate, neck, and ovarian cancers (9–13). Therefore, Jak kinases are attractive therapeutic targets in a variety of disease settings.

Structurally, Jaks have seven regions of sequence similarity named Janus homology (JH) domains (see Fig. 1). The tyrosine kinase domain is localized to the C-terminus in the JH1 domain

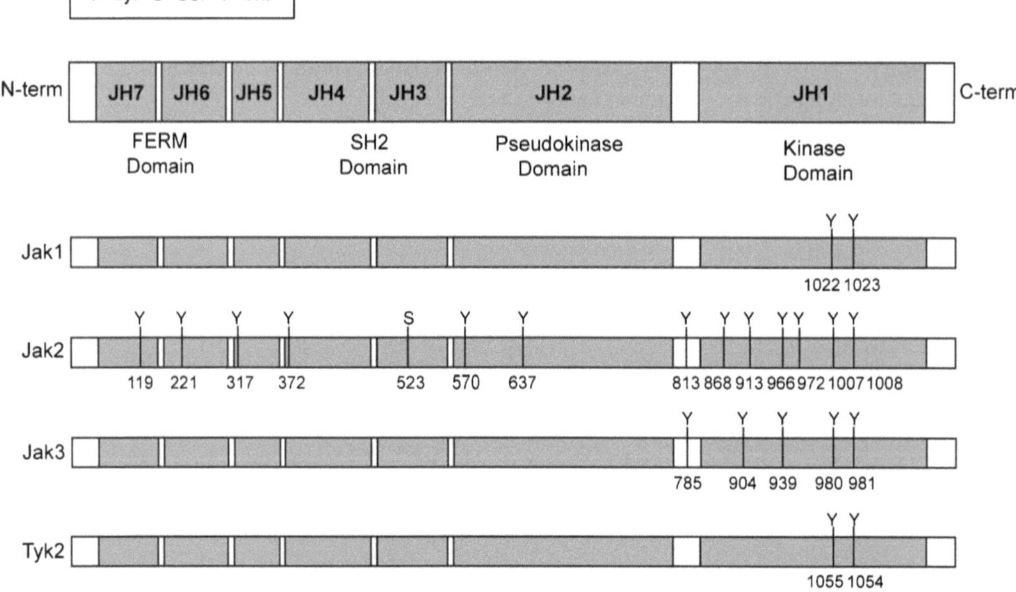

Fig. 1. Schematic representation of Jak structure and location of known phosphorylation sites. Jak proteins share seven regions of Jak homology (JH) domains denoted JH1–JH7. The JH1 domain harbors the tyrosine kinase and conserved Tyr-Tyr (YY)-motif within the autoactivation loop. The JH2 domain contains the pseudokinase that regulates kinase activity and binds substrates. The JH3–JH7 domains are critical for receptor association. *FERM* band 4.1, ezrin, radixin, moesin; *SH2* Src homology 2.

and is the major site of Jak autophosphorylation. For example, Y980 in Jak3 is a positive regulatory site, while Y981 negatively controls its activity (14). Similarly, mutation of the positionally conserved Y1007 to phenylalanine in Jak2 blocked its activation while phenylalanine substitution of Y1008 had no effect (15). Additionally, mutations of the two homologous tyrosines of Y1054 and Y1055 in TYK2 prevented ligand-induced activation of this kinase (16). Although it is not fully understood how phosphorylation of corresponding tyrosines in JAK1 may affect its activity, it has been shown that phosphorylation of Y1023 occurs at higher levels than phosphorylation of Y1022 (17). Recently, our group identified two new tyrosine phosphorylation sites within JH1 domain of Jak3, Y904, and Y939, which positively regulate Jak3 activity upon cytokine-mediated activation (18). Likewise, Y868, Y966, and Y972 have been shown to be required for maximal Jak2 activation, however phosphorylation of Y913 negatively regulates its activation (19, 20). Phosphorylation of Y813 in the linker region between the JH1 and JH2 domain of Jak2 is required for binding to the adaptor protein SH2-Bβ, which further enhances the activity of this enzyme (21). The positionally conserved corresponding Y785 in Jak3 is phosphorylated in response to IL-2 and is also important for binding to SH2-Bβ (22). Jaks are unique in having a catalytically inactive pseudokinase domain (JH2), which has a high degree of sequence similarity to the JH1 domain, however several residues required for kinase activity are altered from the canonical motifs. Y570 and Y637 in the JH2 domain of Jak3 are sites of autophosphorylation that differently regulate its catalytic activity (23, 24). Jaks also have an SH2 domain (JH3–JH4) whose function remains unclear, however phosphorylation of Jak2 on S523 within this domain was shown to function as a negative regulatory mechanism to dampen the activity of this enzyme (25). While little is known about serine phosphorylation in other Jaks, we have observed that Jak3 was serine phosphorylated upon IL-2 stimulation (26). The Jak N-terminal region (JH4–JH7) comprises a FERM (band 4.1, ezrin, radixin, moesin) domain, and mutations identified in this region have established that this domain mediates association with cognate cytokine receptors and regulates kinase activity (27, 28). Indeed, Y119 and Y221 in the FERM domain are sites of Jak2 autophosphorylation that negatively regulate its catalytic activity (23, 29). Additionally, Y317 and Y372 in the JH5 and JH4 domains, respectively, of Jak2 are sites of autophosphorylation that differently regulate its catalytic activity (24, 30).

Jak phosphorylation can be analyzed in a variety of ways, utilizing a range of techniques that provide slightly different types of information. One of the most frequently used techniques is immunoblotting Jak-specific immunoprecipitates from activated cell extracts using anti-phosphotyrosine (α-pY) antibodies. In essence, cells are harvested and lysed in an appropriate buffer and the desired

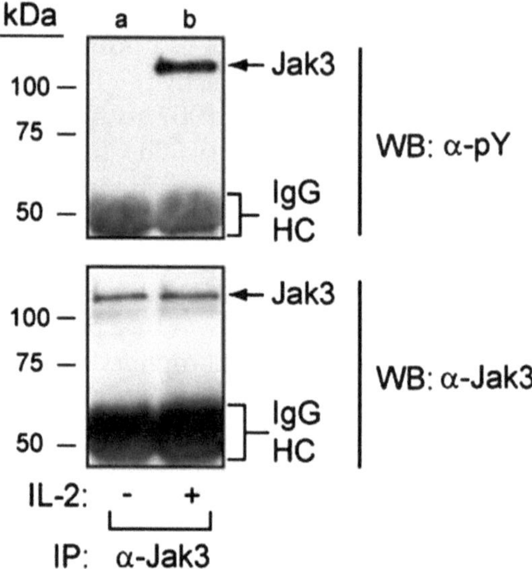

Fig. 2. IL-2-induced tyrosine phosphorylation of Jak3 in primary human lymphocytes. Quiescent PHA activated normal human peripheral blood mononuclear cells were left untreated (*lane a*) or stimulated with IL-2 for 10 min (*lane b*). Cell lysates were immunoprecipitated (IP) with α-Jak3, separated by 7.5% SDS-PAGE and subjected to Western blot analysis as indicated.

Jak immunoprecipitated from the lysate. The immunoprecipitate is then separated by SDS-polyacrylamide gel electrophoresis (SDS-PAGE), transferred to a membrane and the phosphorylated protein detected using phosphotyrosine antibodies by Western blot analysis. Subsequent verification using Jak-specific antibodies to reblot confirms identity of the tyrosine-phosphorylated protein (see Fig. 2). It is also possible to coimmunoprecipitate using this assay. Inducible associations can be identified such as the coimmunoprecipitation of Jak3 with the IL-2Rγ and IL-2Rβ receptor chains after IL-2 stimulation (26). Consequently, receptor pull-down assays are another approach to analyze Jak activation. Autophosphorylation of Jaks can also be analyzed using either radiolabeled "hot" ((γ^{32}P) ATP) or non-radiolabeled "cold" in vitro kinase assays (see Fig. 3). Alternatively, phosphorylation of Jak proteins can be detected by metabolic labeling of cells with (^{32}P)-orthophosphate coupled with autoradiography and phosphoamino acid analysis (PAA) (see Fig. 4). This strategy has the ability of analyzing not only phospho-tyrosine (pY) residues but phospho-serine (pS) and phospho-threonine (pT) residues as well. This is important because there is a building body of evidence that suggests serine phosphorylation of Jaks plays an important role in their regulation and will be an important aspect of future Jak investigations (22, 26).

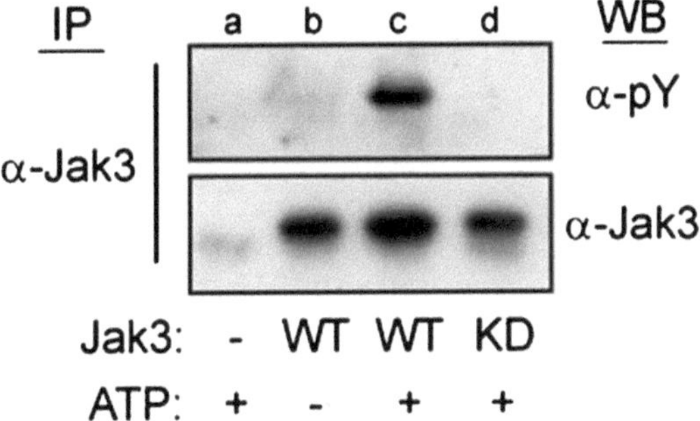

Fig. 3. Analysis of Jak3 activation using a cold in vitro kinase assay. Hek293 cells were transiently transfected with empty (–) (*lane a*), wild-type (WT) (*lanes b and c*), or kinase dead (KD) (*lane d*) Jak3 expressing plasmids. Cells were harvested at 30 h post-transfection, lysed and Jak3 proteins immunoprecipitated (IP). The immunoprecipitates were subjected to an in vitro kinase assay before separation by 7.5% SDS-PAGE and Western blot analysis using the indicated antibodies.

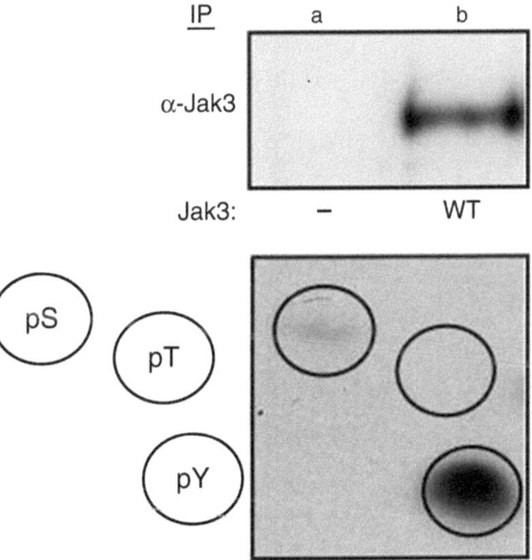

Fig. 4. (^{32}P)-radiolabeling and detection of Jak3 phosphorylation sites using phosphoamino acid analysis. (**a**) HEK293 cells were transiently transfected with empty (–) (*lane a*) or wild-type (WT) (*lane b*) expressing plasmid. Cells were harvested at 30 h post-transfection, lysed and Jak3 proteins immunoprecipitated (IP). The immunoprecipitates were subjected to an in vitro kinase assay in the presence of 100 μM (γ-^{32}P) ATP for 15 min as described in the methods (see Subheading 3.5). The immunoprecipitates were then separated by 7.5% SDS-PAGE, transferred to PVDF membrane and subjected to autoradiography. (**b**) The corresponding Jak3 band was excised and subjected to phosphoamino acid analysis. The position of phospho-serine (pS), -threonine (pT), and -tyrosine (pY) standards were detected by ninhydrin as indicated (*circles*).

This chapter therefore describes four techniques for analysis of Jak phosphorylation and activation: (1) immunoblotting of Jak immunoprecipitates with α-pY, (2) "Hot" and "cold" in vitro kinase assays, (3) metabolic labeling and phosphoamino acid analysis (PAA), and (4) receptor pull-down assays.

2. Materials

2.1. Cell Lysis and Immunoprecipitation

1. Cell lysis buffer: 10 mM Tris–HCl, pH 7.6, 5 mM EDTA, 50 mM NaCl, 30 mM $Na_4P_2O_7$, 50 mM NaF, 1 mM Na_3VO_4, 1% (v/v) Triton X-100, 1 mM phenylmethylsulfonyl fluoride (PMSF), 5 μg/ml aprotinin, 1 μg/ml pepstatin A, and 2 μg/ml leupeptin. Store at 4°C. PMSF, aprotinin, pepstatin A, and leupeptin should be added to lysis buffer directly prior to use (see Note 1).
2. Protein A/Protein G Sepharose slurry: Remove storage buffer by centrifugation at 100×*g* for 5 min at 4°C followed by gentle aspiration of the supernatant. Resuspend the pellet in lysis buffer (at least 5× bead volume). Repeat procedure for a total of three washes. Resuspend the final bead pellet with 2× bead volume of lysis buffer to obtain 50% slurry.
3. 2× SDS-PAGE sample buffer: 20% glycerol, 10% (v/v) 2-mercaptoethanol, 4.6% (w/v) SDS, 0.125 M Tris–HCl, pH 6.8, 0.004% (w/v) bromophenol blue.

2.2. SDS-Polyacrylamide Gel Electrophoresis

1. 4× Resolving gel buffer: 1.5 M Tris–HCl, pH 8.8, 0.4% (w/v) SDS, store at 4°C.
2. 4× Stacking gel buffer: 0.5 M Tris–HCl, pH 6.8, 0.4% (w/v) SDS, store at 4°C.
3. 30% (w/v) Acrylamide:bisacrylamide (29:1) solution.
4. *N*,*N*,*N*,*N*′-Tetramethyl-ethylenediamine (TEMED).
5. 10% (w/v) Ammonium persulfate (APS) solution in water, store at 4°C.
6. 7.5% Separating gel: 1.25 ml 4× Resolving gel buffer, 1.25 ml 30% acrylamide:bisacrylamide, 2.5 ml deionized water, 10 μl APS, 5 μl TEMED. Add APS and TEMED last, right before pouring gel.
7. Stacking gel: 0.625 ml 4× Stacking gel buffer, 0.395 ml 30% acrylamide:bisacrylamide, 1.48 ml deionized water, 15 μl APS, 5 μl TEMED. Add APS and TEMED last, right before pouring gel.
8. 10× Running buffer: 0.25 M Tris–HCl, 1.92 M Glycine, 1.0% (w/v) SDS, pH 8.5–8.8, store at room temperature.
9. Pre-stained molecular weight makers.
10. Capillary end tips (for gel loading).
11. Small gel electrophoresis apparatus with glass plates (0.75 mm spacers) and appropriate combs.

2.3. Transfer

1. 10× Transfer buffer: 1.25 M Tris–HCl, 1.92 M glycine, pH 8.5–8.8, store at room temperature.

2. 1× Transfer buffer: 20% (v/v) methanol, 100 ml 10× Transfer buffer, 700 ml deionized water, store at 4°C. Add methanol last.
3. Wash buffer: 50 mM Tris–HCl, pH 7.6, 200 mM NaCl, 0.25% (v/v) Tween-20.
4. Polyvinylidene difluoride (PVDF) membrane.
5. 3 MM Whatman paper.
6. Semidry transfer apparatus.

2.4. Immunoblot

1. Blocking buffer: Tris-buffered saline (TBS) (50 mM Tris–HCl, pH 7.6, 200 mM NaCl) containing 1% (w/v) bovine serum albumin (BSA) and 0.01% (w/v) sodium azide.
2. TBS-T washing buffer: 50 mM Tris–HCl, pH 7.6, 200 mM NaCl, 0.25% (v/v) Tween-20.
3. Anti-phosphotyrosine (α-pY): clones PY20 or 4G10 (Millipore).
4. Jak-specific antibodies can be obtained from multiple vendors.
5. Horseradish peroxidase-conjugated secondary goat anti-mouse or goat anti-rabbit IgG (depending on the nature of the primary antibody).
6. Enhanced chemiluminescence (ECL) substrate: 1 M Tris–HCl, pH 8.5, 250 mM Luminol, 90 mM *p*-Coumaric acid. Store at 4°C. Immediately before use, add 30 μl H_2O_2 (3.3% v/v) to 10 ml ECL substrate.
7. Stripping buffer: 62.5 mM Tris–HCl, pH 6.7, 100 mM 2-mercaptoethanol, 2% (w/v) SDS, store at room temperature.

2.5. In Vitro Tyrosine Kinase Assay

1. Cell lysis buffer: (see Subheading 2.1, item 1).
2. Kinase buffer: 25 mM HEPES, pH 7.3, 1% (v/v) Triton X-100, 100 mM NaCl, 10 mM $MgCl_2$, 3 mM $MnCl_2$, and 50 μM sodium orthovanadate.
3. Protein A-Sepharose slurry: (see Subheading 2.1, item 2).
4. 2× SDS-PAGE sample buffer: (see Subheading 2.1, item 3).

2.6. Metabolic Labeling

1. Phosphate-free RPMI 1640 medium: Combine 9.3 g RPMI 1640, 2 g Sodium bicarbonate ($NaHCO_3$) in 1 L of sterile deionized water in a pre-sterilized bottle. Do not autoclave RPMI. Store at 4°C. Protect from light.
2. Dialyzed (10,000 Da cutoff) Fetal Bovine Serum (FBS).
3. Cell lysis buffer: (see Subheading 2.1, item 1).
4. Conical base screw cap microcentrifuge tubes. Screw caps with O-rings are preferred.
5. Phosphorus-32 (^{32}P) radionuclide, 10 mCi (370 MBq), Orthophosphoric acid in water, store at −20°C (see Note 2).
6. Coomassie Brilliant Blue R-250 stain.

7. Coomassie Blue destain: 40% (v/v) methanol, 10% (v/v) acetic acid.
8. Filtered pipette tips should be used for all pipetting of (^{32}P) containing products.

2.7. Phosphoamino Acid Analysis

1. pH 1.9 buffer: 25 ml 88% formic acid, 78 ml acetic acid, bring volume up to 1 L using deionized water.
2. pH 3.5 buffer: 50 ml acetic acid, 5 ml pyridine, bring volume up to 1 L with deionized water.
3. PAA standards: 1 mg/ml *O*-phospho-L-serine, 1 mg/ml *O*-phospho-L-threonine, 1 mg/ml *O*-phospho-L-tyrosine.
4. PAA marker dye: 5 mg/ml e-DNP-lysine and 1 mg/ml xylene cyanol solubilized in deionized water containing 30–50% (v/v) of pH 4.7 buffer.
5. 0.25% Ninhydrin: 250 mg ninhydrin dissolved in 100 ml acetone.
6. 20×20 cm TLC plates, 0.1 mm layer thickness.
7. Conical base screw cap microcentrifuge tubes. Screw caps with O-rings are preferable.
8. Thin-layer electrophoresis apparatus.
9. pH 1.9 blotter: Using 3MM Whatman paper, cut a 20 cm×20 cm square and excise blotter holes (see Fig. 5).

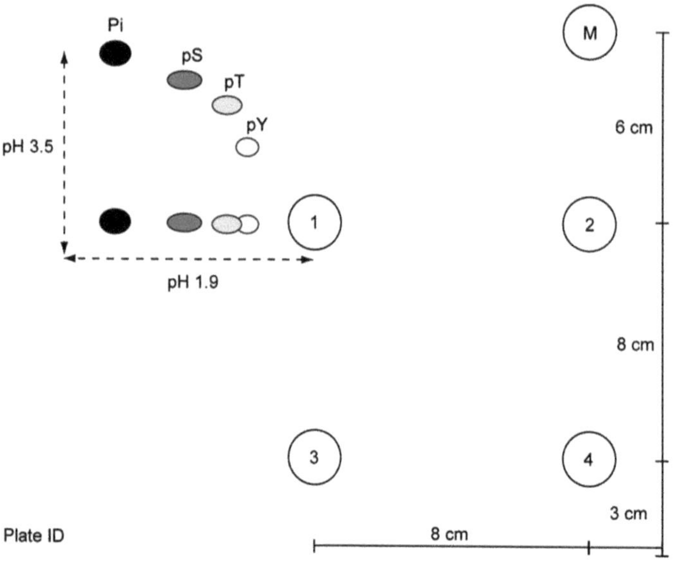

Fig. 5. Schematic of TLC plate setup for two-dimensional phosphoamino acid analysis (2D-PAA). Four samples (*encircled 1–4*) can be analyzed on one TLC plate using the indicated dimensions for spot placement. The PAA marker dye (M) should be spotted in the *top right corner* as indicated. Using a soft lead pencil, the TLC plate can be labeled at the *bottom left corner*. The positions of phosphoserine (pS), phosphothreonine (pT), phosphotyrosine (pY), free phosphate (Pi) post-1D and -2D separation procedures (see Subheading 3.7) are shown for spot one only.

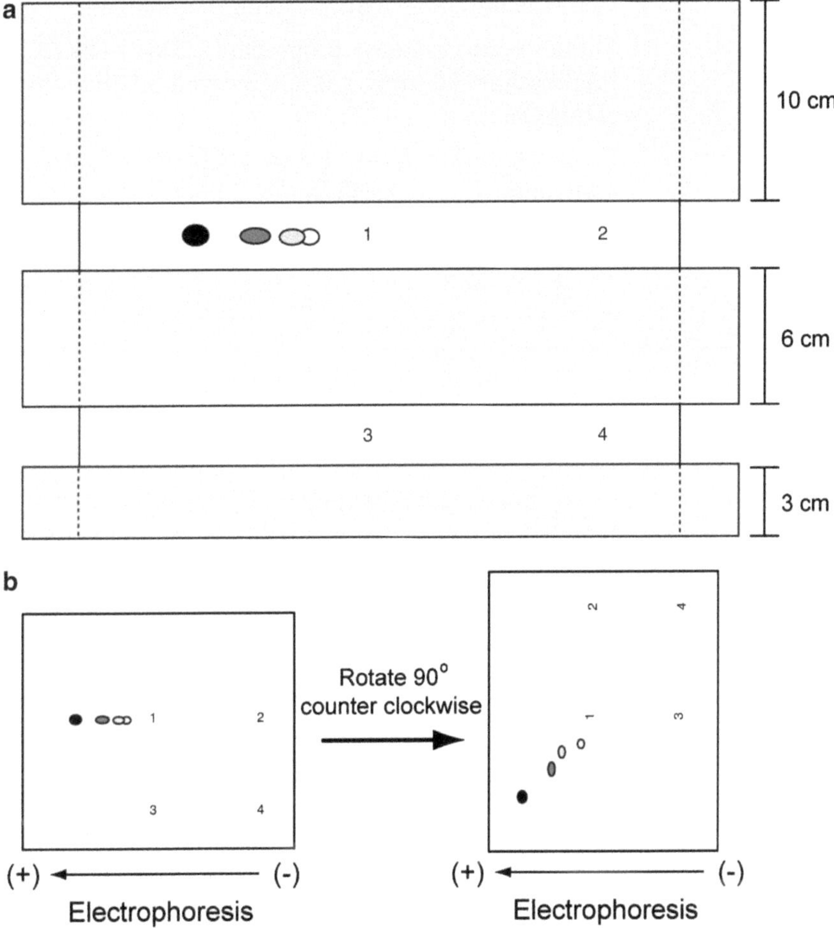

Fig. 6. Preparation of TLC plate for second dimension PAA separation. (**a**) Dampened pH 3.5 blotters should be placed as depicted with the 10 cm wide blotter on the *top*, 6 cm wide blotter in the *middle*, and 3 cm wide blotter on the *bottom* prior to separation in the second dimension. Care should be taken not to lay the blotters directly over the separated samples from the first dimension. (**b**) Following first dimension separation (*left plate*) the TLC plate should be rotated 90 ° counter-clockwise for second dimension separation (*right plate*). The orientation of the anode (+) and cathode (–) are as indicated. The negatively charged PAAs migrate toward the positively charged anode.

10. pH 3.5 blotter: Using 3MM Whatman paper, cut strips for the top (10 cm × 20 cm), middle (6 cm × 20 cm), and bottom (3 cm × 20 cm) (see Fig. 6).
11. Electrophoresis wicks: Using 3 MM Whatman paper, cut a 20 cm × 28 cm strip and fold in half to create one 20 cm × 14 cm wick.
12. Small fan or blow dryer.
13. Spray tool, used to evenly spray 0.25% ninhydrin onto the cellulose plates (Spra-Tool, Aervoe Industries).
14. Speed-Vac centrifuge.

2.8. Receptor Pull Down Assay

1. Cell lysis buffer: (see Subheading 2.1, item 1).
2. Dithiobis(succinimidyl propionate) (DSP) (Pierce Chemical) prepared as 50 mM stock dissolved in dimethyl sulfoxide (DMSO).
3. Blocking buffer: 20 mM Tris–HCl, pH 7.6, 137 mM NaCl containing 1% (w/v) BSA.
4. Wash buffer: 50 mM Tris–HCI, pH 7.6, 200 mM NaCl, 0.05% (v/v) Tween-20.

3. Methods

3.1. Cell Lysis and Immunoprecipitation

1. Following stimulation with or without cytokine/growth factor, pellet the cells by centrifugation (10,000 × *g* for 1 min at 4°C), remove culture media by aspiration and immediately add cold cell lysis buffer (25 μl per one million cells).
2. Briefly vortex to resuspend the cell pellet and rotate for 1 h at 4°C.
3. Pellet insoluble cell debris by centrifugation (12,000 × *g* for 30 min at 4°C).
4. Transfer the cell lysate supernatant to new 1.5 ml eppendorf tubes.
5. Add the appropriate antibody (anti-phosphotyrosine) (5 μg/ml), anti-Jak (4 μg/ml) or normal IgG control (4 μg/ml) to the clarified cell lysate and incubate for 2 h at 4°C with rotation.
6. Add 50 μl of protein A or G-Sepharose slurry to the antibody/cell lysate mixture and rotate for 1 h at 4°C.
7. Wash the immune complex by pelleting the beads by centrifugation (10,000 × *g* for 30 s at 4°C), removing supernatant, resuspending the beads in 1 ml cold lysis buffer and briefly vortexing. Repeat this process for a total of three washes.
8. Elute with 2× SDS sample buffer and boil for 5 min in water bath. Pellet the Sepharose beads by centrifugation (12,000 × *g* for 1 min at room temperature). After cooling to room temperature, the sample is ready for separation by SDS-PAGE. Alternatively, the samples may be stored at −80°C for several months (in aliquots to avoid freeze/thaw cycles).

3.2. SDS-Polyacrylamide Gel Electrophoresis

1. Cast a 7.5% (v/v) acrylamide separating gel. Mix 2.5 ml water, 1.25 ml resolving buffer and 1.25 ml 30% acrylamide in a 15 ml conical tube. Add 10 μl APS and 5 μl TEMED and cast within an 8.2 cm × 10.1 cm × 0.75 mm gel cassette. Allow sufficient space for the stacking gel (two times the comb depth)

and gently overlay with 2-butanol to create a uniform interface and prevent contact with atmospheric oxygen, which inhibits acrylamide polymerization.

2. Once the gel has polymerized (approximately 30 min), pour off the butanol, rinse with deionized water, and cast a stacking gel on top of the separating gel. For a 4.75% stacking gel, mix 1.48 ml water, 625 μl stacking buffer and 395 μl of 30% acrylamide. Add 15 μl APS and 5 μl TEMED to initiate polymerization and immediately overlay on top of the separating gel.
3. Immediately insert the appropriate gel comb into the stacking gel without introducing air bubbles.
4. Once the stacking gel has set (approximately 15 min), place the gel into the electrophoresis unit according to the manufacturer's instructions, add running buffer to the inner and outer chambers of the gel box and carefully remove the comb.
5. Flush out empty wells with running buffer in order to remove any debris that may impede proper current flow.
6. Load protein standard and samples into the wells using gel loading capillary tips (see Note 3).
7. Close the gel box and connect to power supply. Electrophorese at 15 mA/gel until the dye front (from the bromophenol blue) has reached the bottom of the gel.

3.3. Transfer

1. Following electrophoresis, carefully pry the two gel plates apart so that the gel remains intact on one plate. Remove and discard the stacking gel and transfer the remaining resolving gel to a tray containing 1× Transfer buffer.
2. Cut PVDF membrane and six pieces of 3 MM Whatman paper to the size of the gel. Rehydrate the PVDF membrane by soaking in 100% methanol for 30 s, and rinsing in Milli-Q water for 2 min with agitation. Subsequently, equilibrate the membrane for 5 min in 1× Transfer buffer.
3. Briefly submerge one piece of 3 MM Whatman paper in 1× Transfer buffer. Remove excess buffer by tapping several times on the side of the dish before placing onto the cathode of the transfer apparatus, making sure to avoid air bubbles. Repeat with two additional Whatman papers.
4. Carefully place the resolving gel, PVDF membrane, and three additional filter papers. Again, avoid air bubbles between each layer. If necessary, remove air bubbles by rolling a wet test tube over each layer after placing it down (see Note 4).
5. Transfer, using a semidry transfer apparatus, at 150 mA for 1 h.
6. Disassemble the transfer stack, label membrane, and quickly transfer the PVDF membrane to a tray with TBS-T buffer.
7. Wash membrane for 2–5 min before proceeding to Immunoblot.

3.4. Immunoblot

1. Incubate the membrane in blocking buffer on an orbital shaker at room temperature for 1 h (see Note 5).
2. Remove the blocking buffer, add the primary antibody (at appropriate dilution in blocking buffer; start with 1:1,000) and incubate for 1 h on an orbital shaker at room temperature (see Note 6).
3. Remove the primary antibody and wash the membrane with TBS-T three times, for 10 min each, on an orbital shaker at room temperature.
4. Add the horseradish peroxidase-conjugated secondary antibody (at appropriate dilution in blocking buffer; start with 1:2,000) and incubate at room temperature for 30 min on an orbital shaker (see Note 7).
5. Remove the secondary antibody solution and wash three times with TBS-T, for 10 min each, on an orbital shaker at room temperature.
6. Incubate the blot with ECL substrate for 1 min and place the membrane in between two plastic sheets. Maintaining some ECL solution on the membrane substrate depletion during exposure. Expose to X-ray film for 1–10 min to detect signals.
7. To strip the membrane in order to reblot using another antibody, rotate the membrane in stripping buffer at 55°C for 30 min.
8. Wash the membrane three times, for 10 min each, in TBS-T.
9. The membrane is now ready to begin the immunoblot procedure again (see Subheading 3.4, step 1).

3.5. In Vitro Tyrosine Kinase Assay (Hot and Cold)

1. Lyse cells and perform Jak-specific immunoprecipitation according to Subheading 3.1.
2. Wash protein A/G-Sepharose bound Jak/antibody complex three times with cold lysis buffer and once with kinase buffer.
3. Resuspend the bead complex in 50 μl kinase buffer.
4. Initiate the kinase reaction by adding 50 μM cold ATP and incubate at 37°C for 15 min (see Note 8).
5. Stop the reaction by washing the protein A/G-Sepharose bead complex three times with ATP-free kinase buffer and elute the material by boiling in 2× SDS sample buffer for 5 min.
6. Resolve the immunopurified Jak by SDS-PAGE (see Subheading 3.2) and transfer to a PVDF membrane (see Subheading 3.3) (see Note 9).
7. Perform an α-pY and α-Jak immunoblot on the membrane (see Subheading 3.4) (see Note 10).

3.6. Metabolic Labeling

1. Harvest cells and wash three times with phosphate-free RPMI media that has been pre-warmed to 37°C. Pellet the cells by centrifugation at 200 × *g* for 10 min at room temperature. Remove the supernatant and resuspend the cells in phosphate-free RPMI media. Repeat two additional times.
2. Resuspend cells (1×10^7 cells in 0.5 ml) in phosphate-free RPMI containing 2% dialyzed FBS.
3. Resuspend (^{32}P) in phosphate-free RPMI containing 2% dialyzed FBS (see Note 11).
4. Deliver 0.4 μCi of resuspended (^{32}P) to each sample and incubate for 2 h at 37°C. Invert the tubes every 20 min to prevent cells from sedimenting.
5. Stimulate cells with the appropriate cytokine/growth factor at 37°C. Maximum Jak activation is normally achieved with a 10–15 min stimulation.
6. Pellet cells, remove radioactive media from tubes and wash cells one time with cold phosphate-free RPMI medium.
7. Add 0.5 ml cold lysis buffer to each tube and incubate for 45 min at 4°C while rotating.
8. Clarify lysate by centrifugation at 12,000 × *g* for 30 min at 4°C.
9. Perform immunoprecipitation of desired Jak as described above (see Subheading 3.1).
10. Following immunoprecipitation, the sample should be separated by SDS-PAGE (see Subheading 3.2) and transferred to PVDF membrane (see Subheading 3.3).
11. Stain the membrane with Coomassie Blue R250 by incubating for 1 h at room temperature with gentle agitation. Destain the membrane by shaking on an orbital shaker in Coomassie Blue destain buffer. Several changes of the buffer will be required to achieve optimal destain.
12. Use autoradiography to visualize the (^{32}P) incorporated into the Jak proteins.

3.7. Phosphoamino Acid Analysis

1. Utilizing the autoradiograph image (see Subheading 3.6, step 12), locate and excise the Jak bands from the PVDF membrane (see Note 12).
2. In a 1.5 ml screw cap microcentrifuge tube, resuspend the excised band in 50 μl of 6 N HCl, vortex, centrifuge briefly, and incubate at 100°C for 30 min.
3. After hydrolysis, briefly spin down and transfer HCl to a new 1.5 ml screw cap microcentrifuge tube.
4. Dry the samples using a Speed-Vac.

5. Resuspend the hydrolysate in 5–10 μl of pH 1.9 buffer containing 1 μl PAA standards.
6. Centrifuge the samples prior to loading in order to remove any particulate.
7. Load the samples drop by drop (0.5 μl/drop) onto the plate. Allow each drop to completely dry before adding the next drop. Four samples can be analyzed per plate as depicted in the PAA stencil (see Fig. 5).
8. Spot the PAA marker (0.5 μl) at the top of the plate as depicted in Fig. 5.
9. Soak two electrophoresis wicks in pH 1.9 buffer. Set up the electrophoresis apparatus using pH 1.9 buffer according to the manufacturer's instructions.
10. Wet the cellulose plates using a blotter with five holes in it (see Fig. 5) that has been dampened with pH 1.9 buffer (see Note 13).
11. Overlay the pH 1.9 blotter on the cellulose plate by aligning the five holes with the five corresponding spots on the plate. Using your finger, push out the buffer from around each of the five holes as quickly as possible in order to concentrate the spots within each circle. Remove the blotter once each spot is concentrated and the entire plate is wet (see Note 14).
12. Place the plate in the electrophoresis apparatus and run for 25 min at 1.5 kV.
13. Following electrophoresis, remove the plate and dry with a small fan or blow dryer for 20 min (see Note 15).
14. The buffer in the electrophoresis apparatus should be rinsed out and changed to pH 3.5 buffer. The wicks should be replaced with new wicks soaked in pH 3.5 buffer.
15. Dampen the three pH 3.5 blotters (top, middle, bottom) in pH 3.5 buffer containing EDTA (see Note 16).
16. Align the three blotters on the cellulose plate parallel to the first dimension of separation (see Fig. 6).
17. Push out the buffer from the strips by pressing along the blotter edges to wet the plate. It is desired that the buffer from each strip meet the buffer squeezed from the adjacent strip in order to concentrate the sample along a straight line for the second dimension. Remove the blotters once the entire plate is wet and the sample is concentrated. Use a Kimwipe to dab away any excess buffer on the plate.
18. Immediately place the plate in the electrophoresis apparatus, making sure to rotate the plate 90 ° counterclockwise (see Fig. 6).
19. Electrophorese for 16 min at 1.3 kV.

20. Remove the plate and dry with a fan or blow dryer for 20 min.
21. Using a spray-tool, evenly spray the dried plate with 0.25% ninhydrin in acetone (see Note 17).
22. Bake the plate in an oven pre-warmed to 65°C for 15 min to develop the stain (see Note 18).
23. Use autoradiography to detect the amount of (^{32}P) incorporated into each sample.

3.8. Receptor Pull-Down Assay

1. Following stimulation in the presence or absence of cytokine/growth factor, wash the cells with cold phosphate-buffered saline, pH 7.2 containing 1 mM $MgCl_2$ and quickly pellet.
2. Resuspend cells in wash buffer at a concentration of 5×10^7 cells/ml and incubate with the membrane permeable and cleavable cross-linker Dithiobis(succinimidyl propionate) (DSP) at a final concentration of 0.5 mM.
3. Allow the cross-linking reaction to proceed for 15 min on ice and stop by adding 20 mM Tris–HCl, pH 8.5. Wash cells two times and pellet by centrifugation at $12{,}000 \times g$ for 1 min at 4°C (see Note 19).
4. Lyse the cells and perform immunoprecipitation using the anti-cytokine receptors (5 μg/ml), or control IgG (5 μg/ml) as directed in Subheading 3.1.
5. Wash the protein A/G-Sepharose beads three times with lysis buffer and elute with 2× SDS sample buffer.
6. Separate the samples by 7.5% SDS-PAGE (see Subheading 3.2) and subsequently transfer to a PVDF membrane (see Subheading 3.3).
7. For detection of phosphotyrosine proteins, perform an immunoblot using the α-pY antibody (1:1,000) as directed above (see Subheading 3.4).

4. Notes

1. Unless otherwise stated, all solutions should be prepared in distilled deionized water that has a resistivity up to 18.2 MΩ cm.
2. Allow to thaw at room temperature for 30 min prior to use.
3. If loading IP samples, take care not to load the protein A/G Sepharose beads as they will impede the flow of current through the well.
4. The orientation of the transfer stack is dependent on the transfer apparatus used. Remember to place the gel on the side closest to the Cathode electrode plate and the PVDF membrane on the side closest to the Anode electrode plate.

5. Alternatively, blocking of the membrane can be performed overnight at 4°C on an orbital shaker.
6. Alternatively, primary antibody can be used overnight at 4°C on an orbital shaker.
7. Secondary antibody should be diluted to 500 ng/ml in wash buffer containing 1% bovine serum albumin.
8. Alternatively, 30 μCi of (γ-^{32}P) ATP (20 Ci/mmol) may be used for radiolabeled kinase assays.
9. 7.5% SDS-PAGE is acceptable for resolving all Jak family members.
10. For (γ-^{32}P) ATP experiments, autoradiography should be used to visualize isotope-labeled proteins.
11. Volume of RPMI medium used depends on the number of samples being labeled. Generally a concentration of 20 μCi/ml is favorable.
12. PAA should be run from beginning to end in 1 day.
13. Blotter should not be excessively wet.
14. If you notice pools of pH 1.9 buffer on the cellulose plate, remove by dabbing with a Kimwipe. Plates should not be excessively wet.
15. The plate should not smell of acetic acid prior to proceeding.
16. EDTA should be at a final concentration of 0.5 mM and is only added to 200 ml of pH 3.5 buffer used to wet the blotters. Do not use pH 3.5 buffer with EDTA to run the electrophoresis or wet wicks.
17. Sparse amounts of 0.25% ninhydrin are needed to detect phosphoamino acids. Take care not to over saturate the TLC plates.
18. The separated purple spots that appear correspond to the positions of phosphoserine, phosphothreonine, and phosphotyrosine (see Fig. 5).
19. The cells may be thoroughly washed, pelleted, and stored at -80°C until further use. Thaw frozen pellets on ice before proceeding to step 4.

Acknowledgments

This work was supported by grants from the Lizanell and Colbert Coldwell Foundation, Edward N. and Margaret G. Marsh Foundation, and grant G12MD007592 from the National Center on Minority Health and Health Disparities (NCMHD), a component of the National Institutes of Health (NIH).

References

1. Ihle JN, Witthuhn B, Quelle FW, Yamamoto K, Silvennoinen O (1995) Signaling through the hematopoietic cytokine receptors. Annu Rev Immunol 13:369–398
2. Ghoreschi K et al (2009) Janus kinases in immune cell signaling. Immunol Rev 228: 273–287
3. Ihle JN et al (1997) Jaks and Stats in cytokine signaling. Stem Cells 15(Suppl 1):105–111, discussion 112
4. Kawamura M et al (1994) Molecular cloning of L-JAK, a Janus family protein-tyrosine kinase expressed in natural killer cells and activated leukocytes. Proc Natl Acad Sci U S A 91:6374–6378
5. Tortolani PJ et al (1995) Regulation of JAK3 expression and activation in human B cells and B cell malignancies. J Immunol 155:5220–5226
6. Musso T et al (1995) Regulation of JAK3 expression in human monocytes: phosphorylation in response to interleukins 2, 4, and 7. J Exp Med 181:1425–1431
7. Ross JA et al (2007) Regulation of T cell homeostasis by JAKs and STATs. Arch Immunol Ther Exp (Warsz) 55:231–245
8. Leonard WJ, O'Shea JJ (1998) Jaks and STATs: biological implications. Annu Rev Immunol 16:293–322
9. Tam L et al (2007) Expression levels of the JAK/STAT pathway in the transition from hormone-sensitive to hormone-refractory prostate cancer. Br J Cancer 97:378–383
10. Lai SY, Johnson FM (2010) Defining the role of the JAK-STAT pathway in head and neck and thoracic malignancies: implications for future therapeutic approaches. Drug Resist Updat 13:67–78
11. Neilson LM et al (2007) Coactivation of janus tyrosine kinase (Jak)1 positively modulates pro lactin-Jak2 signaling in breast cancer: recruitment of ERK and signal transducer and activator of transcription (Stat)3 and enhancement of Akt and Stat5a/b pathways. Mol Endocrinol 21:2218–2232
12. Gao B et al (2001) Constitutive activation of JAK-STAT3 signaling by BRCA1 in human prostate cancer cells. FEBS Lett 488: 179–184
13. Pesu M et al (2008) Therapeutic targeting of Janus kinases. Immunol Rev 223:132–142
14. Zhou YJ et al (1997) Distinct tyrosine phosphorylation sites in JAK3 kinase domain positively and negatively regulate its enzymatic activity. Proc Natl Acad Sci U S A 94:13850–13855
15. Feng J et al (1997) Activation of Jak2 catalytic activity requires phosphorylation of Y1007 in the kinase activation loop. Mol Cell Biol 17:2497–2501
16. Gauzzi MC et al (1996) Interferon-alpha-dependent activation of Tyk2 requires phosphorylation of positive regulatory tyrosines by another kinase. J Biol Chem 271: 20494–20500
17. Wang R et al (2003) Mechanism of Janus kinase 3-catalyzed phosphorylation of a Janus kinase 1 activation loop peptide. Arch Biochem Biophys 410:7–15
18. Cheng H et al (2008) Phosphorylation of human Jak3 at tyrosines 904 and 939 positively regulates its activity. Mol Cell Biol 28: 2271–2282
19. Argetsinger LS et al (2011) Tyrosines 868, 966, and 972 in the kinase domain of JAK2 are autophosphorylated and required for maximal JAK2 kinase activity. Mol Endocrinol 24: 1062–1076
20. Funakoshi-Tago M et al (2008) Negative regulation of Jak2 by its auto-phosphorylation at tyrosine 913 via the Epo signaling pathway. Cell Signal 20:1995–2001
21. Li Z et al (2007) SH2B1 enhances leptin signaling by both Janus kinase 2 Tyr813 phosphorylation-dependent and -independent mechanisms. Mol Endocrinol 21:2270–2281
22. Kurzer JH et al (2004) Tyrosine 813 is a site of JAK2 autophosphorylation critical for activation of JAK2 by SH2-B beta. Mol Cell Biol 24:4557–4570
23. Argetsinger LS et al (2004) Autophosphorylation of JAK2 on tyrosines 221 and 570 regulates its activity. Mol Cell Biol 24:4955–4967
24. Robertson SA et al (2009) Regulation of Jak2 function by phosphorylation of Tyr317 and Tyr637 during cytokine signaling. Mol Cell Biol 29:3367–3378
25. Mazurkiewicz-Munoz AM et al (2006) Phosphorylation of JAK2 at serine 523: a negative regulator of JAK2 that is stimulated by growth hormone and epidermal growth factor. Mol Cell Biol 26:4052–4062
26. Ross JA et al (2010) Protein phosphatase 2A regulates interleukin-2 receptor complex formation and JAK3/STAT5 activation. J Biol Chem 285:3582–3591
27. Cacalano NA et al (1999) Autosomal SCID caused by a point mutation in the N-terminus of Jak3: mapping of the Jak3-receptor interaction domain. EMBO J 18:1549–1558

28. Zhou YJ et al (2001) Unexpected effects of FERM domain mutations on catalytic activity of Jak3: structural implication for Janus kinases. Mol Cell 8:959–969
29. Funakoshi-Tago M et al (2006) Receptor specific downregulation of cytokine signaling by autophosphorylation in the FERM domain of Jak2. EMBO J 25: 4763–4772
30. Sayyah J et al (2011) Phosphorylation of Y372 is critical for Jak2 tyrosine kinase activation. Cell Signal 23:1806–1815

Chapter 2

Co-immunoprecipitation Protocol to Investigate Cytokine Receptor-Associated Proteins, e.g., Janus Kinases or Other Associated Signaling Proteins

Claude Haan and Serge Haan

Abstract

Jak binding to cytokine receptors has been shown to be a complex and tight interaction. When studying loss-of-function or gain-of-function mutants of the Jaks or cytokine receptors it is often necessary to know if a certain mutant still associates correctly in the context of the signaling complex. The standard technique to show interaction of Jaks with cytokine receptors or other signalling molecules is Co-immunoprecipitation. Here we describe our protocol and discuss different pitfalls that can be encountered during the procedure.

Key words: Co-immunoprecipitation, Cytokine receptors, Janus kinases, Discontinuous SDS-PAGE, Western blotting, Immunodetection

1. Introduction

Due to the lack of structural data, the details of the interaction between cytokine receptors and Janus kinases are unknown. The N-termini of the Jaks consisting of the FERM and the SH2 like domain are involved in binding to cytokine receptors (1–8). Since it has been shown that overexpression of proteins (e.g., Janus kinases) can lead to artifacts in binding behavior (8, 9) expression systems which allow close to endogenous expression of Jaks (7, 10) were developed. The described method allows the detection of co-precipitated endogenous proteins with cytokine receptors and includes different techniques. Parts of the immunoprecipitation-, SDS-PAGE-, Western blotting-, or immunodetection-procedures have been changed (compared to the original protocols) and adapted to lead to higher sensitivity of detection of the co-precipitated proteins.

Sandra E. Nicholson and Nicos A. Nicola (eds.), *JAK-STAT Signalling: Methods and Protocols*, Methods in Molecular Biology, vol. 967, DOI 10.1007/978-1-62703-242-1_2, © Springer Science+Business Media New York 2013

2. Materials

Please check how to securely work with the chemicals that are described in this protocol, before starting the experiment. Prepare all solutions using ultrapure water and analytical grade reagents and follow all waste disposal regulations diligently. All reagents can be stored at room temperature unless otherwise stated.

2.1. Co-immunoprecipitation

1. Lysis buffer: 0.5% IGEPAL-CA630, 20 mM Tris–HCl, pH 7.5, 130 mM NaCl (see Notes 1 and 2). To prepare 1 L of lysis buffer dissolve 2.4 g Tris and 7.6 g NaCl in 900 ml of water and adjust the pH to 7.5 with HCl (see Note 3). Make up with water to 1 L and add 5 ml of IGEPAL-CA630.
2. Washing buffer: 0.1% to 0.5% IGEPAL-CA630, 20 mM Tris–HCl, pH 7.5, 130 mM NaCl, 1 mM sodium vanadate (see Note 4). Prepare this solution as described in the point before and adjust the detergent concentration as needed.
3. 4× Sample buffer: 125 mM Tris–HCl, pH 6.7, 20% glycerol, 10% 2-mercaptoethanol, 4% SDS. Prepare 1 L of 1 M Tris–HCl, pH 6.7 stock solution by dissolving 121.14 g Tris in 900 ml of water and adjust the pH to 6.7 with HCl. Make up to 1 L. To prepare 50 ml of sample buffer use 6.25 ml of 1 M Tris–HCl, pH 6.7, 10 ml glycerol, 10 ml 20% SDS, and 5 ml 2-mercaptoethanol and make up with water to 50 ml. Add very little bromophenol blue to give the solution a blue color.
4. 2× Sample buffer: Dilute 4× sample buffer with water 1:1.
5. Protease inhibitors: We use complete protease inhibitor mix (Roche) (see Note 5).
6. Protein A sepharose: We use Protein A sepharose CL-B4 (GE Healthcare).

2.2. SDS-PAGE

1. 4× separating gel buffer: 1.5 M Tris–HCl, pH 8.8, 0.4% SDS. Dissolve 181.7 g Tris in 800 ml H_2O. Adjust pH to 8.8 with 30% HCl. Make up with water to 1 L and recheck the pH. Add 20 ml of 20% SDS solution.
2. 4× stacking gel buffer: 0.5 M Tris–HCl, pH 6.7, 0.4% SDS. Dissolve 30.3 g Tris in 400 ml H_2O. Adjust pH to 6.7 with 30% HCl. Make up with water to 500 ml and recheck the pH. Add 10 ml of 20% SDS solution.
3. Acrylamide solution 30%, acrylamide-to-bisacrylamide ratio 29:1 (e.g., by Applichem) (see Note 6).
4. *N,N,N′,N′*-tetramethylethylendiamine (TEMED) (see Note 7).
5. Ammonium persulfate (see Note 8).

6. 10× SDS running buffer: 0.25 M Tris-base, 1.92 M glycine, 1% SDS (see Note 9). Dissolve 60.6 g Tris, 288.3 g glycine, and 20 g SDS in 2 L of water. Dilute 10 times before use.

2.3. Western Blotting

1. Anode buffer I: 300 mM Tris-base, 20% methanol. Use 150 ml of a 2 M Tris-base solution and 200 ml methanol and make up to 1 L with water.
2. Anode buffer II: 25 mM Tris-base, 20% methanol. Use 12.5 ml of a 2 M Tris-base solution and 200 ml methanol and make up to 1 L with water.
3. Cathode buffer: 40 mM ε-aminocaproic acid, 20% methanol, 0.01% SDS. Use 40 ml of a 1 M ε-aminocapronic acid solution, 200 ml methanol, and 0.5 ml of a 20% SDS solution and make up to 1 L with water.
4. PVDF membrane (pore size 0,45 μm).
5. Blotting paper (see Note 10) (e.g., Whatman® 3 MM Chr. (0.34 mm thickness)).

2.4. Immunodetection

1. TBS-N: 10 mM Tris–HCl, pH 7.4, 135 mM NaCl, 0.1% IGEPAL CA-630. To prepare 5 L of TBS-N, dissolve 6.1 g Tris and 39.4 g NaCl in 1 L of water and adjust the pH with HCl to 7.5. Add 5 ml of IGEPAL CA-630, mix, and make up with water to 5 L.
2. Blocking buffer: 10% bovine serum albumin (BSA) and 0.01% NaN_3 in TBS-N.
3. pCA-ECL solution: (see Note 11) 100 mM Tris/HCl pH 8.8 (see Note 12), 2.5 mM luminol (IUPAC name: 5-Amino-2,3-dihydro-1,4-phthalazinedione) (see Note 13), 0.2 mM para coumaric acid (pCA, IUPAC name: 3-(4-hydroxyphenyl)-2-propenoic acid), and 2.6 mM hydrogen peroxide (see Note 14). For the pCA-ECL prepare a large batch of ready-to-go solution containing luminol and pCA without the H_2O_2 and keep this at 4°C and in a dark bottle. Prepare 500 ml pCA-ECL by adding 5 ml of 250 mM luminol stock solution (to prepare 250 mM luminol dissolve 1.2 g luminol in 25 ml DMSO) and 1.1 ml of 90 mM pCA stock solution (to prepare 90 mM pCA dissolve 0.15 g of pCA in 10 ml DMSO) in 494 ml 100 mM Tris–HCl, pH 8.8. To activate the pCA-ECL solution for the detection of the Western blots, add 3.5 μl of 30% H_2O_2 (see Note 15) per 10 ml of pCA-ECL solution a few minutes before detection of the Western blot.
4. 4IPBA-ECL solution: 100 mM Tris–HCl, pH 8.8, 1.25 mM luminol, 2 mM 4IPBA (IUPAC name: (4-iodophenyl) boronic acid), 5.3 mM hydrogen peroxide (see Note 16). The 4IPBA-ECL is best prepared fresh from the stocks just before use since longer storage results in increasing background. To prepare

10 ml of 4IBPA-ECL add 50 μl of 250 mM luminol stock, 225 μl of 90 mM 4IPBA stock, and 5 μl of 30% H_2O_2 (to prepare 90 mM 4IPBA stock solution, dissolve 0.22 g of 4-iodophenyl boronic acid in 10 ml DMSO) (see Note 17).

5. Stripping buffer 1 (see Note 18): 2 M glycine, pH 2.5. Add half a teaspoonful of SDS powder (avoid pellets as they take a long time to dissolve and tend to stick to the blot) to 20 ml of buffer when the blot is already covered with buffer. The SDS dissolves during the process (see Note 19). To prepare 2 M glycine, pH 2.5, dissolve 150 g glycine in 800 ml water and adjust the pH to 2.5 using HCl. Make up to 1 L with water and readjust the pH if necessary.
6. Stripping buffer 2 (see Note 20): 60 mM Tris–HCl, pH 6.8, 2% SDS. To prepare 1 L dissolve 7.28 g Tris in 800 ml and adjust the pH to pH 6.8. Add 20 g of SDS and make up with water to 1 L while stirring. Add 80 μl 2-mercaptoethanol per 10 ml of stripping buffer just before use and incubate at 70°C for 20–30 min (see Note 21).

3. Methods

3.1. Cell Lysis and Immunoprecipitation

All steps of cell lysis and immunoprecipitation are performed at 4°C using ice-cold buffers (see Note 22). When showing the interaction between cytokine receptors and Jaks one can either precipitate the receptor and check for co-precipitated Jak or use the inverse approach. However, since cytokine receptors are highly glycosylated, they migrate on SDS-PAGE gels as diffuse bands, while the Janus kinases are visible as sharp bands on Western blots. Probably because it is easier to detect the low levels of co-precipitated protein when this protein does not migrate as a fuzzy band, we have always been more successful in precipitating the cytokine receptor and show the Jak co-precipitation than the other way around. Be sure to validate all antibodies (see Note 23 and Fig. 1) (see Note 24 and Fig. 2) and to include the important controls (see Notes 25, 26 and Fig. 3).

1. Cells are lysed on a 10 cm cell culture dish (placed on ice) (see Note 27) (approx. 5–10 million cells) with 500–1,000 μL of ice-cold lysis buffer containing protease inhibitors. After scraping of the cells from the surface the lysates are transferred to a microfuge tube.
2. Incubate the lysates on ice for 30 min, and then clear the lysates by centrifugation at 12,000×*g* for 15 min in a cooled centrifuge.
3. The cleared lysates are transferred to fresh microfuge tubes and the concentration of the lysates is measured (e.g., by Bradford assay) (see Note 28).

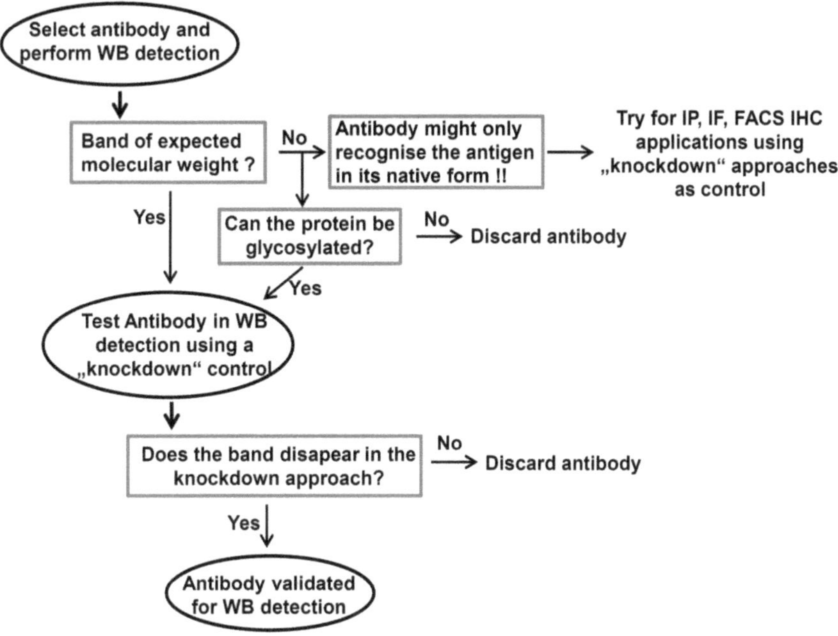

Fig. 1. Possible workflow to test antibody specificity. *WB* Western blot, *IP* immunoprecipitation, *IF* immunofluorescence, *FACS* fluorescence-activated cell sorting, *IHC* immunohistochemistry.

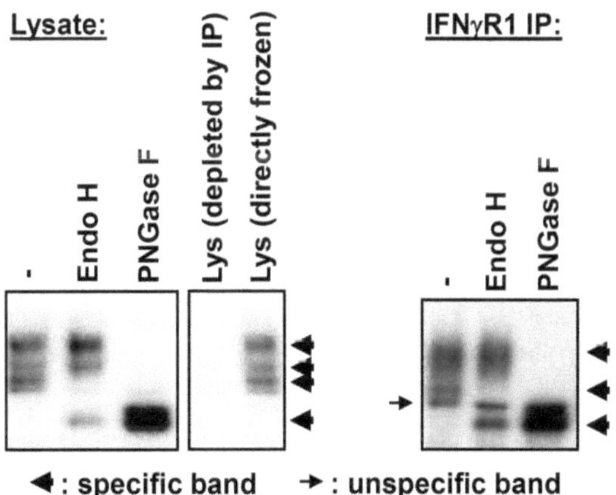

Fig. 2. Lysates and an immunoprecipitation of IFNγR1 were treated with EndoH and PNGaseF (37°C for 1 h following the manufacturer's instructions (New England Biolabs)) to remove the corresponding glycosylation. SDS-PAGE and Western blots were performed and detected with an antibody against IFNγR1. A directly frozen lysate and a lysate depleted by quantitative precipitation of IFNγR1 are also shown as controls. The treatment with EndoH shifts the bands corresponding to the immature ER forms of IFNγR1 towards higher mobility on the Western blot detection. The PNGaseF treatment shifts all glycosylated IFNγR1 forms towards higher mobility. The nonspecific bands are not shifted.

4. Equal amounts of lysate are transferred to fresh microfuge tubes and the volume in each tube is adjusted to 1 ml so that the concentration of protein lysate in each tube is similar. Transfer 90 μl of each lysate to fresh microfuge tubes and add 30 μl of 4× sample buffer before freezing the lysates at −20°C

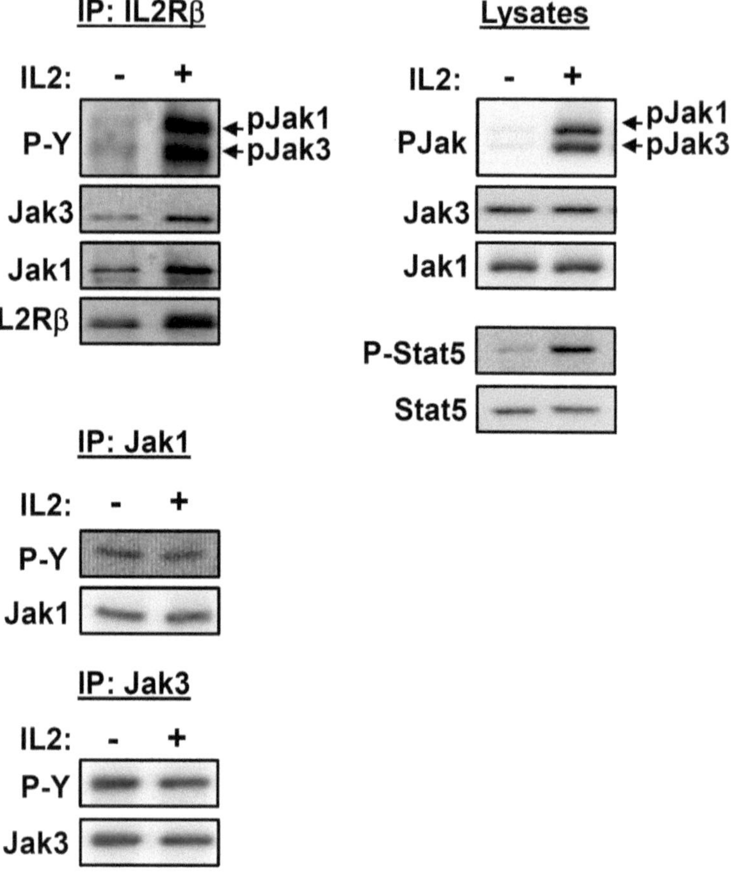

Fig. 3. U4C cells expressing IL2Rβ/γ, Jak1, and Jak3 (10) were stimulated with IL-2. Lysates were prepared and immunoprecipitates with antibodies against IL2Rβ, Jak1, or Jak3 were performed. After SDS-PAGE, Western blots of the lysates and the immunoprecipitates were prepared and detected as indicated. The detection of Jak1 and Jak3 phosphorylation is better from the immunoprecipitates of the IL2Rβ which precipitates the whole signaling complexes than from the immunoprecipitates of Jak1 or Jak3.

(see Note 29). Add 1 μg of precipitating antibody to the remaining lysate and incubate the solution at low speed on an overhead shaker for 2–3 h (or overnight) at 4°C.

5. A protein A sepharose (see Note 30) slurry is prepared in the meantime by equilibrating 2.5 mg protein A sepharose (per IP sample) with 100 μl (per IP sample) lysis buffer for 1 h minimum at 4°C in an overhead shaker (see Notes 31 and 32).
6. 100 μl of protein A sepharose slurry is added to each microfuge tube containing the lysates with the immunocomplexes (see Note 33). The mixture is incubated on the overhead shaker for 1 h at 4°C.
7. The protein A sepharose is collected by centrifugation at 12,000 × *g* for 1 min.

8. The supernatant is discarded (see Note 34) and the pellet containing the sepharose and the immunocomplexes are washed three to five times (see Note 35). 1 ml of washing buffer is added and the immunoprecipitate is re-suspended by vortexing. After a 5-min incubation on ice the sepharose is pelleted by centrifugation at 12,000 × *g* for 1 min. This procedure is repeated 2 to 4 times.
9. After the last washing solution has been removed the sepharose and the immunocomplexes are mixed with 30–50 μl of 2× Sample buffer (depending on the volume of sepharose used, the ratio should roughly be 1:1), vortexed, and stored at −20°C until further use for SDS-PAGE.

3.2. SDS-PAGE

The immunoprecipitates are electrophoretically resolved by mass on a polyacrylamide hydrogel. To achieve better resolution on the gel we recommend discontinuous SDS-PAGE. The sample is first focused on a "stacking gel" and then separated in the second part of the gel, the "separating gel" (see Note 36). There are many types of SDS-PAGE chambers which require slightly different handling when producing the gels or mounting the gel into the electrophoresis chamber, so we do not describe these steps in detail here because they might be specific for our system only (see Note 37).

1. Water, the acrylamide solution, and the 4× separating gel buffer are mixed so that a separating gel of the wanted percentage is achieved (see Note 38). Per 10 ml of solution 5 μl TEMED and 25 μl of 20% APS (see Note 39) are added resulting in final concentrations of both initiators of 0.05% (see Note 40). The mixture is thoroughly mixed and the separating gel is cast between the glass plates. To achieve a level and smooth separating gel surface the polymerizing solution is overlaid with water-saturated isopropanol (see Note 41).
2. When the separating gel is polymerized (see Note 42) remove the isopropanol by inverting and use some water to wash. Remove the water by inverting and aspire the remaining water using a filter paper without touching the gel surface (see Note 43).
3. Water, the acrylamide solution, and the 4× stacking gel buffer are mixed so that a stacking gel of 3% acrylamide is achieved. Per 10 ml of solution add 10 μl TEMED and 30 μl of 20% APS (see Note 44). The mixture is thoroughly mixed and the separating gel is cast between the glass plates. The comb is inserted and the mixture is left to polymerize for 2 h.
4. When the gels are polymerized the comb is removed and the gels mounted in the SDS-PAGE chamber. The running buffer

is filled into the chamber and eventual air bubbles are chased from beneath the gel.

5. The immunoprecipitates and the lysate samples are heat-denatured for 5 min at 95°C.
6. After heating, the samples are centrifuged for 30 s at 13,000 × g (see Note 45). They are now ready to be applied to the gel.
7. The samples are loaded by using a Hamilton syringe or a pipette using very thin tips.
8. The electrophoresis apparatus is now connected to the power supply and the gels are run at 25 mA/gel (see Note 46).
9. When the running front has reached the bottom of the gel the run is stopped, the gel is carefully removed from within the plates, and the stacking gel can be discarded.

3.3. Western Blotting

The transfer of the proteins onto the membrane by Western blotting can be performed in different ways (capillary transfer, electrotransfer). Our method uses PVDF membrane and a semidry Western blotting chamber in combination with a 3 buffer system. The assembly of the Western blotting chamber may be deduced from Fig. 4.

1. During the SDS-PAGE run, 10 pieces of filter paper are cut per gel to be transferred (see Note 47). The PVDF membranes are also cut (see Note 48) and are shortly rinsed with methanol and then put into anode buffer II (see Note 49) for equilibration.
2. The SDS-PAGE separating gels are incubated in cathode buffer for 5 min.

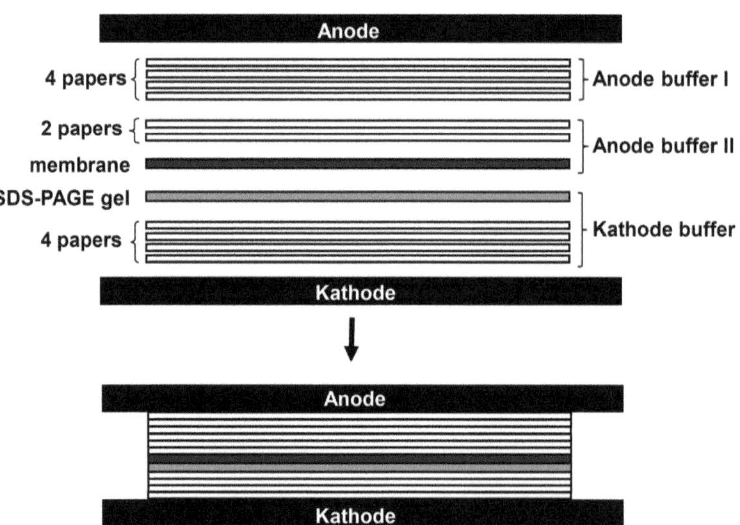

Fig. 4. Typical setup of a semidry Western blotting chamber (3 buffer system).

3. The blotting stack is built now in the Western blotting chamber. We describe it for our chamber which has the cathode as the base of the apparatus (see Note 50) (see also Fig. 4).
4. Filters are incubated in cathode buffer until they are completely wet and are placed on the cathode. Then the gel is placed on the filter papers. After adding the PVDF membrane to the stack two papers incubated with anode buffer II and lastly 4 papers incubated in anode buffer I are added to the stack.
5. Remove eventual residual air bubbles trapped within the stack by rolling a 25 ml pipette (see Note 51) gently over the stack consisting of gel, blotting membrane, and filter papers (move from the middle to the exterior to best chase the air) (see Note 52).
6. Place the top of the chamber carefully (do not displace the stack) on top of the blotting stack and fasten the tightening device according to the make of your blotting chamber (see Notes 53 and 54).
7. Connect the chamber to the power supply and blot for 1 h (see Note 55) at 0.8 mA/cm**2**. The current setting depends on the area of the blotting stack and is kept constant during the run (see Note 56).
8. The blotting membrane is recovered from the stack and immersed in TBS-N buffer in a plastic box (see Note 57). Conserve the blot in TBS-N until proceeding to Western blot detection.

3.4. Immunodetection of Western Blots

1. Block 30 min in blocking buffer (see Note 58).
2. Wash the blot shortly in TBS-N.
3. Add the first antibody solution (1 μg/ml antibody in TBS-N + 0.01–0.1% NaN_3) (see Note 59) for 1 h at room temperature or even better, overnight at 4°C (see Note 60).
4. Wash 3 timcs with TBS-N.
5. Add the HRP-labeled secondary antibody (1:2,500 to 1:5,000) (mind the species of the first antibody!!) (see Note 61).
6. Wash 3 times in TBS-N.
7. Prepare the ECL solution and add H_2O_2 as described in the materials section. Add the solution to the blot for 1 min before detecting the blot.
8. Place the blot between two sheets of plastic foil (see Note 62) and detect the emitted light with any system available to you (see Notes 63 and 64).
9. The Western blots can be stripped and redetected using the so-called stripping buffers. After stripping (scc Subhcading 2) the blot is thoroughly washed and the immunodetection is performed again.

4. Notes

1. We have successfully used different detergents for co-immunoprecipitations. The IGEPAL can also be replaced by 1% Brij 96 or 0.5% Triton X100. We recommend testing different detergents.
2. If phosphorylation events need to be investigated the lysis and washing buffers may also be supplemented with 10 mM $MgCl_2$ and 1 mM sodium vanadate.
3. More diluted solutions of HCl may be used when approaching the target pH to avoid abrupt changes in pH.
4. We always use the same detergent in the lysis and the washing buffer.
5. The protease inhibitors can also be added individually. In this case we use 10 mM PMSF, 1 mM benzamidine, 5 μg/ml aprotinin, 3 μg/ml pepstatin, 5 μg/ml leupeptin, and 1 mM EDTA.
6. We recommend ready-to-use solutions. We have never had problems with these solutions and it avoids working with acrylamide powder, which represents an even greater hazard than the ready-made solutions.
7. The purer the TEMED is the greater the shelf life will be (99% pure is advised). TEMED is subject to oxidation, which causes a gradual loss of TEMED reactivity with time. TEMED is also very hygroscopic and will gradually accumulate water, which will accelerate oxidative decomposition. The latter reason is why we store TEMED at room temperature. If TEMED is stored at 4°C and is opened and handled at room temperature it will more quickly be spoiled due to water condensation. We recommend replacing the TEMED stock half yearly.
8. Ammonium persulfate is also very hygroscopic and additionally decomposes immediately when in contact with water, which results in a rapid loss of reactivity. This is why ammonium persulfate solutions should be prepared fresh daily and the crystalline stock should be replaced half yearly.
9. Never adjust the pH of SDS-running buffer. The added ions might interfere with gel focusing.
10. Blotting paper is alternatively called chromatography paper.
11. We use two self-made ECL solutions, which have been described earlier (11), have good signal intensity, and are more budget friendlier than commercial ECL solutions. The pCA-ECL is a robust product that is easy to handle. The 4IBPA-ECL requires a bit more handling but is interesting in situations where a low background signal has to be achieved. We recommend trying the pCA-ECL (which works just fine for most antibodies) first and switching to 4IPBA-ECL if high background occurs.

12. For both ECL solutions it is important that the pH of the 100 mM Tris–HCl, pH 8.8 is accurate. Often dilutions from 1 M or 1.5 M Tris–HCl, pH 8.6 stocks are performed to get 100 mM Tris–HCl, pH 8.6. pH measurement is only accurate in highly diluted solutions and the pH of a solution shifts to lower pH at dilution. Thus upon dilution of the stock solution the pH might drop well below 8.5. The pH for the HRP/peroxide/luminol reaction is optimal around pH 8.6–8.8; thus make sure to readjust the pH if necessary.
13. Luminol purity grade of 97% is sufficient for this application. However, recrystallization to a higher purity level may still increase the performance of the self-made ECL. We also replace the 30% H_2O_2 stock solution every year.
14. The pCA-ECL has comparable or higher signal intensity compared to the 4IPBA-ECL but shows slightly higher background chemiluminescence. We also obtained higher signal intensity compared to other commercial ECL solutions (11).
15. We replace the 30% H_2O_2 stock solution yearly.
16. The 4IPBA powder is best replaced yearly.
17. The 4IPBA stock solution degrades to boric acid and iodophenol within minutes when it is diluted in the 100 mM Tris–HCl solution. Some people develop headaches when breathing this iodophenol during the detection of the Western blot.
18. We use this stripping buffer when we want to redetect the blot with other phospho-specific antibodies.
19. The solution turns slightly viscous. If the solution is so viscous that the blot does not freely float around, reduce the amount of SDS.
20. We use this buffer to strip blots if we do not plan to redetect with phospho-specific antibodies any more.
21. 2-mercaptoethanol is toxic and volatile. We recommend using tightly sealable boxes (which can be purchased in regular supermarkets) for the stripping procedure if you do not have an oven installed in a hood.
22. It is best to work on ice in a cold room. If no cold room is available it is best to cover the ice container with the samples with a tray so that the samples remain well cooled during the different incubation times.
23. The specificity of the antibody is paramount for the success of the experiment. Different antibodies have to be tested for their suitability in IP and Western blot. Figure 1 provides an overview of the workflow for validating antibodies.
24. Cytokine receptors are heavily glycosylated which results in highly shifted (in comparison to the expected molecular weight), multiple, and diffuse bands in Western blot detections.

Thus it is hard to interpret if the antibody of choice which is used to detect the cytokine receptor is specific. Alternatively or additionally to an siRNA approach to validate the antibody, the receptor can be de-glycosylated by treatment of the lysates with different glycosidases (see Fig. 2). Endoglycosidase H (EndoH) is a recombinant glycosidase which cleaves the chitobiose core of high mannose and some hybrid oligosaccharides from N-linked glycoproteins but does not release complex oligosaccharides. When proteins are correctly processed through the endoplasmic reticulum (ER) and Golgi, they become resistant to EndoH. Thus treatment of the cytokine receptor immunoprecipitates or of the lysates with EndoH will reveal which protein bands on the Western blot have not yet been processed beyond the ER. PNGaseF (also called N-glycosidase F) is an amidase which cleaves between the innermost GlcNAc and asparagine residues of high mannose, hybrid, and complex oligosaccharides from *N*-linked glycoproteins. Thus treatment of lysates or immunoprecipitates with PNGase F will reveal the native polypeptide size of a protein. The molecular mass of the un-glycosylated cytokine receptor band can now also be compared to the theoretical expected value. All Western blot bands that are not shifted upon the two treatments are nonspecific bands.

25. Control samples showing the specificity of the co-immunoprecipitation should also be included. One important control is to perform the IP protocol with lysate without precipitating antibody. This shows if the putatively "co-precipitated" protein is in fact nonspecifically interacting with the sepharose. A lysate in which the co-precipitated protein is knocked down (e.g., by an siRNA approach) should also be performed. This control will reveal whether the detection antibody (the one which is used in the Western blot detection) is unspecifically detecting another protein of similar migration behavior. This seemingly improbable situation is surprisingly common and has occurred several times in our lab. One explanation for this being so common might be that antibody-providing companies often test their antibodies in the presence of excess peptide (against which the antibody has been generated) as a negative control (personal communication of many sales representatives of different companies). Of course using this peptide in any assay (Western blot detection, immunoprecipitation, or immunofluorescence) as a negative control will compete with the specific and the unspecific signal generated by the antibody (and thus is no real control). Obviously antibodies detecting proteins of the right molecular weight in Western blots will be chosen for sale even though there is unrecognized, nonspecific detection of a protein migrating with a similar mobility (only experiments in cells not expressing this protein of choice can yield information about specificity versus unspecificity of the

antibody). Thus we strongly recommend using knockdown experiments to validate the precipitating and the detecting antibodies. Fortunately, for Janus kinases, negative control cell lines lacking each of the Jaks exist. Jak3 expression is restricted to the hematopoietic system (12) and the fibrosarcoma cell lines U4C, γ2A, and U1D lack Jak1, Jak2, or Tyk2, respectively (4, 13, 14). If a protein cannot be knocked down efficiently it is a good idea to verify co-precipitation with more than one detecting antibody. Last, but not least, it is important to perform Western blots of the lysates from which the immunoprecipitates have been generated in order to compare the lysate expression levels of the same proteins detected in the immunoprecipitation. This might not only reveal differences of expression but also possible degradation problems which might occur in the lysates during immunoprecipitation and which might account for unreproducible results.

26. Janus kinase phosphorylation upon cytokine stimulation can also be determined from the co-immunoprecipitates for some robust cytokine receptor antibodies (e.g., IL2Rβ immunoprecipitates; see Fig. 3). This can be of advantage since many antibodies used to precipitate the Jaks (to investigate their phosphorylation) lead to a strong background phosphorylation of the Jaks upon IP so that the cytokine-induced phosphorylation can barely be seen anymore, although it can clearly and specifically be detected from the lysates (see Fig. 3 and (10)). These problems in detecting activated phosphorylated Jaks from Jak immunoprecipitations might result from autoactivation of the Jaks upon aggregation on the sepharose or it might be that the majority of precipitated Jaks do not associate with the cytokine receptor through which the stimulus has occurred (the four Jaks bind to a much greater variety of cytokine receptors). Moreover a proportion of the Jaks which have been precipitated might not have been bound to cytokine receptors present at the cell surface (as some of the cytokine receptors are present in the ER or Golgi as part of their maturation process and in which Jak binding is involved (reviewed in ref. (8)). However, when precipitating the cytokine receptor complex, the associated Jak is sure to be activated after stimulation through this cytokine receptor. This latter situation reflects the lysate situation much better if the signal intensity is strong enough to be detected reproducibly (which is more difficult in this Co-IP situation). Much depends on the quality of the antibodies used in this case.

27. This only applies to adherent cells. If working with non-adherent cells, pellet the wanted amount of cells by centrifugation at 200 rpm for 5 min. Remove the medium and resuspend the cell pellet in ice-cold lysis buffer.

28. Any method available can be used here.
29. The total amount of lysate should not be used for the immunoprecipitation here, since a gel with the lysates should always be run in parallel to the IP to make sure that expression levels of the different proteins investigated by IP can also be checked in the lysate. This also enables you to check if a given treatment (e.g., cytokine stimulation) was successful, for example, by detecting activated signalling proteins in the lysate by Western blot. This 120 μl of lysate is enough to produce 4–5 Western blots.
30. Protein A fixed to magnetic beads can of course also be used. In this case follow the protocol of the magnetic bead provider to precipitate the immunocomplexes.
31. Take care to always prepare protein A sepharose slurry for a few more samples than you actually have.
32. Protein A binds to Fc-fragments of antibodies. Unfortunately it does not bind all isotypes equally well and even binds some very inefficiently. If you use a precipitating antibody isotype that does not bind protein A, you can still check if protein G (which has a different binding specificity for the different isotypes of antibody) is able to precipitate your antibody. If this also does not work a "bridging" antibody can be used. Antibodies recognizing a certain isotype of antibody (this would be the isotype of your precipitating antibody) can be bought from different providers. Such an antibody can be used to saturate the protein A sepharose (3 μg can be used for this). After washing 3 times with lysis buffer this antibody-bound sepharose A can be used to precipitate the immunocomplexes.
33. Before adding the slurry to the samples shorten the pipette tip so that the orifice is widened somewhat. With some tips the opening is so small that sepharose particle aggregates might block the tip, which leads to unequal amounts of sepharose added, which in turn yields unreproducible results.
34. When performing the experiment for the first time you can keep an aliquot of this supernatant to check if the IP was quantitative. This can be checked by directly performing a Western blot of the supernatant or by performing a second IP with this supernatant. In this way the amount of antibody can be optimized if wanted.
35. The strength of the binding of the co-immunoprecipitated protein and the background of the Western blot detection determine the washing steps. If the background of the blot is high more washing steps are needed. However if the co-precipitated signal is weak the washing steps might better be reduced in case the complex is not very stable. In the same line of thought a higher concentration of detergent in the washing

buffer reduces background while less detergent might enhance the signal intensity of the co-precipitated protein. This is why the composition of the washing buffer might need to be optimized. We recommend testing the lysis buffer as washing buffer first as this works for most situations.

36. Acrylamide is toxic and so are PAGE gels. Also polymerization reactions are not accomplished to 100% and some highly toxic monomers are embedded in the gel matrix. Do not touch PAGE gels without wearing gloves.
37. However, special care has to be taken to ascertain the best possible migration behavior of the samples, e.g., some systems have a tendency to accumulate air bubbles under the gel when mounting the gel. Little things like this can have massive repercussions on the resolution of the protein bands in the gel.
38. This depends on the proteins which should be resolved on the gel. 7.5–10% gels are best for most Janus kinases and cytokine receptors.
39. Best prepare the 20% ammonium persulfate shortly before use.
40. As little ammonium persulfate and TEMED as possible should be used. Excess of either can lead to oxidation of sample proteins or changes in buffer pH, which can affect the resolution of the proteins on the gel. Polymerization should occur within 15–20 min but the time one should allow for the polymerization to be complete is about 2 h. If the gelation takes longer than 20 min to start, the inhibitory effects of atmospheric oxygen will begin to appear and the amounts of peroxide and TEMED should be increased. The gels should be cast at room temperature (23–25°C is optimal for polymerization). It is best to have all materials and solutions at room temperature before starting. Oxygen can serve as an inhibitor since it can function as a free radical trap. This is why degassing the solutions used to produce the gel can help if there are problems of reproducibility.
41. This also protects the polymerizing gel from atmospheric oxygen.
42. Usually the gel is polymerized enough after 15–20 min to remove the isopropanol and then to cast the stacking gel on top of the separating gel. The two gels have polymerized for 1–2 h by the time the whole procedure is finished. Thus one does not have to wait for 2 h before casting the stacking gel.
43. This is a precaution because the presence of isopropanol could lead to precipitation of some proteins.
44. Stacking gels have more contact with oxygen than the isopropanol-overlaid separating gel. This is why we use more initiators for stacking gels.

45. This is important since it removes insoluble debris which can lead to poorer resolution of the samples.
46. Depending on the buffer system the gels are run under very different conditions. This only applies to our Laemmli-type system.
47. Always cut the paper so that the paper area exceeds the gel area so that you do not "lose" lanes when the gel slightly extends during handling. This can happen with gels of low percentage.
48. Handle the membrane only with tweezers or gloves and try to touch it as little as possible. Also take care to label your membranes in one way or another so that blots do not get mixed up. We label our membranes in one corner with a regular pen. However one has to take care that the writing does not dissolve when incubating the membrane with methanol.
49. Always wear gloves when handling the aqueous buffers. All buffers contain 20% methanol.
50. Western blot chambers of a different make can have the anode at the base and the cathode as lid. The stack is then built in the inverse order as described here.
51. Plastic 25 ml pipettes can be shortened (to 15–20 cm depending on the size of the blotting chamber) quite easily to fit this purpose. Unshortened pipettes often hit the outer rim of the chamber resulting in inefficient purging of the air.
52. Be careful not to press too much so that you do not purge the buffer from the stack. If the stack is too dry the gel will shrink during the procedure which will have an impact on the aspect of the bands in the Western blot detection and which will also lead to an inefficient transfer.
53. Always try to blot more than 1 gel to ensure that the top is placed flat on the stack surface.
54. If the chamber does not have a tightening mechanism (e.g., screws) to apply some pressure during the blotting procedure it is best to place a weight on top of the chamber. This serves to counteract the possibility that the gaseous reaction products (O_2 and H_2 which form at the anode and at the cathode, respectively) may interfere with the maintenance of a homogenous electric field across the blotting stack during the blotting procedure. If no pressure is applied, over time these gases may increasingly form bubbles between the layers of the blotting stack.
55. For different proteins optimal blotting times might be different. For high-percentage gels the transfer is slower. Small proteins are transferred faster.
56. Also here alternative settings might work as well.
57. The membrane must stay completely wet at all times!

58. 5% skim milk powder can also be used and is best prepared fresh each day.
59. We reuse these antibody solutions for up to 3 months, depending on the quality of the antibody of course. This does not work for all antibodies. Some are too unstable in high dilution to be used for long periods of time. This has to be determined for each antibody.
60. An overnight incubation gives better results, especially for the detection of the co-precipitated proteins.
61. We do not add NaN_3 (it inhibits the HRP) to the secondary antibody solution and also do not reuse the secondary antibody solution.
62. This guarantees that the blot stays wet during the procedure which is important if you want to strip and re-probe the blot again.
63. Depending on the means of light detection the data are more or less fit for quantitation. Quantitation of ECL signals from exposed films by densitometry has a very low linear range and should really be avoided. Quantitation of CCD camera-generated files can be reliable and the range of linear fit is much greater (15). The activity of, e.g., HRP (the enzyme catalyzing the ECL reaction) changes over time during the detection. Thus the linear range is limited as it is in other assays using enzymes (e.g., ELISA) and thus a standard has to be run on every blot to perform an accurate quantitation. This standard can be a series of dilutions of the detected lysate or IP (e.g., a mix of stimulated versus unstimulated lysate) so that at least a relative quantitation can be performed. Care has to be taken at the start of the experiment to ensure that there is enough material to perform the standard. Furthermore there are many ECL providers and they use different reagents so that different chemical reactions lead to the generation of the light and the linearity and duration of light emission varies from product to product. This leads to inter-laboratory differences that are suboptimal for quantitative data comparison. Thus, ECL is unlikely to become the standard method for Western quantitation in the future.
64. The blots can also be detected using fluorescently labeled secondary antibodies. Blocking buffers and washing steps might have to be optimized for this. Fluorescence detection is the method of choice for quantitation of Western blots as the linear range is far greater than that for ECL. Far infrared fluorophores, which do not interfere with auto-fluorescence of biomolecules (in contrast to many "visible-spectrum" fluorophores), are best suited to weak signals such as those associated with the detection of co-precipitated proteins.

Acknowledgements

This work was supported by the University of Luxembourg grants R1F105L01 and R1F107L01.

References

1. Richter MF, Dumenil G, Uze G, Fellous M, Pellegrini S (1998) Specific contribution of Tyk2 JH regions to the binding and the expression of the interferon alpha/beta receptor component IFNAR1. J Biol Chem 273: 24723–24729
2. Zhao Y, Wagner F, Frank SJ, Kraft AS (1995) The amino-terminal portion of the JAK2 protein kinase is necessary for binding and phosphorylation of the granulocyte-macrophage colony-stimulating factor receptor beta c chain. J Biol Chem 270:13814–13818
3. Chen M, Cheng A, Chen YQ, Hymel A, Hanson EP, Kimmel L, Minami Y, Taniguchi T, Changelian PS, O'Shea JJ (1997) The amino terminus of JAK3 is necessary and sufficient for binding to the common gamma chain and confers the ability to transmit interleukin 2-mediated signals. Proc Natl Acad Sci U S A 94: 6910–6915
4. Kohlhuber F, Rogers NC, Watling D, Feng J, Guschin D, Briscoe J, Witthuhn BA, Kotenko SV, Pestka S, Stark GR, Ihle JN, Kerr IM (1997) A JAK1/JAK2 chimera can sustain alpha and gamma interferon responses. Mol Cell Biol 17:695–706
5. Hilkens CM, Is'harc H, Lillemeier BF, Strobl B, Bates PA, Behrmann I, Kerr IM (2001) A region encompassing the FERM domain of Jak1 is necessary for binding to the cytokine receptor gp130. FEBS Lett 505:87–91
6. Haan C, Is'harc H, Hermanns HM, Schmitz-Van De Leur H, Kerr IM, Heinrich PC, Grotzinger J, Behrmann I (2001) Mapping of a region within the N terminus of Jak1 involved in cytokine receptor interaction. J Biol Chem 276:37451–37458
7. Haan S, Margue C, Engrand A, Rolvering C, Schmitz-Van de Leur H, Heinrich PC, Behrmann I, Haan C (2008) Dual role of the Jak1 FERM and kinase domains in cytokine receptor binding and in stimulation-dependent Jak activation. J Immunol 180:998–1007
8. Haan C, Kreis S, Margue C, Behrmann I (2006) Jaks and cytokine receptors—an intimate relationship. Biochem Pharmacol 72: 1538–1546
9. Haan C, Heinrich PC, Behrmann I (2002) Structural requirements of the interleukin-6 signal transducer gp130 for its interaction with Janus kinase 1: the receptor is crucial for kinase activation. Biochem J 361:105–111
10. Haan C, Behrmann I, Haan S (2010) Perspectives for the use of structural information and chemical genetics to develop inhibitors of Janus kinases. JCMM 14(3):504–527
11. Haan C, Behrmann I (2007) A cost effective non-commercial ECL-solution for Western blot detections yielding strong signals and low background. J Immunol Methods 318:11–19
12. Johnston JA, Kawamura M, Kirken RA, Chen YQ, Blake TB, Shibuya K, Ortaldo JR, McVicar DW, O'Shea JJ (1994) Phosphorylation and activation of the Jak-3 Janus kinase in response to interleukin-2. Nature 370:151–153
13. Guschin D, Rogers N, Briscoe J, Witthuhn B, Watling D, Horn F, Pellegrini S, Yasukawa K, Heinrich P, Stark GR et al (1995) A major role for the protein tyrosine kinase JAK1 in the JAK/STAT signal transduction pathway in response to interleukin-6. EMBO J 14:1421–1429
14. Pellegrini S, Dusanter-Fourt I (1997) The structure, regulation and function of the Janus kinases (JAKs) and the signal transducers and activators of transcription (STATs). Eur J Biochem 248:615–633
15. Dickinson J, Fowler SJ (2002) Quantification of proteins on western blots using ECL. In: Walker JM (ed) The protein protocols handbook, 2nd edn. Humana Press Inc., Totowa, pp 429–437

Chapter 3

In Vitro JAK Kinase Activity and Inhibition Assays

Jeffrey J. Babon and James M. Murphy

Abstract

The discovery that a range of myeloproliferative diseases and leukemias are associated with Janus Kinase (JAK) mutations has highlighted the importance of JAK/STAT signalling in disease and sparked a renewed interest in developing JAK inhibitors. In vitro kinase assays are the most direct and quantitative method to assess mutant forms of JAK for altered enzymatic properties as well as verifying and quantifying the affinity and efficacy of potential inhibitors. Here, we describe protocols for heterologous expression and purification of JAK kinases from insect cells, assays to determine the activity of these purified kinases, and finally inhibition assays to determine the effectiveness of potential inhibitors.

Key words: JAK, Kinase, Enzyme, Inhibition assay, Activation, Enzyme kinetics, JAK1, JAK2, JAK3, TYK2

1. Introduction

Cytokine binding to a specific, cognate cell-surface receptor initiates an intracellular signalling cascade that is driven by activation of a family of receptor-bound tyrosine kinases known as Janus Kinases (JAKs) (1). Once activated, JAKs phosphorylate Signal Transducers and Activators of Transcription (STATs), transcription factors normally sequestered in the cytoplasm (2). Activated STATs dimerize and translocate into the nucleus, where they upregulate transcription of a suite of cytokine-responsive genes (3–5).

There are four mammalian JAKs, JAK1-3, and TYK2, each consisting of four domains (6–8). The N-terminal FERM domain binds to the appropriate membrane-bound receptor whilst the C-terminal kinase (catalytic) domain phosphorylates substrate proteins. Between these are a noncanonical SH2 domain and a pseudokinase domain. Although full-length JAKs are difficult to express using recombinant systems, the kinase domain can be easily expressed and purified using the baculovirus expression system in *Spodoptera frugiperda* (*Sf*21 or *Sf*9) insect cells.

Sandra E. Nicholson and Nicos A. Nicola (eds.), *JAK-STAT Signalling: Methods and Protocols*, Methods in Molecular Biology, vol. 967, DOI 10.1007/978-1-62703-242-1_3, © Springer Science+Business Media New York 2013

A number of different leukemias and myeloproliferative disorders are caused by JAK mutations (9–21). A hallmark of these diseases is that JAK is mutated in such a way as to render it constitutively active. This can be either via *point mutation* or by *oncogenic fusion* of JAK. The end result is that the cell's normal cytokine signalling pathways are activated even in the absence of cytokine (22). This aberrant activation then leads to dysregulated proliferation and disease. Given this, there is intense interest in characterizing the enzymatic properties of mutant forms of JAK (turnover rate, substrate affinity, and ATP affinity) as well as developing and quantifying small-molecule inhibitors of JAKs.

In vitro kinase assays are a valuable method to gain mechanistic insights into both kinase activity and, where inhibitors exist, inhibition. These assays are the only method available to quantify enzymatic parameters such as K_M and k_{cat}. Here, we describe protocols to measure kinase activity (and inhibition) using a radiometric assay. We find this assay to be the most robust and quantitative method to measure kinase activity and inhibition.

2. Materials

1. Plasmid pFastBac HTb (Life Technologies, Carlsbad, CA, USA).
2. Equipment and reagents for PCR (thermocycler, deoxynucleotide triphosphates, high-fidelity and Taq polymerases, DNA templates, oligonucleotide primers).
3. Restriction enzymes (BamHI, NotI, alkaline phosphatase).
4. Equipment for DNA electrophoresis (Tris–acetate buffer, agarose, gel tanks, ethidium bromide, UV transilluminator).
5. Competent *Escherichia coli* cells for cloning (DH10B or TOP10).
6. Luria broth (LB)-agar plates containing 100 μg/mL ampicillin.
7. LB media containing 100 μg/mL ampicillin.
8. DNA miniprep kit.
9. Big Dye Terminator sequencing reagents (Life Technologies/ Applied Biosystems) and oligonucleotide primers for sequencing.
10. DH10Bac chemically competent cells (Life Technologies).
11. 15 mL Round-bottomed polypropylene and polystyrene tubes (BD).
12. Super broth (35 g tryptone, 20 g yeast extract, 5 g NaCl, 2.5 mL 2 M NaOH to 1 L with deionized water).
13. LB-Bac agar plates (LB-agar containing 50 μg/mL Kanamycin, 10 μg/mL Tetracycline, 7 μg/mL Gentamicin, 0.1 μg/mL BluoGal, 40 μg/mL IPTG).

14. LB-Bac media (Luria-Bertani broth containing 50 μg/mL Kanamycin, 10 μg/mL Tetracycline, 7 μg/mL Gentamicin).
15. Solution I (15 mM Tris–HCl pH 8, 10 mM EDTA pH 8, 100 μg/mL RNase A).
16. Solution II (0.2 M NaOH, 1% v/v SDS).
17. Solution III (3 M sodium acetate–acetic acid pH 5.5).
18. Isopropanol.
19. 70% ethanol.
20. TE: 10 mM Tris–HCl, 1 mM EDTA, pH 8.
21. *Spodoptera frugiperda* (*Sf*9 or *Sf*21) insect cells (Life Technologies).
22. Sf900 II SFM insect cell media (Life Technologies).
23. 1 L Schott bottles and 2.8 L Fernbach flasks.
24. Humidified 27°C incubators: stationary and shaking.
25. Equipment for cell culture (Class II Biosafety cabinet, microscope for cell inspection and counting, hemocytometer, slide, and coverslip).
26. Tissue culture-treated 6-well plates (e.g., Corning catalog number 3516).
27. Insect cell transfection reagent (e.g., Life Technologies CellFectin II).
28. 50 mL Polypropylene Falcon tubes.
29. Vented cap 75 and 150 cm^2 cell culture flasks.
30. Cytobuster reagent (Merck, Whitehouse Station, NJ, USA).
31. Equipment for sodium dodecyl sulfate-polyacrylamide gel electrophoresis (SDS-PAGE).
32. Equipment and reagents for Western blotting (transfer apparatus, PVDF membrane, milk powder, anti-His primary antibody, horseradish peroxidase-conjugated secondary antibody, chemiluminescent reagent, device for imaging blot or film, and developer).
33. Lysis Buffer A: 200 mM NaCl, 20 mM HEPES–HCL pH7.5, 5 mM imidazole–HCl buffer pH 7.5, 5% v/v glycerol, 5 mM 2-mercaptoethanol.
34. Phenylmethylsulfonyl fluoride (PMSF) and EDTA-free Complete Protease Inhibitor cocktail tablets (Roche Applied Science, Penzberg, Germany).
35. Wash Buffer B: 200 mM NaCl, 20 mM HEPES–HCl pH 7.5, 35 mM imidazole–HCl pH 7.5, 5% v/v glycerol, 5 mM 2-mercaptoethanol.

36. Elution Buffer C: 200 mM NaCl, 20 mM HEPES–HCl pH 7.5, 250 mM imidazole–HCl pH 7.5, 5% v/v glycerol, 5 mM 2-mercaptoethanol.
37. Sonicator.
38. Peristaltic pump.
39. 1 mL Ni^{2+} immobilized metal affinity chromatography (IMAC) cartridge.
40. 10× kinase buffer: 200 mM Tris pH 8.0, 1 M NaCl, 1 mM DTT, 40 mM $MgCl_2$, 1 mg/mL BSA.
41. 3× kinase buffer (60 mM Tris pH 8.0, 300 mM NaCl, 0.3 mM DTT, 3 mM ATP, 12 mM $MgCl_2$, 0.3 mg/mL BSA, 3 mM peptide substrate).
42. γ-^{32}P-ATP (3000 Ci/mmol, 10 mCi/mL).
43. Tyrosine kinase substrate peptide (e.g., STAT5b: RRAKAADGYVKPQIKQVV).
44. P81 phosphocellulose paper (Whatman, GE Healthcare).
45. Phosphorimager device.
46. Liquid scintillation counter.

3. Methods

The methods described below outline (1) expression of JAK kinase in insect (*Spodoptera frugiperda*) cells, (2) affinity purification of the expressed JAK kinases, (3) assay of JAK activity and (4) inhibition assay of potential JAK inhibitors.

3.1. Expression of JAK in Sf21 Cells

This section describes a convenient procedure for cloning each JAK into the pFastBac HTb vector (Life Technologies) by PCR, restriction digestion, and ligation.

3.1.1. Preparation of Expression Construct

1. The kinase domain boundaries used for each of the four JAKs are as follows (see Note 1):

   JAK2—836-1132 (Genbank: Protein, AAH54807; cDNA, BC054807).

   TYK2—875-1183 (Genbank: Protein, AAH94240; cDNA, BC094240).

   JAK1—861-1153 (Genbank: Protein, EDL30874; cDNA, AK141210).

   JAK3—807-1124 (Genbank: Protein, AAC50950; cDNA, U09607).

Each kinase domain can be conveniently cloned into pFastBac HTb via 5′ BamHI and 3′ NotI restriction sites, as encoded by the following oligonucleotides (mismatches in uppercase):

JAK1 5′ (BamHI site underlined) CGCGGATcCacaacagaggtggaccccactc.

JAK1 3′ (NotI site underlined) ATAAGAATGCGGCCGCttattttaaaagtgcttcaaatccttc.

JAK2 5′ (BamHI site underlined) CGCggATcCtttgaagacagggaccctacacag.

JAK2 3′ (NotI site underlined) ATAAGAATGCGGCCGCtcacgcagctatactgtcccgg.

JAK3 5′ (BamHI site underlined) CGCggATcccagctctatgcctgccaagac.

JAK3 3′ (NotI site underlined) ATAAGAATGCGGCCGctatgaaaaggacagggagtggtg.

TYK2 5′ (BamHI site underlined) CGCggATcCtcggctgtgaactcagactcacc.

TYK2 3′ (NotI site underlined) ATAAGAATGCGGCCGCtcagcacacgctgaacacggaag.

Standard PCR protocols should be employed to amplify each cDNA from its respective template using a high-fidelity DNA polymerase (such as Phusion Hot Start, Finnzymes) and then cloned into the pFastBac HTb vector using standard procedures. Insert DNA sequences should be verified by Big Dye Terminator sequencing before proceeding.

3.1.2. Preparation of Bacmids

This section describes the transformation of the pFastBac expression construct into DH10Bac (Life Technologies) or DH10MultiBac (ATG Biosynthetics, Merzhausen, Germany) chemically competent cells to generate bacmids by recombination. Further details of procedures are described in the Bac-to-Bac (Life Technologies) or MultiBac Turbo (ATG Biosynthetics) product manuals.

1. Aliquot 20–100 μL DH10Bac or DH10MultiBac chemically competent cells, thawed on ice, into precooled 15 mL round-bottomed polypropylene tubes, and incubate with 1 μL of 1 ng/μL pFastBac HTb:JAK miniprep DNA on ice for 30 min.
2. Heat shock cells for 45 s in a 42°C water bath before returning to ice for 2 min.
3. Add 0.8 mL Super broth to transformed cells and recover for 4–5 h at 37°C, 220 rpm before plating 0.3 mL on LB-Bac plates. Culture plates for 36–48 h at 37°C to allow sufficient time for blue pigment to develop in bacmid recombination-negative colonies.

4. Streak selected white colonies on fresh LB-Bac plates and incubate for 36–48 h to ensure that chosen colonies are not contaminated with blue colonies.
5. Inoculate 2.5 mL LB-Bac media with an isolated white colony and culture for 16 h at 37°C, 220 rpm.
6. Transfer 1.5 mL of culture to a microfuge tube and pellet cells at 13,000 × *g*, 1 min using a benchtop microfuge.
7. Completely remove supernatant and resuspend pellet in 0.3 mL ice-cold Solution I by gently pipetting up and down. It is crucial to avoid vortexing at any stage during bacmid purification, since the large bacmid DNA is prone to fracturing.
8. Lyse cells by adding 0.3 mL room-temperature Solution II. Mix by gentle inversion and incubate for <5 min at room temperature.
9. Add 0.3 mL ice-cold Solution III. Mix by inversion and incubate on ice for 10 min.
10. Centrifuge in a benchtop centrifuge (maximum speed, 10 min) to eliminate insoluble material.
11. Pipette 0.75 mL of lysate supernatant into a precooled microfuge tube containing 0.75 mL isopropanol. Mix by inversion and rest on ice for 10–15 min.
12. Pellet bacmid DNA by centrifugation in a benchtop microfuge at 4°C (maximum speed, 30 min).
13. Remove and discard supernatant using a pipette (rather than decanting), taking care not to disturb the DNA pellet.
14. Add 0.5 mL ice-cold 70% ethanol to the tube. Invert tube gently before centrifuging for 5 min in a benchtop microfuge.
15. Using a pipette, remove supernatant, again taking care to avoid disturbing the DNA pellet.
16. Repeat 0.5 mL 70% ethanol wash and spin. Carefully remove supernatant using a pipette again. Spin briefly to collect residue and remove by pipette.
17. Dry DNA pellet at room temperature for 10–15 min. Add 100 μL 1× TE pH 8.0. Since the bacmid is prone to shear damage, do not pipette up and down, but allow pellet to dissolve in TE at room temperature. Typically, 1.5 mL of culture will yield 100 μL of 300–800 ng/μL bacmid DNA. Bacmid DNA should be stored at 4°C or aliquoted and stored at −80°C to avoid DNA fractures that may arise from freeze–thaw cycles.
18. Verify the presence of the JAK cDNA in the bacmid by performing an analytical PCR reaction using 0.5 μL bacmid DNA as the template with the M13 reverse primer and a JAK-specific forward primer.

3.1.3. Preparation of Baculoviruses

This section describes the introduction of bacmid DNA into *Spodoptera frugiperda* (*Sf*9 or *Sf*21) cells and the subsequent preparation of high-infectivity baculoviruses.

1. *Spodoptera frugiperda Sf*9 or *Sf*21 cells can be purchased from Life Technologies and maintained in commercially sourced, complete, serum-free culture media, such as Sf900-II SFM (Life Technologies) or ESF921 (Expression Systems) media. Insect cells are grown in the absence of antibiotics and should be manipulated only in a Biosafety Cabinet under sterile conditions with strict adherence to aseptic technique. All vessels and apparatus used for insect cell culture should be sterile.
2. 50–300 mL *Sf*21 cell cultures, at a cell density of $0.3\text{–}4 \times 10^6$ cells/mL, should be grown in 1 L Schott bottles (or equivalent) in a humidified 27°C incubator shaking at 130 rpm. Typically, a culture at 0.3×10^6 cells/mL will reach a density of 3×10^6 cells/mL after 3 days.
3. For *Sf*21 transfection, 0.9×10^6 cells in a 2 mL volume of Sf900-II SFM should be adhered to each well of a tissue culture-treated 6-well plate (such as Corning Costar catalog number 3516) for 25–30 min in a humidified 27°C incubator.
4. To prepare the bacmid for transfection, add 1 μg of Bacmid DNA to 100 μL of Sf-900 II SFM medium in a 15 mL polystyrene tube. For each transfection, mix 6 μL of an appropriate transfection reagent, such as CellFectin II (Life Technologies) or GeneJuice (Merck), with 100 μL of Sf-900 II SFM medium in a separate 15 mL polystyrene tube, and add 106 μL to each DNA–medium mix. Tap tube gently to mix and incubate at room temperature for 30–45 min.
5. Add 0.8 mL Sf900-II SFM media to each bacmid:transfection reagent mix.
6. Well by well, aspirate media from each well of a 6-well plate and replace with a bacmid mix. It is advisable to add the bacmid mix by slowly dropping on to the cell monolayer to avoid detaching the adhered cells.
7. After 5–6 h in a humidified 27°C incubator, aspirate the supernatant from each cell monolayer and replace with 2 mL Sf900-II SFM. This should be done well by well to avoid cells drying out. Care should be taken to add the fresh media dropwise to avoid detaching cells from the monolayer. Most importantly, care should be taken to avoid cross-contamination between wells.
8. Incubate for 4 days in a humidified 27°C incubator.
9. Harvest P1 virus by pipetting the supernatant from each well of the 6-well plate to labeled 15 mL polystyrene tubes. Centrifuge at $500 \times g$, 5 min at room temperature to remove debris. The supernatant should be transferred to a new 15 mL

polystyrene tube and stored in the dark at 4°C. Under these storage conditions, the P1 virus will retain infectivity for >6 months.

10. High-infectivity virus should be prepared by two amplification passages. Firstly, P2 virus should be prepared by adhering 10^7 *Sf*21 cells (in a 20 mL volume of Sf 900-II SFM media) to a T75 vented cell culture flask (such as Corning) for 25–30 min at 27°C before adding 100 μL P1 virus.
11. After 3 days in a humidified incubator at 27°C, decant supernatant from T75 flask into a 50 mL Falcon tube. Centrifuge at 500×*g* for 5 min at room temperature to eliminate debris. Decant supernatant into a fresh 50 mL Falcon tube and store at 4°C in the absence of light.
12. Prepare P3 virus by adhering 2×10^7 *Sf*21 cells to T150 vented cell culture flasks (such as Corning) in 30 mL Sf 900-II SFM media for 20–30 min at 27°C. To each flask, add a further 20 mL Sf 900-II SFM and 50 μL P2 virus. Return to humidified 27°C incubator.
13. Harvest P3 virus after 3 days by decanting supernatants from T150 flasks into 50 mL Falcon tubes and centrifuging for 5 min at 500×*g* to eliminate debris. Transfer supernatants to sterile vessels for storage at 4°C and wrap in aluminum foil to exclude light. P3 virus stored under these conditions will retain infectivity for >6 months.

3.1.4. Determination of Optimal Conditions for Protein Expression

This section describes a procedure to empirically determine the volumetric ratio of P3 virus to the number of *Sf*21 cells for optimal protein expression.

1. P3 virus titrations are performed in 6-well plates (such as Costar catalog number 3516). 10^6 *Sf*21 cells in 2 mL Sf900-II SFM are adhered for 25–30 min at 27°C.
2. Titrate P3 virus by adding 5, 20, 50, and 100 μL P3 virus to successive wells. Place plate in a humidified 27°C incubator.
3. After 48 h, aspirate supernatant and lyse infected monolayer with 0.3 mL ice-cold Cytobuster reagent (Novagen). Place 6-well plate on ice and pipette reagent up and down over cell monolayer to ensure complete lysis.
4. Transfer lysates to 1.5 mL microfuge tubes and spin at 4°C, 13,000 × *g* to pellet insoluble material.
5. Prepare 10 μL samples of lysate supernatants and pellets (resuspended in 0.3 mL water) and resolve proteins using standard reducing SDS-PAGE.
6. Transfer proteins from SDS-PAGE gel to PVDF membrane using standard techniques and perform Western blot analysis with an anti-His tag antibody. Typically, we observe

optimum soluble JAK yields for infections using 50 μL P3 virus per 10^6 *Sf*21 cells.

3.1.5. Protein Expression

1. Culture 400–500 mL *Sf*21 cells to a density of $2.5–3.5 \times 10^6$ cells/mL in a 2.8 L Fernbach flask at 85–90 rpm, 27°C (see Note 2).
2. Based on the ratio of P3 baculovirus per 10^6 *Sf*21 cells determined to give optimal protein expression in Subheading 3.1.4, calculate and add an appropriate volume of P3 baculovirus to the Fernbach flask.
3. Allow infection and JAK expression to occur for 46–50 h at 27°C in a humidified shaking (8–90 rpm) incubator.
4. Pellet cells by centrifugation at $500 \times g$ for 5 min.
5. Snap-freeze pellets in liquid nitrogen and store at −80°C until use.

3.2. Purification of JAK from Sf21 Cells

Note: The cloning strategy outlined in Subheading 3.1.1 results in an N-terminal His_6 tag. This section describes the purification of His_6-tagged JAKs (see Note 3).

1. Thaw cell pellets in a water bath at room temperature. Resuspend in Lysis Buffer A (supplemented with 1 mM PMSF and EDTA-free Complete Protease Inhibitor cocktail (Roche)), using 5 mL of lysis buffer for every pellet from 100 mL *Sf*21 culture. Pool cell suspension in a single 50 mL Falcon tube.
2. Lyse the cell suspension by sonication using 6×10s bursts with a 10-s rest between each burst. Use a moderate power level and perform in an ice/water bath to ensure that the temperature of the lysate is maintained at 4°C.
3. Pellet the insoluble material by centrifugation ($>20{,}000 \times g$, 30 min, 4°C). Clarify supernatant by passing through a syringe-driven 0.45 μm filter and collecting the filtered material.
4. Connect a 1 mL Ni^{2+} IMAC cartridge to a peristaltic pump, such as a Gilson Minipuls3, and set up in a cold room (see Note 4). Wash cartridge with 5–10 mL Lysis Buffer A at a flow rate of 1 mL/min. Suitable Ni^{2+} cartridges include HiTrap Chelating SP (GE Healthcare) charged with 0.1 M $NiCl_2$, HisTrap HP (GE Healthcare), or NiMAC (Novagen).
5. Pass clarified lysate over Ni^{2+} IMAC cartridge via peristaltic pump at a flow rate of ~1 mL/min.
6. Wash the Ni^{2+} IMAC cartridge with 7–10 mL of Lysis Buffer A via peristaltic pump at 1 mL/min flow rate.
7. Repeat wash with 7–10 mL Wash Buffer B.
8. Elute His_6-JAK by passing 5–6 mL of Elution Buffer C through Ni^{2+} IMAC cartridge via peristaltic pump at a flow rate of

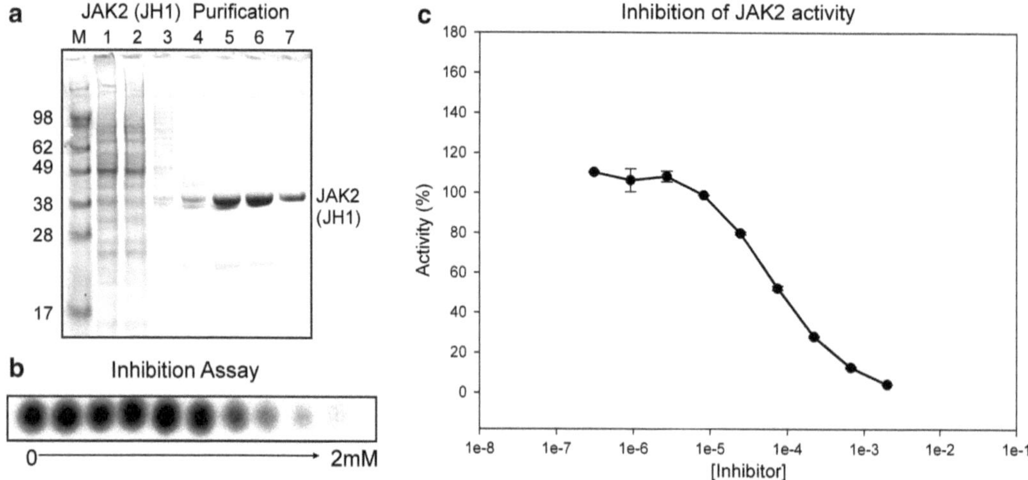

Fig. 1. Purification of JAK2 kinase domain and its use in in vitro inhibition assays. (**a**) SDS-PAGE analysis of JAK2 (JH1 domain) purified from insect cells. *Lanes 1–4*: Whole cell lysate, soluble cell lysate, imidazole wash, and imidazole elution fractions. *Lanes 5–7* Gel filtration fractions. (**b**) In vitro Inhibition assay of JAK2 using ADP performed according to Subheading 3.4. The ADP concentration was varied from 0 to 2 mM (left to right). (**c**) Analysis of the Inhibition assay performed in (**b**). The IC_{50} of the inhibitor used (ADP) can be read off the plot as the concentration required to give 50% inhibition (100 μM in this case).

~1 mL/min and collecting eluate in a 50 mL Falcon tube containing 25 mL Lysis Buffer A (see Note 5).

9. Concentrate the supernatant to 0.5 mL by ultrafiltration using a 10 kDa MWCO spin concentrator (see Note 6).
10. Further purify by gel filtration using a Superdex 200 10/300 column. Use Tris-Buffered Saline (pH 7.5) as the running buffer (see Note 7).
11. Analyze fractions by SDS-PAGE, pool desired fractions, snap-freeze in liquid nitrogen, and store at −80°C until required for use (see Fig. 1a).

3.3. JAK Activity Assays

The first protocol describes the determination of K_M and V_{max} for the purified kinases. These assays use a synthetic peptide that contains only a single tyrosine as a substrate. Ideally this will be based on known JAK substrates (for example the STAT5b(693-708) peptide: RRAKAADGYVKPQIKQVV). See Note 8. These reactions contain two substrates: ATP and peptide. To determine $K_M^{peptide}$ and $V_{max}^{peptide}$ the ATP concentration will be held constant and in excess (2 mM) whilst the peptide concentration is varied (covered in Subheading 3.3.1). Conversely, to determine K_M^{ATP} and V_{max}^{ATP} the peptide concentration will be held constant at (4 mM) and in excess whilst the ATP concentration is varied (covered in Subheading 3.3.2).

3.3.1. JAK Activity Assay (To determine $K_m^{peptide}$ and $V_{max}^{peptide}$)

This protocol results in ten individual reactions, each with a different peptide substrate concentration. Each reaction will be performed in a single well of a microtiter plate in a total volume of 15 μl with the aid of a multichannel pipette. Therefore, care must be taken with all pipetting steps and the reactions designed such that 5 μl is the lowest volume pipetted into the final reaction. Even with these precautions, and a state-of-the-art multichannel pipette, volumetric errors are likely to be the biggest source of deviation. It is recommended that these reactions be performed in duplicate so that any errors can be estimated. The final composition of each reaction will be as follows:

30 mM Tris–HCl pH 8.0.

100 mM NaCl.

0.1 mg/mL BSA.

2 mM ATP (see Note 9).

4 mM $MgCl_2$ (see Note 10).

1 μCi γ-^{32}P-ATP.

0.1 mM DTT.

20 nM JAK enzyme.

0–4 mM peptide substrate.

1. Prepare 10× kinase assay buffer.
2. Prepare 1 ml of 30 mM ATP using ATP-Na_2 salt (powder) stored at −20°C. This must be performed immediately prior to the assay setup.
3. Prepare 100 μl of 12 mM peptide by dissolving lyophilized peptide in 30 mM Tris–HCl, pH 8.0. Ensure that the pH of the final solution is 7–9 (see Note 11).
4. Using a 96-well microtiter plate, make a series of ten twofold serial dilutions (using 30 mM Tris–HCl, pH 8.0) of peptide substrate. The final volume in each well should be 5 μl (see Note 12).
5. Prepare sufficient mastermix for 12 reactions (18 μl 10× kinase buffer, 12 μl 30 mM ATP, 1.2 μl γ-^{32}P-ATP, 28.8 μl H_2O). From this step you must exercise radioactive safety precautions and perform all manipulations behind a perspex screen. Add 5 μl of mastermix to peptide substrate.
6. The reaction is started by the addition of enzyme (see Note 13). Therefore enzyme (JAK, 60 nM) should be pre-aliquoted into ten empty wells in the microtiter plate and allowed to equilibrate to room temperature. Start all reactions simultaneously by transferring 5 μl of JAK into wells containing substrate/mastermix with a multichannel pipette. Mix well by pipetting up and down ten times. Cover plate with clear tape to prevent evaporation.

7. Allow the reaction to proceed for 10 min, then, using a multi-channel pipette, spot 4 μl of each reaction onto P81 phosphocellulose paper, and immediately place paper into 100 ml 5% H_3PO_4. The paper should be marked so that the location of each spot will be known.
8. Take a second time-point at 20 min by repeating step 9 (see Note 14).
9. Wash P81 paper three more times using 100 ml 5% H_3PO_4, 20 min each time. Perform a final 10-s wash in Acetone and allow paper to air-dry.
10. Expose P81 paper to a phosphorimager plate overnight and scan.
11. Excise each spot individually and measure incorporated radioactivity by scintillation counting (see Note 15).
12. Plot Incorporated radioactivity vs. (Substrate) and fit to the Michaelis–Menten equation using software such as Sigmaplot.
13. Measure total radioactivity by scintillation counting of 4 μl of a 1:100 dilution of the remaining reaction mixture. The equation

$$\frac{\text{Incorporated radioactivity}}{\text{Total radioactivity}(\times 100)}[\text{ATP}](2\times 10^{-3}\,\text{M})$$
$$\text{Reaction volume}(15\times 10^{-6}\,\text{L})$$

yields the total amount of phosphoproduct formed (in moles, see Note 16). Dividing this number by the number of moles of JAK enzyme present in the reaction and again by reaction time, in seconds, yields the absolute turnover rate (s^{-1}) or k_{cat} when the substrate concentration is saturating.

3.3.2. JAK Activity Assay (To determine K_M^{ATP} and V_{max}^{ATP})

This is essentially identical to Subheading 3.3.1 except that the ATP concentration is varied (0–2 mM) whilst the substrate peptide concentration is held constant (4 mM). Note that the specific activity of ATP (in terms of Ci/mmol) per reaction remains constant.

1. Follow Subheading 3.3.1, steps 1–4.
2. Prepare 20 μl of 6 mM ATP (containing 1 μl of γ-^{32}P-ATP) and make a series of twofold serial dilutions using a 96-well microtiter plate. The final dilution should contain 20 mM Tris–HCl pH 8.0 and 0 mM ATP. From this step you must exercise radioactive safety precautions and perform all manipulations behind a perspex screen.
3. Follow Subheading 3.3.1, step 6.

4. Prepare sufficient mastermix for 12 reactions (18 μl 10× kinase buffer, 42 μl 17 mM peptide). Add 5 μl of mastermix to wells containing substrate.
5. Follow Subheading 3.3.1, steps 8–13.
6. Plot Incorporated radioactivity vs. (Substrate) and fit to the Michaelis–Menten equation using software such as Sigmaplot.
7. Follow Subheading 3.3.1, step 15.

3.4. JAK Inhibition Assays

This protocol describes the determination of IC_{50} for potential JAK inhibitors (see Note 17; Fig. 1b, c). These assays are similar to those detailed in the previous sections except that each reaction is performed using a constant concentration of ATP and substrate and a variable concentration of inhibitor. The final composition of each reaction will be as follows:

30 mM Tris–HCl pH = 8.0.

100 mM NaCl.

0.1 mg/mL BSA.

2 mM ATP (see Note 18).

4 mM $MgCl_2$.

1 μCi γ-^{32}P-ATP.

0.1 mM DTT.

20 nM JAK enzyme.

1 mM peptide substrate.

Variable inhibitor concentration.

1. Add 0.1 μl γ-^{32}P-ATP per reaction to 3× kinase buffer (see Notes 19 and 20).
2. Using a 96-well microtiter plate, make a series of ten twofold serial dilutions (using 30 mM Tris–HCl, pH 8.0) of inhibitor. The final volume in each well should be 5 μl (see Note 21).
3. Add 5 μl of kinase buffer from step 1 to each well.
4. The reaction is started by the addition of enzyme. Therefore enzyme (JAK, 60 nM) should be pre-aliquoted into ten empty wells in the microtiter plate and allowed to equilibrate to room temperature. Start all reactions simultaneously by transferring 5 μl of JAK into wells containing substrate/mastermix with a multichannel pipette. Mix well by pipetting up and down ten times. Cover plate with clear tape to prevent evaporation.
5. Follow Subheading 3.3.1, steps 9–12.
6. Plot incorporated radioactivity vs. log (inhibitor concentration). The inflection point of the curve yields IC_{50}.

4. Notes

1. We routinely prepare the kinase domains of all four JAKs using the described method. The poor expression and low yields of soluble, full-length JAKs preclude their preparation by the same method at present.
2. Expressing active JAK results in relatively poor yields (ca. 0.5 mg/L). A standard method to increase yield is to add a JAK inhibitor to the expression media, for example the ATP-competitive inhibitor CMP-6 (Merck). Whilst this increases yield by tenfold it is, in our hands, impossible to remove post-purification and will therefore interfere with downstream assays.
3. A modified vector that encodes an N-terminal Glutathione-S transferase (GST) can also be used and the GST-JAK expressed under identical conditions to those described above. In that case use standard GST purification protocols to purify the GST-JAK fusion protein.
4. A column bed can be prepared from Ni-NTA resin in place of a prepacked cartridge. Care should be taken to avoid exposure of the His_6-JAK to the high imidazole concentration of Elution Buffer C for long periods of time, such as when gravity flow is used for elution.
5. We have observed that high concentrations of imidazole may compromise JAK solubility and immediate dilution of the eluate limits protein precipitation.
6. ATP, like ATP-competitive inhibitors, helps to stabilize JAK by binding to the active site and inducing a more "closed" conformation. In the absence of ATP, JAK is liable to aggregate upon concentration. We find this especially problematic with JAK1. Hence, at this point, ATP and $MgCl_2$ can be added to JAK (final concentrations 1 mM and 2 mM, respectively) to help stabilize it.
7. After gel-filtration chromatography JAK should be >95% pure. If further purification is required then anion-exchange chromatography using a Mono-Q column can be employed. In this case Tris–HCl pH 8.5 should be used as the running buffer and the enzyme eluted with a gradient of 0–500 mM NaCl.
8. The two arginine residues are placed at the N-terminus so that the peptide will be positively charged at acidic pH and thus easily separated from nucleotides and free phosphate by adsorption onto P81 phosphocellulose paper. They can also be placed at the C-terminus if preferred.
9. We have observed that JAK begins to lose activity in concentrations of ATP greater than 2 mM.

10. It is likely that JAK binds ATP concurrently with two magnesium ions; therefore the ATP:Mg^{2+} ratio should be kept at 1:2.
11. Be careful to ensure that the pH is correct; we observe significant residual Trifluoroacetic acid and Acetonitrile in peptides from a variety of suppliers which will inhibit JAK at high concentrations. Lyophilization of the peptide from purified water will usually resolve this problem.
12. In our hands the use of a microtiter plate and multichannel pipette yields the greatest accuracy and allows the assay to be performed in a high-throughput fashion. However, many other formats also yield acceptable results.
13. Thaw and dilute JAK immediately prior to performing the assay(s). We see significant loss of activity over 24–48 h.
14. Performing two time-points allows the researcher to determine whether activity is linearly proportional to time. This condition is necessary to determine kinetic parameters such as K_M and k_{cat}. It is not strictly necessary for determining the IC_{50} of inhibitors.
15. Scintillation counting allows the incorporation of radioactivity to be correlated to an absolute quantitation of product formation; however this information is only required for determination of k_{cat} and not K_M and can hence be omitted if desired.
16. The ×100 factor is as a result of diluting the 4 μl of reaction mix by 100 before measuring via scintillation counting.
17. Proteins can be tested for inhibitory activity using this assay, as well as small molecules. We assay the inhibitory activity of Suppressors of Cytokine signalling (SOCS) proteins using this methodology (23). In this case the assay is a useful technique to determine whether a protein is a direct inhibitor of JAK.
18. Do not exceed 2 mM ATP if 4 mM $MgCl_2$ is used. We observe inhibition of enzyme activity when higher ATP concentrations are used. This may be due to ATP rather than MgATP binding to the active site which forms a nonproductive complex.
19. If the JAK inhibitor being tested is ATP competitive then a lower concentration of ATP (100 μM) can be used which will yield more sensitive results.
20. By keeping the ATP and substrate concentrations saturating and ensuring that time-points are taken where product formation is linear with time the IC50 values obtained using this methodology will be a good approximation of the true Ki value, at least for a noncompetitive inhibitor. However in many instances an IC_{50} determination will be sufficient and the concentrations of ATP and substrate can be varied at will.
21. The presence of DMSO is very often necessary to ensure that small molecule inhibitors are soluble. If this is required ensure that the concentration of DMSO in the final reaction is <5%.

Acknowledgements

This work was made possible through Victorian State Government Operational Infrastructure Support and the Australian Government NHMRC IRIISS. This research was supported by an NHMRC Program Grant (461219), NIH Grant (CA022556), NHMRC CDA (JJB, 516777), ARC Future Fellowship (JMM, FT100 100100), and NHMRC Project Grant (1011804).

References

1. Wilks AF, Harpur AG (1994) Cytokine signal transduction and the JAK family of protein tyrosine kinases. Bioessays 16:313–320
2. Shuai K, Ziemiecki A, Wilks AF, Harpur AG, Sadowski HB, Gilman MZ, Darnell JE (1993) Polypeptide signalling to the nucleus through tyrosine phosphorylation of Jak and Stat proteins. Nature 366:580–583
3. Schindler C, Shuai K, Prezioso VR, Darnell JE Jr (1992) Interferon-dependent tyrosine phosphorylation of a latent cytoplasmic transcription factor. Science 257:809–813
4. Shuai K, Schindler C, Prezioso VR, Darnell JE Jr (1992) Activation of transcription by IFN-gamma: tyrosine phosphorylation of a 91-kD DNA binding protein. Science 258: 1808–1812
5. Shuai K, Horvath CM, Huang LH, Qureshi SA, Cowburn D, Darnell JE Jr (1994) Interferon activation of the transcription factor Stat91 involves dimerization through SH2-phosphotyrosyl peptide interactions. Cell 76:821–828
6. Wilks AF, Harpur AG, Kurban RR, Ralph SJ, Zurcher G, Ziemiecki A (1991) Two novel protein-tyrosine kinases, each with a second phosphotransferase-related catalytic domain, define a new class of protein kinase. Mol Cell Biol 11:2057–2065
7. Harpur AG, Andres AC, Ziemiecki A, Aston RR, Wilks AF (1992) JAK2, a third member of the JAK family of protein tyrosine kinases. Oncogene 7:1347–1353
8. Velazquez L, Fellous M, Stark GR, Pellegrini S (1992) A protein tyrosine kinase in the interferon alpha/beta signaling pathway. Cell 70:313–322
9. James C, Ugo V, Le Couedic JP, Staerk J, Delhommeau F, Lacout C, Garcon L, Raslova H, Berger R, Bennaceur-Griscelli A, Villeval JL, Constantinescu SN, Casadevall N, Vainchenker W (2005) A unique clonal JAK2 mutation leading to constitutive signalling causes polycythaemia vera. Nature 434: 1144–1148
10. Scott LM, Tong W, Levine RL, Scott MA, Beer PA, Stratton MR, Futreal PA, Erber WN, McMullin MF, Harrison CN, Warren AJ, Gilliland DG, Lodish HF, Green AR (2007) JAK2 exon 12 mutations in polycythemia vera and idiopathic erythrocytosis. N Engl J Med 356:459–468
11. Bercovich D, Ganmore I, Scott LM, Wainreb G, Birger Y, Elimelech A, Shochat C, Cazzaniga G, Biondi A, Basso G, Cario G, Schrappe M, Stanulla M, Strehl S, Haas OA, Mann G, Binder V, Borkhardt A, Kempski H, Trka J, Bielorei B, Avigad S, Stark B, Smith O, Dastugue N, Bourquin JP, Tal NB, Green AR, Izraeli S (2008) Mutations of JAK2 in acute lymphoblastic leukaemias associated with Down's syndrome. Lancet 372:1484–1492
12. Campbell PJ, Scott LM, Buck G, Wheatley K, East CL, Marsden JT, Duffy A, Boyd EM, Bench AJ, Scott MA, Vassiliou GS, Milligan DW, Smith SR, Erber WN, Bareford D, Wilkins BS, Reilly JT, Harrison CN, Green AR (2005) Definition of subtypes of essential thrombocythaemia and relation to polycythaemia vera based on JAK2 V617F mutation status: a prospective study. Lancet 366:1945–1953
13. Baxter EJ, Scott LM, Campbell PJ, East C, Fourouclas N, Swanton S, Vassiliou GS, Bench AJ, Boyd EM, Curtin N, Scott MA, Erber WN, Green AR (2005) Acquired mutation of the tyrosine kinase JAK2 in human myeloproliferative disorders. Lancet 365: 1054–1061
14. Kralovics R, Passamonti F, Buser AS, Teo SS, Tiedt R, Passweg JR, Tichelli A, Cazzola M, Skoda RC (2005) A gain-of-function mutation of JAK2 in myeloproliferative disorders. N Engl J Med 352:1779–1790
15. Levine RL, Wadleigh M, Cools J, Ebert BL, Wernig G, Huntly BJ, Boggon TJ, Wlodarska I,

Clark JJ, Moore S, Adelsperger J, Koo S, Lee JC, Gabriel S, Mercher T, D'Andrea A, Frohling S, Dohner K, Marynen P, Vandenberghe P, Mesa RA, Tefferi A, Griffin JD, Eck MJ, Sellers WR, Meyerson M, Golub TR, Lee SJ, Gilliland DG (2005) Activating mutation in the tyrosine kinase JAK2 in polycythemia vera, essential thrombocythemia, and myeloid metaplasia with myelofibrosis. Cancer Cell 7:387–397

16. Mullighan CG, Zhang J, Harvey RC, Collins-Underwood JR, Schulman BA, Phillips LA, Tasian SK, Loh ML, Su X, Liu W, Devidas M, Atlas SR, Chen IM, Clifford RJ, Gerhard DS, Carroll WL, Reaman GH, Smith M, Downing JR, Hunger SP, Willman CL (2009) JAK mutations in high-risk childhood acute lymphoblastic leukemia. Proc Natl Acad Sci U S A 106:9414–9418
17. Walters DK, Mercher T, Gu TL, O'Hare T, Tyner JW, Loriaux M, Goss VL, Lee KA, Eide CA, Wong MJ, Stoffregen EP, McGreevey L, Nardone J, Moore SA, Crispino J, Boggon TJ, Heinrich MC, Deininger MW, Polakiewicz RD, Gilliland DG, Druker BJ (2006) Activating alleles of JAK3 in acute megakaryoblastic leukemia. Cancer Cell 10:65–75
18. Cornejo MG, Kharas MG, Werneck MB, Le Bras S, Moore SA, Ball B, Beylot-Barry M, Rodig SJ, Aster JC, Lee BH, Cantor H, Merlio JP, Gilliland DG, Mercher T (2009) Constitutive JAK3 activation induces lymphoproliferative syndromes in murine bone marrow transplantation models. Blood 113:2746–2754
19. Lacronique V, Boureux A, Valle VD, Poirel H, Quang CT, Mauchauffe M, Berthou C, Lessard M, Berger R, Ghysdael J, Bernard OA (1997) A TEL-JAK2 fusion protein with constitutive kinase activity in human leukemia. Science 278:1309–1312
20. Reiter A, Walz C, Watmore A, Schoch C, Blau I, Schlegelberger B, Berger U, Telford N, Aruliah S, Yin JA, Vanstraelen D, Barker HF, Taylor PC, O'Driscoll A, Benedetti F, Rudolph C, Kolb HJ, Hochhaus A, Hehlmann R, Chase A, Cross NC (2005) The t(8;9)(p22;p24) is a recurrent abnormality in chronic and acute leukemia that fuses PCM1 to JAK2. Cancer Res 65:2662–2667
21. Bousquet M, Quelen C, De Mas V, Duchayne E, Roquefeuil B, Delsol G, Laurent G, Dastugue N, Brousset P (2005) The t(8;9)(p22;p24) translocation in atypical chronic myeloid leukaemia yields a new PCM1-JAK2 fusion gene. Oncogene 24:7248–7252
22. Dusa A, Mouton C, Pecquet C, Herman M, Constantinescu SN (2010) JAK2 V617F constitutive activation requires JH2 residue F595: a pseudokinase domain target for specific inhibitors. PLoS One 5:e11157
23. Babon JJ, Kershaw NJ, Murphy JM, Varghese L, Laktyushin A, Lucet IS, Norton RS, Nicola NA (2012) Immunity 36:239–250

Chapter 4

Quantitative Analysis of JAK Binding Using Isothermal Titration Calorimetry and Surface Plasmon Resonance

Jeffrey J. Babon

Abstract

Janus Kinases (JAKs) are the key effector kinases that initiate intracellular signalling cascades in response to cytokines and growth factors. As such, a large number of cytoplasmic proteins interact with JAKs both as substrates and as components of regulatory machinery designed to ensure correct activation and termination of signalling. In vitro techniques such as Isothermal Titration Calorimety and Surface Plasmon Resonance are valuable methods to verify and quantify the interaction between JAK and potential binding partners or substrates. Here we describe protocols that exploit both of these in vitro techniques in order to more fully understand the intracellular JAK signalling network.

Key words: Isothermal titration calorimetry, Surface plasma resonance, JAK, JH1, Binding

1. Introduction

Cytokine signalling cascades are driven intracellularly by the activation of receptor bound Janus Kinases (JAKs) (1–3). By virtue of their catalytic kinase domain, JAKs are able to tyrosine phosphorylate intracellular substrates (4, 5). The major substrates of JAK-catalyzed phosphorylation are the intracellular domains to which the JAKs are scaffolded and subsequently, the Signal Transducer and Activator of Transcription (STAT) transcription factors that dock onto the cytokine receptors once they are phosphorylated. STATs are the direct downstream targets of activated JAKs and, once activated, are able to translocate into the nucleus and drive the appropriate biological response by up-regulating the transcription of cytokine-responsive target genes (6). In addition to these well-characterized components of the JAK/STAT pathway there have also been a large number of other intracellular proteins implicated as substrates for JAK-catalyzed phosphorylation (7–9). Enzyme–substrate interactions usually display fast-on/fast-off-type kinetics

Sandra E. Nicholson and Nicos A. Nicola (eds.), *JAK-STAT Signalling: Methods and Protocols*, Methods in Molecular Biology, vol. 967, DOI 10.1007/978-1-62703-242-1_4, © Springer Science+Business Media New York 2013

and hence most proteins that are substrates of JAK phosphorylation would not be expected to form a stable, long-lived complex with the kinase. Another class of proteins that engage in transient interactions with JAK are phosphatases which inactivate JAK by dephosphorylating its activation loop. These include SHP-1, SHP-2, PTB-1B, TCPTP, and CD45 (10–13). In these situations JAK is acting as the substrate, rather than the enzyme; however once again a stable interaction is not required for catalysis to occur.

When identifying and validating components of the JAK interactome, a distinction needs to be made between proteins that are substrates of JAK-catalyzed phosphorylation, such as those listed above, and those that engage in a stable interaction with the kinase. This protocol is designed for use in the discovery and validation of this latter class. Examples of proteins that form a stable interaction with JAK include regulatory proteins such as the Suppressors of Cytokine Signalling (SOCS) family and the cytokine receptor family. SOCS proteins bind JAK and then either inhibit their activity or catalyze their ubiquitination (14–16). Importantly, they are not usually substrates for JAK-catalyzed phosphorylation. Cytokine receptors on the other hand represent an example of a class of proteins that are both substrates and binding partners. These receptors act as scaffolds to facilitate JAK auto-activation (in *trans*); however they are also the first targets of JAK-catalyzed phosphorylation following cytokine stimulation (17).

When identifying JAK-binding proteins, standard approaches such as co-immunoprecipitation are a useful first step. However such methodologies cannot distinguish between proteins that bind directly to JAK and those that bind via a third party. Due to the large number of proteins which bind the intracellular domain of cytokine receptors this has resulted in an artificially large set of JAK interacting proteins being identified in the literature. In vitro analysis of the interaction between JAK and a potential binding partner, using purified proteins, is one of the few ways that this question can be resolved. Of the large number of potential in vitro methodologies that can be exploited for this purpose, two of the most powerful are Isothermal Titration Calorimety (ITC) and Surface Plasmon Resonance (SPR). These techniques are capable of both verifying that the interaction is direct and quantifying the affinity of the interaction. ITC is an equilibrium analysis (18, 19) and can directly measure the affinity, stoichiometry, and enthalpy of an interaction and indirectly measure its entropy. In contrast, SPR is a kinetic analysis (20). It can also measure affinity and in addition measure the on- and off-rates of complex formation. When used in combination, ITC and SPR provide a nearly complete description of a bimolecular interaction.

This chapter provides protocols for characterizing a potential JAK interaction using both ITC and SPR.

2. Materials

The most important requirements for these protocols are access to ITC and SPR platforms. We use a Microcal omega VP-ITC microcalorimeter (MicroCal Inc., Northampton, MA) and a Biacore 3000 Biosensor (GE Healthcare), respectively, for these applications and hence these protocols are written with those particular platforms in mind.

ITC Buffer: 150 mM NaCl, 20 mM HEPES pH 7.5, 5 mM 2-mercaptoethanol (see Note 1).

1. 10 kDa MWCO dialysis tubing.
2. 5% Deconex detergent solution.
3. Purified JAK protein expressed as a GST-fusion (see Note 2).
4. Purified or semi-purified "test" protein.
5. Biacore CM5 sensor chip.
6. Anti-GST antibody.
7. *N*-Hydroxysuccinimide (*NHS*) (Sigma).
8. Ethyl(dimethylaminopropyl) carbodiimide (*EDC*) (Sigma).
9. Ethanolamine (Sigma).
10. Biacore Running Buffer: 150 mM NaCl, 20 mM HEPES pH 7.5, 0.5 mM EDTA, 0.01% Tween-20.
11. Biacore Regeneration Buffer (10 mM Glycine pH 2.0).

3. Methods

The methods described below outline (1) Equilibrium Analysis of JAK binding by ITC and (2) Kinetic Analysis of JAK binding by SPR. We also include tips on analysis of the raw data from these methodologies.

3.1. Equilibrium Analysis of JAK Binding by Isothermal Titration Calorimetry

This experiment requires 2 mL of 10 μM purified JAK (JH1 domain only, see Note 2) and 500 μl of 100 μM purified binding protein (ligand); hence this technique uses large amounts of protein compared to SPR. The most important factor in a successful ITC experiment is that the JAK and ligand must be in identical buffers as any pH difference can give rise to a large heat of neutralization which will obscure the binding signal. The most robust method of ensuring that the buffers are identical is to dialyze the JAK and ligand in separate dialysis bags in the *same* beaker of buffer overnight (see Note 3). The other key requirement for a successful ITC experiment is accurate quantitation of the concentration of both JAK and ligand (see Notes 4 and 5).

1. Prepare 3 L of ITC buffer.
2. Place 2.5 mL of 10 μM JAK into a 10 kDa MWCO dialysis bag. Fasten tightly with dialysis clips and place into the 3 L of buffer.
3. Place 0.7 mL of 100 μM ligand (test protein) into a dialysis bag of the appropriate MWCO. Fasten tightly with dialysis clips and place into the 3 L of buffer (see Note 6).
4. Allow the proteins to dialyze overnight at 4 °C with constant stirring.
5. Remove samples from dialysis bags and place into separate micro-centrifuge tubes.
6. Centrifuge at 13,000 × *g* for 10 min to remove any precipitate.
7. De-gas JAK and ligand under vacuum for 10 min with stirring (see Note 7).
8. Set calorimeter thermostat to 25 °C.
9. Place approximately 2 mL of JAK into cell chamber using a long-needled syringe. Remove any excess material that is above the neck of the cell chamber.
10. Fill the ITC syringe with ligand solution. Direct the ITC software to purge and refill syringe to remove any trapped air bubbles.
11. Place syringe (containing ligand) into cell (containing JAK).
12. Direct software to equilibrate with a syringe stirring speed of 330 rpm and then perform an initial 5 μl injection of ligand into the cell. This injection will be ignored in the final data analysis (see Note 8). A further 30 × 10 μl injections of ligand into the cell, with 3-min intervals, is then performed to yield the final raw data.
13. Steps 7–12 should be repeated using ITC buffer instead of JAK in the cell; this is the buffer control run and acts as a baseline for any buffer differences or heat of dilution effects that may affect the signal.

3.1.1. Analysis of ITC Data

1. Data is best analyzed using the Origin™ Software (MicroCal).
2. Use Origin™ to calculate the binding isotherm from the raw data; it may be necessary to subtract the signal from each of the injections in the blank (buffer control) run if the latter gave a significant signal.
3. Use the least-squares fitting module in the origin software to fit the data to a one- or multisite analysis as appropriate. The software will output the K_d, stoichiometry, enthalpy, and entropy of the interaction.

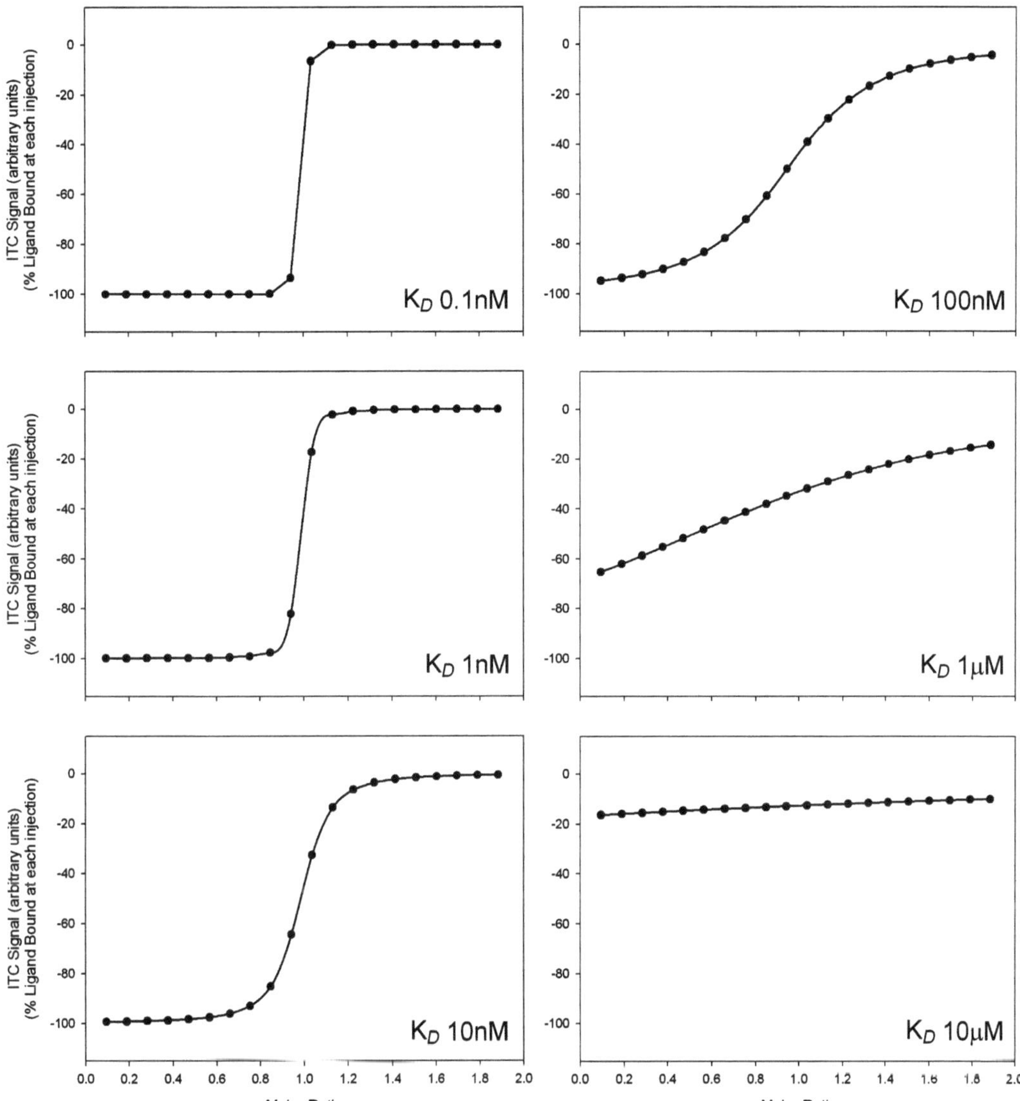

Fig. 1. Simulated ITC data for interactions with a range of affinities. ITC curves were simulated for a range of different affinities that cover most biological interactions (10^{-10} to 10^{-5} M). Interactions weaker than ~1 μM or stronger than ~10 nM would not be accurately quantifiable unless the concentrations of protein and ligand were increased or decreased, respectively. The concentrations of protein and ligand used in this simulation were 10 and 100 micromolar respectively.

4. Examine both the outputted K_d data and the binding isotherm (see Note 9). This method of data analysis is only accurate when the binding isotherm is obviously sigmoidal (S-shaped, see Fig. 1). If this is not the case then the experiment should be repeated with tenfold higher concentration of both JAK and ligand (see Note 10). For best results the equation is $K_d \times C = 10–50$ (where C is the concentration of JAK). Simulated ITC results for several binding affinities are shown in Fig. 2.

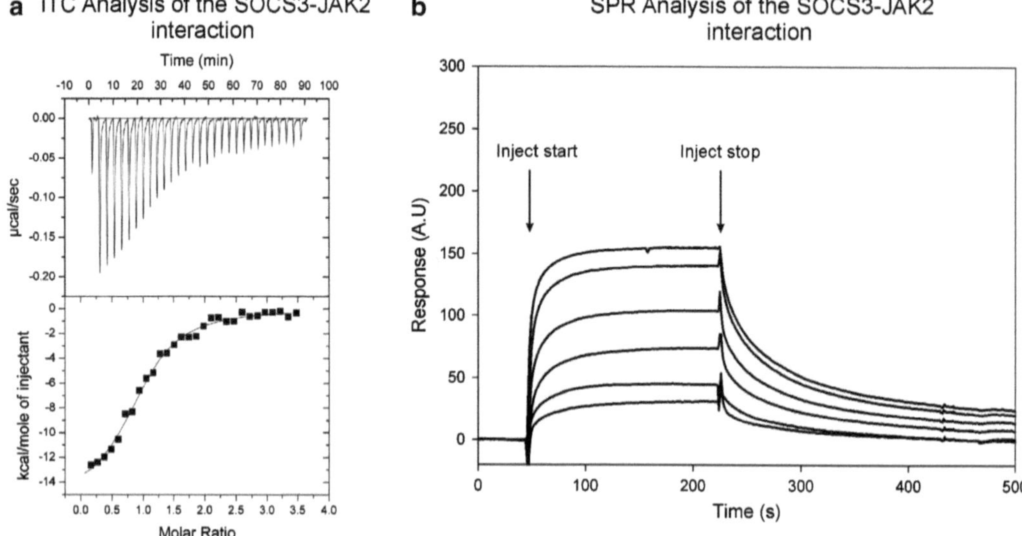

Fig. 2. Example of an ITC and SPR experiment used to quantify the JAK2-SOCS3 interaction. (**a**) ITC analysis of the SOCS3–JAK2 interaction was performed according to Subheading 3.1. 10 μM JAK2 was present in the ITC cell, into which 100 μM SOCS3 was titrated in thirty 10 μl aliquots. (**b**) SPR analysis of the SOCS3–JAK2 interaction was performed according to Subheading 3.2. GST-JAK2 was bound to the biacore chip and series of dilutions (2 μM–62.5 nM) of SOCS3 was passed over. The curvature of the region between the injection starting and stopping is used to measure the on-rate whilst the curvature of the region after the injection has stopped is used to measure off-rate. Data from both analyses could be fit to a single site binding model with a K_d of approximately 1 μM.

5. Examine the outputted stoichiometry value. The majority of biological interactions are 1:1; however 2:1 (or 1:2) interactions are relatively common. Outputted results suggesting non-integer values such as 0.7 or 1.3 are most likely due to errors in protein concentration estimation.
6. The equation used to fit the binding isotherm will output values for K_d, stoichiometry, and enthalpy. These are essentially calculated from the shape of the curve, the inflection point of the curve, and the *y*-intercept, respectively. A value for the entropy of the interaction is also given. This is the most indirect quantity obtained by ITC analysis and should be treated with care (see Note 11).

3.2. Kinetic Analysis of JAK Binding by Surface Plasmon Resonance

This experiment requires approximately 0.5 mL of 10 μg/mL purified GST-JAK and 0.5 mL of 2 μM purified binding protein (ligand) (see Note 12) and hence is a more "protein-friendly" method to quantify binding when compared to ITC. This protocol covers preparation of the GST-JAK-loaded biacore chip and analysis of the JAK–ligand interaction by passing a series of different concentrations of ligand over the immobilized GST-JAK.

1. Covalently bind anti-GST antibody to the CM5 biacore chip using amine coupling by following steps 2–4 (see Note 13).

2. Activate the surface by injecting 50 μl of 0.2 M EDC/ 0.05 M NHS in water with a flow rate of 5 μl/min for 10 min.
3. Couple anti-GST antibody to the activated chip using 50 μl of 10 μg/mL antibody in 10 mM sodium acetate pH 5.0 with a flow rate of 5 μl/min for 10 min.
4. Deactivate excess groups by injecting 35 μl of 1 M ethanolamine–HCl pH 8.5 over the chip at a flow rate of 5 μl/min for 7 min.
5. Allow biacore buffer to flow over the chip for at least 2 h at a flow rate of 20 μl/min (see Note 14).
6. Prepare 500 μl of 10 μg/mL GST-JAK in biacore buffer (see Note 15).
7. Prepare 500 μl of 2 μM ligand (binding protein) and use this to make a series of twofold dilutions of ligand (2, 1, 0.5, 0.25, 0.125, 0.06, 0 μM).
8. Use the biacore control software to set up a method to pass 15 μl of GST-JAK over the chip (flow rate: 5 μl/min, 3 min) followed by 60 μl of 2 μM ligand (flow rate: 20 μl/min, 3 min). Allow 3 min between injections and then allow ligand binding to be monitored for 5 min after the injection ceases.
9. Regenerate the chip surface by injecting 100 μl of Regeneration Buffer (50 μl/min, 2 min).
10. Repeat steps 8–9 six times using the six remaining ligand dilutions.

3.2.1. Analysis of SPR Data

1. Biacore data is best analyzed using the BIAEVALUTION™ software (GE Healthcare).
2. Overlay the raw data from the seven experiments and use the BIAEVALUTION™ software to fit the data using a 1:1 Langmuir binding model.
3. Deriving a binding constant from Biacore data uses three independent variables:
   (a) k_{on} (the association rate constant) is derived from the shape of the association curve as ligand is passed over the chip. This rate is dependent upon concentration and the shape of the curve will differ between each dilution of ligand.
   (b) k_{off} (the dissociation rate constant) is derived from the shape of the dissociation curve as the ligand dissociates from the chip after the injection is complete. This rate is independent of concentration and the shape of the curve will be roughly equal for each dilution of ligand.
   (c) The total response (in RU) as a function of ligand concentration. This value can be plotted as a function of ligand concentration and fitted to the equation

$$\mathrm{RU} = \mathrm{RU}_{\mathrm{max}} \times (\mathrm{ligand})/(K_{\mathrm{d}} + (\mathrm{ligand})).$$

4. The Langmuir binding model in the BIAEVALUTION™ software uses all the available information detailed above (k_{on}, k_{off}, RU) to determine K_d and this is the most robust method of estimating affinity (see Note 16).

4. Notes

1. The presence of reducing agent is important to help stabilize JAK; however DTT is usually avoided when performing ITC reactions as it is unstable in aqueous solution and its heat of decomposition may obscure the real signal. In practice the DTT signal is rarely a concern; however the use of 2-mercaptoethanol is advised to avoid this potential problem.
2. We have successfully purified and tested the JH1 domain of all four JAKs using this procedure and have also successfully tested larger JAK2 fragments that incorporate the JH2 domain.
3. Other successful methods of buffer exchange include Gel filtration of both components, lyophilization from a volatile buffer, and resuspension or the use of centrifugal concentrator. However, extended dialysis is our preferred method of buffer exchange.
4. We routinely use absorption at 280 nm with a calculated molar extinction coefficient to quantify our protein concentrations. However, this method is most accurate when the protein is completely unfolded. Therefore it is suggested that a small aliquot of protein is diluted 1:10 in 6 M Guanidine–HCl and the absorbance measured.
5. Errors in protein concentration impact most heavily upon the measured stoichiometry and enthalpy values. For example overestimating the ligand concentration by 50% will result in a 50% underestimation of ΔH and a 50% error in stoichiometry.
6. The optimal concentration of JAK used in these experiments depends on the strength of the interaction and the enthalpy of the interaction. If the interaction is tight ($K_d < 10^{-8}$ M) then the JAK concentration can generally be lowered to 2 μM. If the interaction is weak ($K_d > 10^{-6}$ M) then the JAK concentration may need to be increased significantly in order for quantifiable binding to be observed. The concentration of the test protein needs to be correspondingly scaled up or down along with JAK. Note that the relative concentrations of JAK and test protein are relatively inflexible. A ratio of 1:8 to 1:12 is appropriate if the stoichiometry is expected to be 1:1.

7. This step is important as the syringe will be stirring over the course of the experiment which will result any trapped air being released as bubbles which can release a very large amount of heat as they collide with the solvent and walls of the vessel.
8. This first injection is performed as there is often a small amount of trapped air at the syringe outlet which results in an incorrect volume being expelled during the first injection.
9. Too high an affinity results in a step-function that cannot be fit to a continuous equation. If this is the case then the affinity is likely underestimated by the software. Too low an affinity results in a curve with a low curvature and no obvious inflection point which does not approach zero by the final injection. Such a curve can only be examined qualitatively (to indicate that low-affinity binding is occurring) but cannot be accurately quantified.
10. As a rule of thumb, increasing the concentration of both proteins by tenfold gives a similar shaped curve to that obtained if the concentration was unchanged but the affinity was tenfold higher. For example an interaction that resulted in an ITC curve such as the fifth panel in Fig. 1 (i.e., K_d 1 μM) would be difficult to quantitate accurately. However, if the concentration of both JAK and ligand were increased tenfold then the curve would resemble that shown in the fourth panel which is easily quantifiable. The same procedure can be used if the affinity is too high to be easily measured; in this case one could *decrease* the concentration of both proteins by tenfold to make the curve less "step-like" and more sigmoidal. However as the enthalpy signal detected with each injection is directly proportional to the concentration of ligand then the sensitivity of the equipment is often limiting if the proteins are too dilute.
11. The entropy value is an indirect quantity as it is derived from the equation

    $$\mathbf{RT\ln(}K_d\mathbf{)} = \mathbf{\Delta H} - \mathrm{T}\Delta\mathrm{S},$$

    where the elements in bold are known after analysis of the ITC data. Therefore the entropy value must be treated with caution as it is both indirect and heavily dependent upon the buffer used if protonation/deprotonation of amino-acid side chains occur as part of the interaction. Moreover it can be heavily temperature dependant.
12. We have successfully tested SOCS2, SOCS3, SOCS4, and SOCS5 as the "ligand" using this procedure.
13. It may be possible to covalently bind JAK to the chip surface (instead of requiring anti GST); however we prefer to use the procedure outlined in this protocol.
14. During this period watch to see that the baseline has stopped drifting and is stable.

15. JAK is unstable in the absence of reducing agent; however the presence of reducing agent quickly destroys the anti-GST antibody bound to the chip. We have observed that storing GST-JAK in 1 mM DTT and then diluting it 1,000-fold immediately prior to loading is the most effective method of maintaining JAK and the chip surface.
16. Multistep binding mechanisms and nonspecific binding to the chip surface will often render an interaction un-fittable using the standard algorithms. We have found low-level nonspecific binding to be especially problematic. This can manifest itself as the trace not returning to baseline after the ligand injection is stopped and results in the failure of a kinetic analysis. If this occurs, the affinity of the interaction can still be estimated by plotting the maximum response of each injection (*y*-axis) vs. ligand concentration (*x*-axis) and fitting this to $RU = RU_{max} \times (\text{ligand})/(K_d + (\text{ligand}))$ to yield K_d.

Acknowledgements

This work was made possible through Victorian State Government Operational Infrastructure Support and the Australian Government NHMRC IRIISS. This research was supported by an NHMRC Program Grant (461219), NIH Grant (CA022556), NHMRC CDA (516777), and NHMRC Project Grant (1011804).

References

1. Wilks AF, Harpur AG (1994) Bioessays 16: 313–320
2. Wilks AF, Oates AC (1996) Cancer Surv 27: 139–163
3. Ihle JN, Witthuhn BA, Quelle FW, Yamamoto K, Silvennoinen O (1995) Annu Rev Immunol 13:369–398
4. Wilks AF (1989) Proc Natl Acad Sci U S A 86: 1603–1607
5. Wilks AF, Harpur AG, Kurban RR, Ralph SJ, Zurcher G, Ziemiecki A (1991) Mol Cell Biol 11:2057–2065
6. Shuai K, Ziemiecki A, Wilks AF, Harpur AG, Sadowski HB, Gilman MZ, Darnell JE (1993) Nature 366:580–583
7. Ma XY, Sayeski PP (2007) Biochemistry 46: 7153–7162
8. Rui LY, Mathews LS, Hotta K, Gustafson TA, CarterSu C (1997) Mol Cell Biol 17: 6633–6644
9. Cengel KA, Freund GG (1999) J Biol Chem 274:27969–27974
10. Alexander DR (2002) Curr Biol 12: R288–R290
11. Jiao HY, Berrada K, Yang WT, Tabrizi M, Platanias LC, Yi TL (1996) Mol Cell Biol 16: 6985–6992
12. Klingmuller U, Lorenz U, Cantley LC, Neel BG, Lodish HF (1995) Cell 80:729–738
13. Ihle JN, Witthuhn BA, Quelle FW, Silvennoinen O Tang B, Yi T (1994) Blood Cells 20:65–80
14. Hilton DJ, Richardson RT, Alexander WS, Viney EM, Willson TA, Sprigg NS, Starr R, Nicholson SE, Metcalf D, Nicola NA (1998) Proc Natl Acad Sci U S A 95:114–119
15. Hilton DJ (1999) Cell Mol Life Sci 55: 1568–1577
16. Kile BT, Schulman BA, Alexander WS, Nicola NA, Martin HM, Hilton DJ (2002) Trends Biochem Sci 27:235–241

17. Ihle JN, Witthuhn BA, Quelle FW, Yamamoto K, Thierfelder WE, Kreider B, Silvennoinen O (1994) Trends Biochem Sci 19:222–227
18. Tame JRH, O'Brien R, Ladbury JE (1998) Isothermal titration calorimetry of biomolecules. In: Ladbury JE, Chowdhry BZ (eds) Biocalorimetry: applications of calorimetry in the biological sciences. John Wiley & Sons Ltd, Chichester, NY, pp 27–38
19. Ladbury JE, Chowdhry BZ (1996) Chem Biol 3:791–801
20. McDonnell JM (2001) Curr Opin Chem Biol 5:572–577

Chapter 5

Determination of Protein Turnover Rates in the JAK/STAT Pathway Using a Radioactive Pulse-Chase Approach

Anna Dittrich, Elmar Siewert, and Fred Schaper

Abstract

The turnover rate of different protein species in a signal transduction network strongly affects the impact of the given species on the outcome of a stimulus. Whereas stable, long-lived proteins mainly account for the transmission of a signal, unstable short-lived species often comprise regulatory functions. Here, we describe a method to determine the half-lives of proteins of the JAK/STAT pathway by a pulse-chase approach in cell culture. First, radioactive labeling with ^{35}S-methionine is carried out to label newly synthesized proteins (pulse). Subsequently, the dynamics of the decay of these proteins is monitored in the absence of labeled amino acids over a defined time period (chase). For this purpose the protein of interest is isolated by immunoprecipitation from total cell lysates, separated on an SDS-polyacrylamide gel, and subsequently visualized by autoradiography.

Key words: Protein dynamics, Half-life, Turnover rate, ^{35}S-methionine, Autoradiography, Pulse-chase

Abbreviations

SOCS	Suppressor of cytokine signalling
SHP2	SH2 containing phosphatase 1
JAK	Janus kinase
TYK	Tyrsoine kinase
gp130	glycoprotein 130 (CD130)
gp80	glycoprotein 80 (IL-6R alpha ,CD126)
STAT	signal transducer and activator of transcription

1. Introduction

JAK-STAT signaling is a very rapid process. Activation (phosphorylation) of receptor chains is detectable within minutes after binding of the ligand to the specific cytokine- or interferon receptor. STAT phosphorylation peaks about half an hour after stimulation,

Sandra E. Nicholson and Nicos A. Nicola (eds.), *JAK-STAT Signalling: Methods and Protocols*, Methods in Molecular Biology, vol. 967, DOI 10.1007/978-1-62703-242-1_5,

immediately followed by accumulation of the activated STATs in the nucleus. In line with these fast kinetics, gene induction of target mRNA can be demonstrated as early as 30 min after stimulation. Interestingly, both STAT activation and gene induction are transient due to the activity of negative regulators of signaling such as inducible feedback kinase inhibitors (e.g., SOCS proteins) or phosphatases within the receptor complex (e.g., SHP2) or within the nucleus (e.g., TC-PTP).

However, negative regulation not only limits response intensities but also more importantly reduces response times. A system with negative autoregulation or feedback mechanisms responds more rapidly to a stimulus without increasing the steady state of the signaling molecules. In contrast, accelerating the response time of an ordinarily regulated process is associated with an increased steady-state level.

The turnover of a signaling component influences the dynamics of signal transduction. Under conditions of steady state, the concentration of a protein is determined by the ratio of the rate of induction (β) and the rate of degradation (α). The response time to a defined stimulus ($R_{1/2}$) comprises the time required to increase the expression level to 50% of the steady-state expression. Whereas the rate of induction positively influences the steady state, the rate of degradation affects the response time inversely ($R_{1/2} = \log(2)/\alpha$). Thus, it is obvious that signaling components with a rapid rate of degradation (α), as realized by rapid turnover and short half-lives ($T_{1/2}$) are important components to speed up signal transduction and regulate signal transduction.

In addition to regulating the expression level of a protein, the availability or its specific function (e.g., enzymatic activity) can generally be regulated by posttranslational modifications, such as phosphorylation or acetylation. Stable proteins with long half-lives are ready at hand. They are not restricted in availability but are rather regulated at the posttranslational level. The expression of rapidly degraded proteins is often induced by a certain stimulus. They exert their major functions mainly without further posttranslational modifications. Typical examples for rapidly induced, short-lived signaling components of the JAK/STAT pathway are regulatory SOCS proteins, with typical half-lives of 1–2 h (e.g., refs. 1–5). In contrast long-living proteins with half-lives between 4 and 18 h, such as STAT factors or SHP2, are often regulated by posttranslational modifications (e.g., refs. 4, 6–8). A systems biology-based analysis of protein half-lives in mammalian signal transduction networks mathematically confirmed these two classes of signaling species. The authors addressed long-lived stable transmitter proteins and flexible negative feedback regulators with short half-lives (9).

The biochemical analysis of protein half-lives can thereby assign a protein to one of these classes and can give evidence of its physiological function. Furthermore, a growing number of quantitative computational models of signal transduction pathways calls for the exact determination of protein half-lives in different cellular systems.

Here, we demonstrate a radioactive pulse-chase approach to determine the half-lives of specific signaling components of the JAK/STAT pathway. Cells are incubated with ^{35}S methionine for 30 min. During this pulse the radioactive amino acids are incorporated into newly synthesized proteins. Subsequently, the dynamics of the decay of these radioactively labeled proteins is monitored in the absence of labeled amino acids over a defined time period (chase). For this purpose whole cell lysates are prepared at different time points. Proteins of interest are isolated by immunoprecipitation, separated on an SDS-polyacrylamide gel and visualized by autoradiography. The decay of radioactively labeled proteins as detected by autoradiography is then used to calculate the half-lives of the protein of interest. This method is widely used for the determination of half-lives of both stable and unstable proteins (see, for example, refs. 2, 4, 10.

Stimulating the cells with cytokines or inhibitors after the labeling period allows determination of whether the half-life of the protein of interest is affected by the individual treatment.

To enhance the amount of newly expressed proteins during the pulse, the cells can be transfected with expression vectors for the protein of interest. This also enables the analysis of the influence of mutations on protein stability and the isolation of proteins via a protein tag if no appropriate antibodies are available. It is noteworthy that (moderate) overexpression of the proteins seems not to influence their turnover rates compared to endogenous protein expression; however, this must be controlled for each individual protein and expression system by comparing the half-lives of endogenous and exogenous proteins.

Alternatively, it is possible to transiently induce expression of the protein of interest via a tetracyclin-dependent pTET off vector to avoid radioactive labeling. The synthesis of the protein of interest is induced by a coexpressed transactivator in the absence of doxycycline. As soon as doxycyline is added to the medium, it binds and inhibits the transactivator and the promoter is silenced. To discriminate endogenous and transiently induced exogenous expression of the protein of interest, the latter needs to be tagged. Subsequently the decay of the exogenous protein can be detected by immunoprecipitation of the tagged proteins, SDS-PAGE and subsequent Western blotting (see, e.g., ref. 11). Fluorescent tags enable us to follow the decay of the protein of interest in living cells by microscopy (12).

Finally, protein turnover of unstable proteins can be analyzed by blocking new protein synthesis with cycloheximide and subsequent detection of the decay of the remaining protein using e.g. Western blotting and immunostaining (see, e.g., refs. 13–15). Although this method circumvents radioactive labeling, it is not feasible to analyze longer periods of time (stable proteins) due to the general toxicity of cycloheximide. Furthermore, this approach may produce artifacts since cycloheximide may additionally interfere with the synthesis of regulatory proteins involved in the decay of the protein of interest.

2. Materials

2.1. Consumables

1. Lipofectamin 2000 (Life Technologies, Carlsbad, CA, USA).
2. TRAN^{35}S-LABEL (MP Biomedicals, Santa Ana, CA, USA).
3. Protein A Sepharose (GE Healthcare, Little Chalfont, UK).
4. Radiographic film (Curix Ortho HT-U film, Agfa).

2.2. Equipment

1. Vertical electrophoresis system.
2. Radiographic film processor.
3. PhosphorImager (Storm 840, GE Healthcare).

2.3. Buffers

1. PBS: 200 mM NaCl, 2.5 mM KCl, 8 mM Na_2HPO_4, 1.5 mM KH_2PO_4, pH 7.4.
2. RIPA lysis buffer: 50 mM Tris–HCl pH 7.4, 150 mM NaCl, 1 mM EDTA, 0.5% Nonidet P-40, 15% glycerol, 1 mM NaF, 1 mM phenylmethanesulfonylfluoride, 1 μg/ml leupeptin, 1 μg/ml pepstatin, 5 μg/ml aprotinin.
3. Denaturing lysis buffer: 1% SDS, 10 mM Tris–HCl (pH 7.4).
4. Immunoprecipitation buffer: 5 mM Tris–HCl pH 7.4, 75 mM NaCl, 0.5 mM EDTA, 0.5 mM EGTA, 0.5% Triton X-100, 0.25% Nonidet P-40, 0.1 mM phenylmethanesulfonylfluoride.
5. RIPA wash buffer: 50 mM Tris–HCl pH 7.4, 100 mM NaCl, 1 mM EDTA, 0.1% Nonidet P-40, 15% glycerol, 1 mM NaF, 1 mM phenylmethanesulfonylfluoride, 1 μg/ml leupeptin, 1 μg/ml pepstatin, 5 μg/ml aprotinin.
6. SDS/PAGE loading buffer: 62 mM Tris–HCl pH 6.8, 10% glycerol, 2% SDS, 5% mercaptoethanol.
7. Separating gel (7%, 10%, 12.5%): 7.34 ml, 5.9 ml, 4.7 ml distilled water, 3.36 ml, 4.8 ml, 6 ml 30% acrylamide/bis (29:1), 3.8 ml 1.5 M Tris–HCl pH 8.8, 75 μl 20% SDS, 15 μl TEMED, 75 μl 20% ammonium persulfate (APS).
8. Stacking gel: 4 ml distilled water, 635 μl 30% acrylamide (29:1), 313 μl 2 M Tris–HCl pH6.8, 25 μl 20% SDS, 5 μl TEMED, 40 μl 20% APS.
9. SDS running buffer: 25 mM Tris, 192 mM glycine, 0.1% SDS.

10. Gel fixation buffer I: 40% methanol, 10% acetic acid, 50% distilled water?
11. Gel fixation buffer II: 1 M sodium salicylate.

2.4. Media

1. Standard medium: DMEM (Gibco), supplemented with 10% fetal bovine serum, 100 mg/l streptomycin, 60 mg/l penicillin.
2. OptiMEM (Gibco).
3. Labeling medium: Eagle's minimal essential medium (MP Biomedical) supplemented with 300 mg/l glutamine and 0.2% bovine serum albumin.

2.5. Antibodies for Immunoprecipitation

Flag (Sigma), gp130 (M20, Santa Cruz), JAK1 (New England Biolabs), JAK2 (06-255, Upstate), SHP2 (C18, Santa Cruz), STAT1 (N-terminal, BD Transduction Lab), STAT3 (C20, Santa Cruz), STAT5 (BD Transduction Lab), Tyk2 (BD Transduction Lab) (see Note 1).

2.6. Plasmids

pSVL (Amersham Bioscience): eukaryotic expression vector containing coding sequences of Flag-SOCS1, Flag-SOCS3, gp130, JAK1, JAK2, SHP2, STAT1, STAT3, TYK2.

3. Methods

All steps can be performed with adherent cell lines or primary cells. However, it may be necessary to adapt media and steps related to cell culture and transfection to suitable conditions. All steps described were adjusted for COS-7 cells and have also been applied to HepG2 cells to check for half-lives of endogenous proteins.

3.1. Transfection of Cells (See Notes 2–4)

1. Cells are grown in standard medium supplemented with fetal bovine serum and streptomycin/penicillin up to 60% density at 37°C, 5% CO_2 in a water-saturated atmosphere.
2. Dilute 1 μg of pSVL expression vector/6 cm cell culture dish in 100 μl OptiMEM. Add 3 μl Lipofectamine 2000/6 cm cell culture dish to 200 μl OptiMEM. Allow both solutions to equilibrate to room temperature for 5 min and subsequently mix them. Do not vortex. Let the mixture rest for 15 min at room temperature without shaking.
3. Replace the culture medium by OptiMEM. Use the least possible volume of medium to guarantee a high concentration of expression vectors.
4. Drip the transfection mix into the medium and mix by gently shaking the cell culture dish.

5. Cultivate the cells for 6 h at 37°C, 5% CO_2 in a water-saturated atmosphere.
6. Replace the culture medium with standard medium and cultivate the cells for 36 h at 37°C, 5% CO_2 in a water-saturated atmosphere (see Note 5).

3.2. Metabolic Labeling

1. 36 h after transfection, wash the cells with pre-warmed PBS (37°C) to remove residual unlabeled amino acids. Replace PBS by pre-warmed MEM, supplemented with arginine, leucine, glucose, inositol, 0.2% BSA, and 10 mM HEPES (pH 7.4). Cultivate cells for 15 min at 37°C, 5% CO_2 in a water-saturated atmosphere (see Notes 6 and 7).
2. Add 4.3 MBq TRAN^{35}S-LABEL to the medium and cultivate cells for 30 min at 37°C, 5% CO_2 in a water-saturated atmosphere (see Notes 8 and 9).
3. Wash the cells with pre-warmed PBS and change medium to standard medium supplemented with 5% FBS (see Note 10).

3.3. Cell Lysis

Cells are harvested after various periods of time (chase). Duration of the chase depends on the half-life of the protein of interest and needs to be adjusted (see Note 11).

1. Aspirate the medium and wash the cells two times in ice cold PBS (see Note 12).
2. Aspirate residual PBS and freeze the whole cell culture dish at −80°C (see Note 13).
3. Subsequent to harvesting cells at the final time point place the frozen cell culture dishes on ice and cover the cells with 500 μl RIPA lysis buffer (see Notes 14–16). Incubate the cells for 1 h on ice.
4. Centrifuge the lysate for 10 min at max. speed in a microfuge at 4°C.
5. Transfer the supernatant into a fresh reaction tube and discard the pellet.

3.4. Immunoprecipitation

1. Incubate the lysate with 1 μg of antibody specific to your protein of interest overnight at 4°C on a rotating shaker (see Notes 17 and 18).
2. Simultaneously soak 4 mg Protein-A-sepharose/sample in 100 μl RIPA-lysis buffer overnight at 4°C on a rotating shaker (see Note 19).
3. Add 100 μl of resuspended Protein-A-Sepharose beads to each of the samples containing the lysate and antibodies. Incubate for 4 h at 4°C on a rotating shaker.
4. Centrifuge the suspension 5–10 s at maximum speed in a microfuge to separate Protein-A-Sepharose–antibody–protein

complexes from residual lysate. Carefully aspirate the lysate and wash the Protein-A-Sepharose/protein complexes four times with 400 μl RIPA wash buffer (see Notes 20 and 21).

5. Subsequent to the last washing step centrifuge again for 10–15 s without washing. Carefully aspirate as much residual washing buffer as possible.
6. Mix the Protein-A-Sepharose/protein complexes with 30 μl 2 × SDS loading buffer by gentle vortexing and boil them for 10 min at 89°C. Cool the sample on ice.
7. Centrifuge for 10 min at max. speed and collect the supernatant in a fresh tube. Avoid transferring any Sepharose.
8. Centrifuge again for 10 min at max. speed. The supernatant is ready to be analyzed by SDS-PAGE.

3.5. SDS/PAGE

Proteins are separated on a standard discontinuous SDS gel. Depending on the size of your protein of interest you have to adapt the acrylamide concentration of the gel.

1. Prepare an SDS-polyacrylamide gel and load the whole supernatant (approx. 25 μl). Use a protein marker with appropriate molecular weight standards. Radioactively labeled protein standards are useful.
2. Run the SDS-polyacrylamide gel.
3. Detach one glass plate and transfer the second plate carrying the gel into a plain dish. Incubate the gel in fixation buffer I for 30 min at room temperature (see Note 22).
4. Wash the gel for 30 min in H_2O.
5. Incubate the gel in 1 M sodium salicylate for 30 min (see Note 23).
6. Aspirate as much solution as possible. Put a sheet of 3MM paper on the gel. Carefully remove the glass plate. The gel will stick to the paper. Cover the gel with Saran wrap. Take care that no Saran covers the bottom side of the 3MM paper (see Note 24).
7. Dry the gel at 70°C in a vacuum dryer until it is completely free of water.

3.6. Detection of Radioactivity (See Note 25)

1. Detection by autoradiography: Lay the dried gel on a radiographic film and expose it in a developing cassette. Expose the film for up to 2 days and develop it using a standard radiographic film processor or manually. Do not forget to mark the sizes of the markers as they do not appear on the film if nonradioactive protein standards have been used.
2. Detection by phosphorimager: Place the gel on the phospho imager and run detection mode.

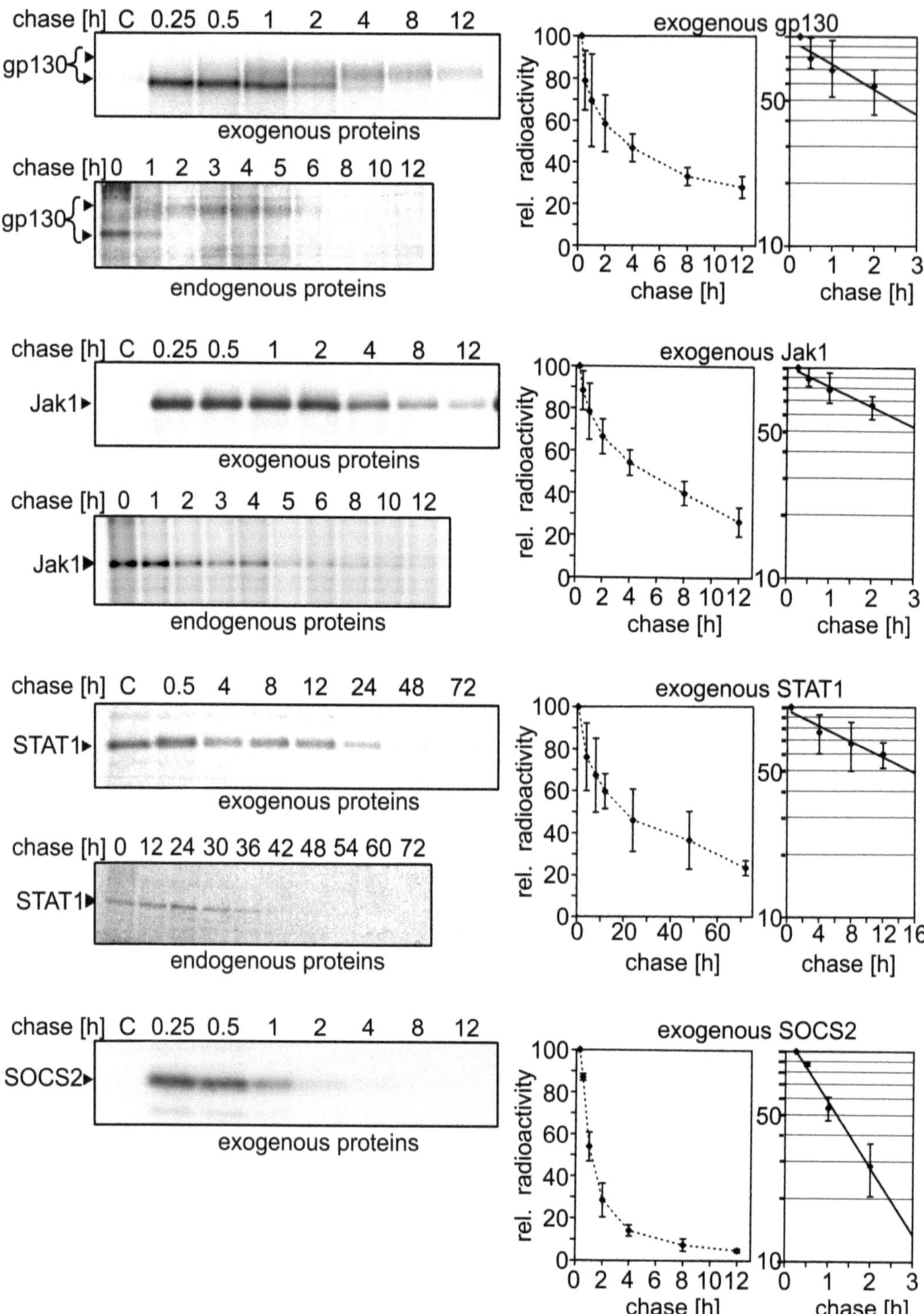

Fig. 1. Half-lives of selected proteins involved in JAK/STAT signaling. The cDNAs encoding the various signaling molecules were transfected into COS7 cells (exogenous proteins). Endogenous proteins were analyzed in HepG2 cells. The cells were pulse-labeled with (^{35}S)-methionine and chased for the times indicated. Cellular extracts were used for immunoprecipitation with corresponding antibodies and subjected to SDS-PAGE and autoradiography. Data are given as one representative

Table 1
Half-lives of proteins involved in JAK/STAT signaling

Protein	gp130	JAK1	JAK2	JAK2	SOCS1	SOCS2	SOCS3	STAT3	STAT1	SHP2
Half-life (h)	2.5	3.2	1.9	2	1.5	1	1.6	8.5	16	18–20

The table was reproduced from an article previously published by us in the FEBS Journal (former European Journal of Biochemistry) (4) with kind permission by John Wiley & Sons, Inc. (License Number 2620180991497)

3.7. Analysis of Half-Lives

1. Quantify signal intensity at each time point.
2. Set the first (maximal) value to 100% and plot your data in a half-logarithmic graph where y (logarithmic scale) is the signal intensity and x is the time. Perform an exponential regression analysis. In this depiction the regression curve should be linear so you can easily calculate the half-life of your protein of interest (i.e., the time when 50% of the initial signal remains). See Fig. 1 and Table 1 for example illustrations (see Note 26).

4. Notes

1. The choice of antibody depends on your cellular system. These antibodies worked well in COS-7 and HepG2 cells. For other cells and species other antibodies might be more suitable.
2. Transfection can also be performed using electroporation or magnetic transfection. Use the method best-suited for your cells.
3. You can use this method to compare the half-lives of different mutants of your protein. If your immunoprecipitating antibody does not recognize the mutant form you can also fuse a protein tag to your protein and use this tag for immunoprecipitation. In addition, we used tagged proteins if there was no good antibody for immunoprecipitation available. Be aware that a tag might influence protein stability.
4. Include untransfected control cells to confirm overexpression of the protein of interest. However, endogenous proteins might be detectable in this control (see Fig. 1, STAT1).
5. The time period after transfection is flexible and can be shortened if necessary. If you prolong the time to gain more exog-

Fig. 1. (continued) of several experiments. For quantification of radioactive proteins, gels were analyzed using a phosphorimager (Storm 840). The initial maximal radioactivity was defined as 100%. Data represent the mean ± SEM of three to seven experiments. Correlation coefficient is >0.90. Note the logarithmic scale of the y-axis in the *right panels*. The figures and data are reproduced from an article previously published by us in the FEBS Journal (former European Journal of Biochemistry) (4) with kind permission by John Wiley & Sons, Inc. (License Number 2620180991497).

enous protein consider that the expression vector may become degraded or diluted through cell proliferation.

6. Labeling of endogenous proteins is possible if they are produced fast enough to incorporate sufficient amounts of radioactive amino acid during the pulse. In our hands, this worked in HepG2 cells for gp130, STAT1, and JAK1. You can also induce expression of a protein by an exogenous stimulus, e.g., endogenous SOCS proteins can be induced by stimulation with a cytokine.
7. This preincubation in minimal essential medium eagle without TRAN^{35}S Label is necessary to reduce any traces of residual (non-labeled) amino acids. The cells will consume these amino acids.
8. Be careful working with radioactivity. Radioactive sulfur is volatile. If you have a laminar flow hood for radioactive work use it. Otherwise, work under an extractor hood. This may produce some problems with sterility. You can therefore add antibiotics to the medium.
9. We place the cell culture dishes in a plastic box with activated charcoal. The charcoal binds the volatile sulfur and you avoid contamination of the incubator. Do not seal the box completely since this will not allow proper equilibration of temperature, CO_2 content and humidity with that of the incubator.
10. Radioactive fluids need to be disposed of with regard to the rules of your lab.
11. If you want to analyze the influence of a certain stimulus (e.g., inhibitor, cytokine) on protein degradation you can add it right after the pulse.
12. Quite often very early time points of the chase tend to fluctuate. This is due to ongoing protein synthesis. You can avoid this by waiting half an hour before starting the chase. As the protein is degraded continuously this will not influence your analysis.
13. Freezing of the plates allows you to perform lysis of all plates at one time under the same conditions. This is of great importance for longer chases.
14. Add the protease and phosphatase inhibitors just prior to use in the lysis buffer. If you store the buffer too long the inhibitors lose their potency.
15. In the methods section, we describe a native lysis of the cells. However some antibodies do not work for immunoprecipitation under native conditions. In our hands, this occurs for antibodies raised against the JAKs. To circumvent this problems, we use a denaturing protocol:
    - Pipette 500 μl hot (80°C) denaturing lysis buffer on the cells and harvest them with a rubber scraper. Transfer

the lysate into a reaction tube and boil it for 10 min at 80°C. Drill a small hole in the cap of your reaction tube. This avoids bursting of the reaction tube due to high temperatures.

- Eliminate cell debris by centrifugation.
- Because the SDS in the lysis buffer will impair immunoprecipitation, you have to dilute the lysate 1:10 in immunoprecipitation buffer. Subsequently, use the whole volume for immunoprecipitation.

16. If you have enough cells you can use more lysis buffer (e.g., 800 μl) and divide it in two parts. Thereby you can analyze two proteins from one sample. Be aware that this requires a high-level endogenous expression rate of your proteins of interest.
17. As most reaction tubes do not seal completely we wrap them in parafilm and additionally store them in a second closed vessel. This circumvents radioactive contamination.
18. The exact amount of antibody needs to be titrated for your conditions. We gained good efficiency of immunoprecipitation with 1–3 μg of antibody.
19. Of course it is also possible to perform immunoprecipitation with magnetic protein-A/G beads as described by the suppliers. In addition, you have to keep in mind that Protein-A does not bind every antibody isotype, so you may have to include a bridging antibody. Preincubation of Sepharose with rabbit-anti-mouse antibodies usually enhances efficiency of immunoprecipitation with monoclonal mouse antibodies.
20. As Sepharose is not easily visible we use a 10 μl pipette tip attached to a Pasteur pipette to aspirate the supernatant. This reduces the suction power and thereby the loss of Sepharose.
21. Pipette washing buffer in one rapid stream on the Sepharose. Do not mix by pipetting because too much Sepharose will stick to the pipette tip. Instead of pipetting briefly vortex the mixture.
22. If you want to interrupt the experiment you can fix the gel overnight in fixation buffer I.
23. Incubation of the gel in sodium salicylate enhances the detection of radioactive signals (16).
24. Saran wrap should be only on the gel side, otherwise the gel will not dry on the vacuum dryer.
25. Radioactivity can either be detected by autoradiography or by a phosphorimager. In our hands, the first generates superior pictures, whereas the latter is superior for quantification.
26. Some proteins may shift their mobility due to posttranslational modifications, such as glycosylation of receptor proteins (see Fig. 1, gp130).

References

1. Chen XP, Losman JA, Cowan S, Donahue E, Fay S, Vuong BQ, Nawijn MC, Capece D, Cohan VL, Rothman P (2002) Pim serine/threonine kinases regulate the stability of Socs-1 protein. Proc Natl Acad Sci U S A 99: 2175–2180
2. Haan S, Ferguson P, Sommer U, Hiremath M, McVicar DW, Heinrich PC, Johnston JA, Cacalano NA (2003) Tyrosine phosphorylation disrupts elongin interaction and accelerates SOCS3 degradation. J Biol Chem 278: 31972–31979
3. Kario E, Marmor MD, Adamsky K, Citri A, Amit I, Amariglio N, Rechavi G, Yarden Y (2005) Suppressors of cytokine signaling 4 and 5 regulate epidermal growth factor receptor signaling. J Biol Chem 280:7038–7048
4. Siewert E, Müller-Esterl W, Starr R, Heinrich PC, Schaper F (1999) Different protein turnover of interleukin-6-type cytokine signalling components. Eur J Biochem 265:251–257
5. Vuong BQ, Arenzana TL, Showalter BM, Losman J, Chen XP, Mostecki J, Banks AS, Limnander A, Fernandez N, Rothman PB (2004) SOCS-1 localizes to the microtubule organizing complex-associated 20S proteasome. Mol Cell Biol 24:9092–9101
6. Blesofsky WA, Mowen K, Arduini RM, Baker DP, Murphy MA, Bowtell DD, David M (2001) Regulation of STAT protein synthesis by c-Cbl. Oncogene 20:7326–7333
7. Lee CK, Bluyssen HA, Levy DE (1997) Regulation of interferon-alpha responsiveness by the duration of Janus kinase activity. J Biol Chem 272:21872–21877
8. Tam NW, Ishii T, Li S, Wong AH, Cuddihy AR, Koromilas AE (1999) Upregulation of STAT1 protein in cells lacking or expressing mutants of the double-stranded RNA-dependent protein kinase PKR. Eur J Biochem 262:149–154
9. Legewie S, Herzel H, Westerhoff HV, Bluthgen N (2008) Recurrent design patterns in the feedback regulation of the mammalian signalling network. Mol Syst Biol 4:190
10. Frantsve J, Schwaller J, Sternberg DW, Kutok J, Gilliland DG (2001) Socs-1 inhibits TEL-JAK2-mediated transformation of hematopoietic cells through inhibition of JAK2 kinase activity and induction of proteasome-mediated degradation. Mol Cell Biol 21:3547–3557
11. Dittrich A, Khouri C, Dutton Sackett S, Ehlting C, Böhmer O, Albrecht U, Bode JG, Trautwein C, Schaper F (2012) Glucocorticoids increase interleukin-6 dependent gene induction by interfering with the expression of the SOCS3 feedback inhibitor. Hepatology 55(1):256–266
12. Köster M, Hauser H (1999) Dynamic redistribution of STAT1 protein in IFN signaling visualized by GFP fusion proteins. Eur J Biochem 260:137–144
13. Kamura T, Sato S, Haque D, Liu L, Kaelin WG Jr, Conaway RC, Conaway JW (1998) The Elongin BC complex interacts with the conserved SOCS-box motif present in members of the SOCS, ras, WD-40 repeat, and ankyrin repeat families. Genes Dev 12:3872–3881
14. Kim H, Hawley TS, Hawley RG, Baumann H (1998) Protein tyrosine phosphatase 2 (SHP-2) moderates signaling by gp130 but is not required for the induction of acute-phase plasma protein genes in hepatic cells. Mol Cell Biol 18:1525–1533
15. Sasaki A, Inagaki-Ohara K, Yoshida T, Yamanaka A, Sasaki M, Yasukawa H, Koromilas AE, Yoshimura A (2003) The N-terminal truncated isoform of SOCS3 translated from an alternative initiation AUG codon under stress conditions is stable due to the lack of a major ubiquitination site, Lys-6. J Biol Chem 278:2432–2436
16. Manteuffel R, Weber E (1983) Fluorographic detection of tritium-labelled proteins in immunoelectropherograms with the water-soluble fluor, sodium salicylate. J Biochem Biophys Methods 7:293–297

Chapter 6

Designing RNAi Screens to Identify JAK/STAT Pathway Components

Katherine H. Fisher, Stephen Brown, and Martin P. Zeidler

Abstract

The JAK/STAT signaling pathway has essential roles in multiple developmental processes, including stem cell maintenance, immune responses, and cellular proliferation. As a result, it has been extensively studied in both vertebrate systems and lower complexity models, such as *Drosophila*. Given its connection with such a wide range of biological functions, it is no surprise that pathway misregulation is frequently associated with multiple human diseases including cancer. While the core components of the pathway, and a number of negative regulators, are well known and conserved in many organisms, more subtle levels of regulation and inter-pathway crosstalk are less well understood. With the emergence of RNA interference (RNAi) as a tool to knock down gene expression and so evaluate protein function, high-throughput screens have been developed to identify pathway regulators on a genome-wide scale. Here we discuss the approaches and methods employed thus far for identification of pathway regulators using RNAi in *Drosophila*. Furthermore, we discuss possible approaches for future screens and the significant potential for applying RNAi technology in vertebrate models.

Key words: RNAi, siRNA, JAK/STAT signaling, *Drosophila*, High-throughput screening

1. Introduction

Signal transduction pathways form part of a complex communication system that allows cellular function to occur under tight spatiotemporal control. The appropriate cellular response to cues within the microenvironment is essential for the coordination of basic functions, such as proliferation and apoptosis, as well as more complex actions such as immune responses, morphogenesis, tissue development and repair. Aberrant signaling can lead to numerous human diseases, including immunodeficiencies and both solid and haematopoietic tumors (1–3). Significant research effort has been dedicated to the dissection of signal transduction cascades, leading to the identification of core pathway components and elucidation

Sandra E. Nicholson and Nicos A. Nicola (eds.), *JAK-STAT Signalling: Methods and Protocols*, Methods in Molecular Biology, vol. 967, DOI 10.1007/978-1-62703-242-1_6, © Springer Science+Business Media New York 2013

of their functions in vivo. However, complex questions regarding crosstalk of signaling pathways, more subtle levels of regulation, and combinatorial "systems" approaches, are having an increasing influence on signal transduction research. Recent advances in "omic" technologies have greatly expanded the scientist's toolkit and have provided a wealth of opportunities to investigate these issues on a much larger scale than previously possible. Indeed, whole genome analysis of pathway components and the analysis of their interconnections are resulting in significant fundamental insights and may ultimately assist in the identification of many more potential therapeutic targets.

Representing an innate defense against viral infection, RNA interference (RNAi) is a conserved gene-silencing process where the introduction of double-stranded (ds) RNAs causes endogenous mRNA degradation in a sequence-specific manner. The RNAi effect was first described in *Caenorhabditis elegans* where both injection (4) and soaking of dsRNA (5), as well as feeding *Escherichia coli* which express dsRNA (6), has proven to be a powerful and effective tool (7, 8). Injecting dsRNA into *Drosophila* embryos (9), as well as bathing cultured *Drosophila* cells in dsRNA (10) have also proven to be highly effective approaches (11–13). In both *C. elegans* and *Drosophila*, long dsRNAs are taken up into cells and processed by the Dicer machinery into short 21 bp fragments (14, 15). These short-interfering (si) RNAs then act as templates for the RISC complex which use the complementarity between the 21 bp antisense strand and cellular mRNAs to provide targeting specificity and ultimately catalyses the destruction of the transcript (Fig. 1a) (15–17).

Although the Dicer-mediated response to dsRNA is conserved in mammalian cells, the introduction of long dsRNA (>27 bp) also induces a parallel antiviral response, mediated by interferon production and cytotoxicity (18, 19). However, such interferon-induced responses are not triggered by shorter, 21 bp siRNAs, which are sufficient to form effective RISC complexes and thus mediate the desired RNAi effect (Fig. 1b) (20). Similarly, inducible plasmid-based short hairpin (sh) RNAs that utilize the micro-RNA processing mechanism have also been developed (21) with genome-wide libraries now available for both vertebrate cells and *Drosophila* in vivo systems (Fig. 1c) (22).

Since the initial identification of 21 bp siRNAs as a viable tool for gene knockdown in vertebrate cells (20, 23) significant progress has been made in the chemistry of siRNA synthesis, bioinformatic prediction of siRNA efficacy, specificity (24, 25), and delivery (26). These developments have been taken up and developed by commercial producers of siRNAs such that gene knockdown in vertebrate cells is now both efficient and an increasingly routine tool. Despite these technical advances, the costs associated with large-scale siRNA-mediated screening, the need to transfect siRNAs into mammalian

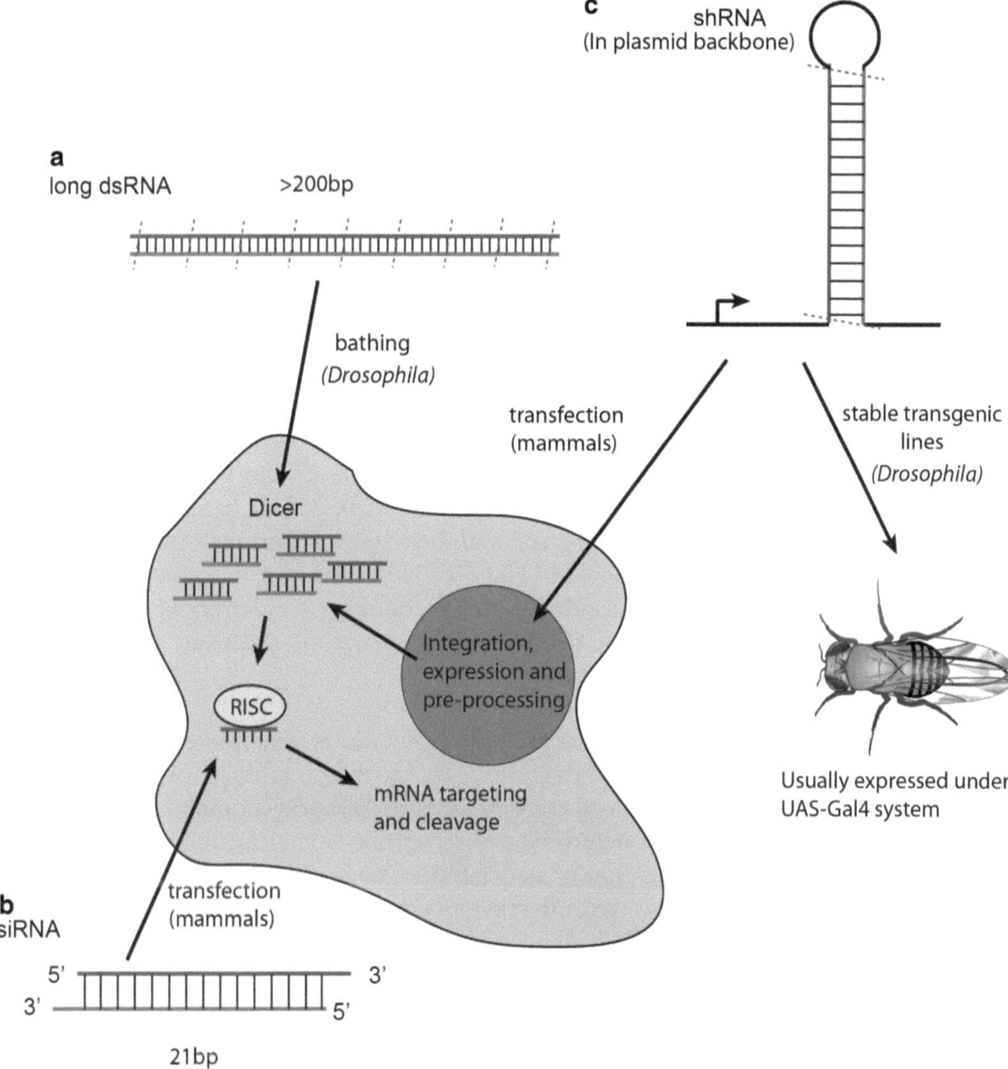

Fig. 1. Approaches to RNAi knockdown in *Drosophila* and mammalian cell culture. In *Drosophila* cell culture, long dsRNAs (**a**) are incubated in serum-free media to induce uptake and are then processed by the cellular Dicer machinery to generate 21 bp siRNAs, which interact with the RISC complex to induce target knockdown. Shorter siRNAs (**b**) transfected into mammalian cells are able to associate directly with the RISC complex and induce mRNA destruction. Plasmids encoding shRNAs (**c**) represent a third potential approach that can be transfected into cells, or injected to produce transgenic flies. shRNAs are expressed and processed into siRNAs within a cell to induce RNAi-mediated silencing.

cell lines to trigger the response and the inherent complexity and redundancy of the vertebrate genome, make large-scale screens in vertebrate cells a challenging and expensive undertaking.

By contrast, dsRNAs (usually 100–500 bp in length) can be readily designed to target *Drosophila* transcripts. They can be amplified from genomic DNA by PCR and *in vitro* transcribed to cost-effectively generate large quantities of dsRNA (27).

Furthermore, many cultured *Drosophila* cell lines readily take up dsRNAs added to the culture media, thus avoiding the need for expensive transfection reagents. In addition, the *Drosophila* genome is fully sequenced, well annotated, and has relatively low complexity with many of the primary cellular mechanisms and signaling pathway components present in vertebrates being conserved. These factors have resulted in extensive use of *Drosophila* cell lines for genome-scale high-throughput RNAi screens over recent years. Indeed, bioinformatic advances, designed to minimize off-target effects, have been used to generate second- and even third-generation RNAi libraries (28, 29) which are now widely available via facilities such as the Harvard "*Drosophila* RNAi Screening Center" (http://www.flyrnai.org/) and the University of Sheffield based "Sheffield RNAi Screening Facility" (http://www.rnai.group.shef.ac.uk/). As a result, large-scale screens have been carried out to identify novel genes involved in a wide range of cellular processes including cell viability, cell division, cell death, and signal transduction (11–13, 30, 31). Here we describe how we, and others, have utilized this technology to identify components and regulators of the JAK/STAT signaling pathway, and discuss the potential for future screens.

1.1. JAK/STAT Signaling

As described elsewhere in this volume, a simplified, linear model of JAK/STAT signalling begins with the binding of an extracellular cytokine to a transmembrane receptor complex, composed of homo- or heteromers of receptors, triggering autophosphorylation and activation of associated Janus Kinases (JAKs). The JAKs phosphorylate tyrosine residues on the receptor, forming docking sites for Signal Transduction and Activator of Transcription (STAT) proteins, which are then themselves tyrosine phosphorylated by JAKs. Phosphorylated STATs can then dissociate, dimerise, and translocate into the nucleus, where they bind to palindromic DNA binding sites and induce transcription of target genes (32–34). In humans, over fifty cytokines can signal through four JAKs and seven STATs (35), with homo- and hetero-dimerization of receptors, JAKs and STATs thought to confer tissue specificity. By contrast to the complexities of the vertebrate system, *Drosophila* JAK/STAT signaling is initiated by just three ligands (Upd, Upd2, Upd3) and transduced via one receptor (Dome), to one JAK (Hop) and one STAT (STAT92E) (Fig. 2). The fly pathway maintains high structural and functional conservation to vertebrate JAK/STAT signaling (reviewed in ref. 32) and so makes *Drosophila* an attractive, low-complexity model organism in which to study this pathway. In addition to core pathway components, negative regulators, such as SOCS-family proteins have also been identified in both systems (36–38).

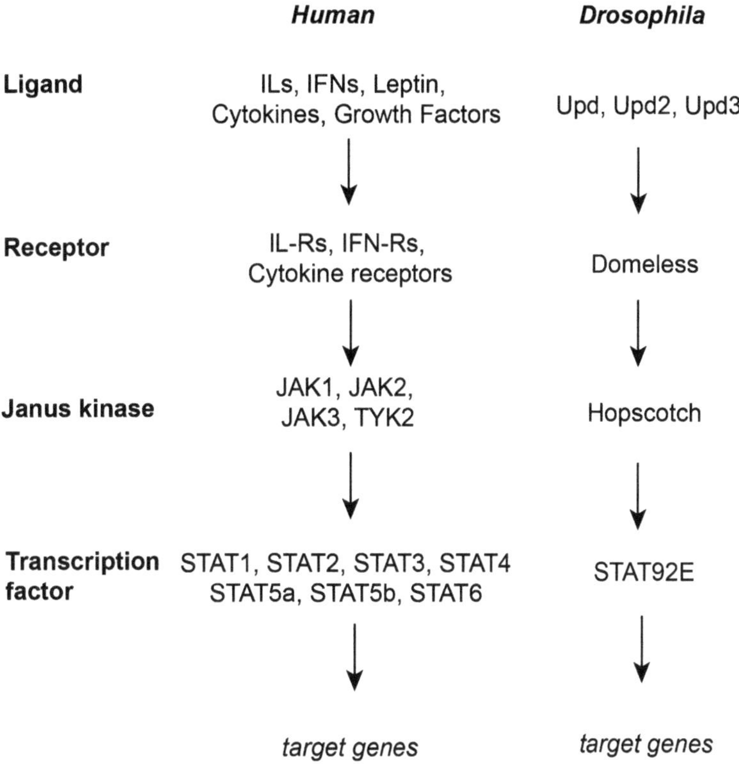

Fig. 2. Evolutionary conservation of the JAK/STAT signaling cascade. In both humans and *Drosophila*, extracellular ligands bind to a dimerized receptor, which triggers a phosphorylation cascade catalyzed by the associated JAK kinase(s). The STAT transcription factors are in turn phosphorylated, and subsequently dimerize and translocate into the nucleus to activate target gene expression. While the human pathway consists of multiple components at each stage, the *Drosophila* cascade comprises just three ligands, and one each of the receptor, kinase, and transcription factor.

2. Materials

The key requirement for undertaking genome-wide RNAi screens in *Drosophila* cells is a dsRNA library. First generation libraries including the Heidelberg HD1 collection, the Harvard *Drosophila* RNAi Screening Center Genome-wide RNAi Library version 1.0 (DRSC 1.0) and the *Silencer®* library offered commercially by Ambion have been available for some time (reviewed in ref. 28) In recent years, advances in our understanding of the biology underlying the RNAi pathway and the development of sophisticated bioinformatic tools, such as E-RNA*i* (29), have allowed the generation of significantly improved second generation libraries. The resulting Heidelberg HD2 and DRSC 2.0 libraries contain 13,900 (DRSC 2.0) to 14,587 (HD2) individual dsRNAs targeting >90% of the transcripts annotated in Flybase Genome annotation release 5.24. The dsRNAs have been designed to improve efficacy and avoid off-target matches of >19 nucleotides and CAN repeats where possible. In cases where no unique dsRNAs avoiding potential off-target effects can be designed, multiple independent dsRNAs designed to the same primary target but different potential off-targets have been developed (28). A recent study has used the JAK/STAT signaling pathway as a case study for analyzing the HD2 library. They found a significant reduction in the off-target effects observed over the earlier library version (39).

Screening itself can be undertaken either in-house using home-made dsRNAs designed using tools such as E-RNA*i* or bought in libraries (the HD2 library for example being available from the Sheffield RNAi Screening Facility (SRSF)). Alternatively, many groups have undertaken successful screens at dedicated, publically available, facilities such as the SRSF (www.rnai.group.shef.ac.uk) or the Harvard DRSC (www.flyrnai.org). In both facilities, external screeners are able to apply to undertake screens in a dedicated, state-of-the-art facility with access to both the equipment and expertise required to develop assays, screen the available dsRNA libraries and analyze their results.

3. Methods

3.1. Design Principles of RNAi Screens for JAK/STAT Components

Given the wide range of potential assays that can be developed to measure aspects of JAK/STAT signaling, it is not possible to provide a single "protocol" for the design of genome-wide RNAi screens. However, very valuable generalized methods of RNAi screening have been described elsewhere (27, 40) and our own approaches are discussed (see Notes 1–11). Previous screens as well as current and potential future screening strategies are therefore discussed below as a guide to the possibilities and potential pitfalls involved.

3.2. Transcriptional Assays

One of the most direct approaches to identify regulators of signaling cascades is to assay transcriptional activity. In 2005, two genome-scale RNAi screens were published describing the identification of JAK/STAT pathway components and regulators in *Drosophila* cells using similar, but different transcriptional reporters and dsRNA libraries (41, 42). Each of these screens used a pathway reporter consisting of multiple copies of a STAT92E-binding sequence, from the promoter of a known target gene, namely *Draf* (43) or *socs36E* (44, 45). These binding sites were placed upstream of a reporter gene encoding *Firefly* luciferase, a protein whose expression level can be accurately and easily quantified using a straightforward luminometric enzyme activity assay, and a plate reading luminometer (46). In addition, both groups used a constitutively expressed *Renilla* luciferase to normalize for transfection efficiency and cell viability (Fig. 3). As a consequence, it is not surprising that

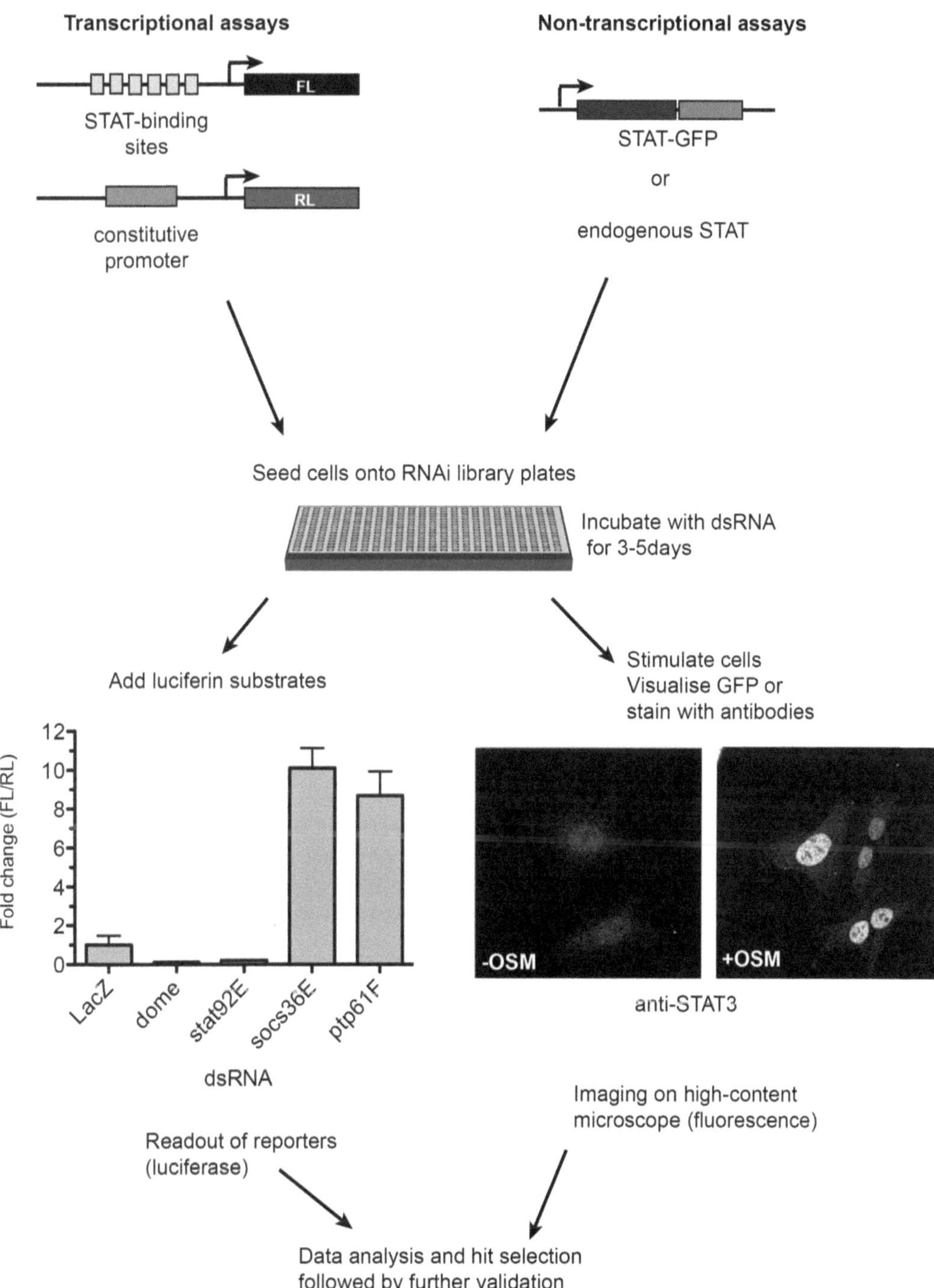

Fig. 3. Potential workflows of JAK/STAT RNAi screens. Cells containing either transcriptional assays (*left*) or non-transcriptional assays (*right*) are batch transfected with any reporter plasmids, then added to 384-well microtiter plates containing a pre-aliquotted RNAi library. Gene silencing is allowed to proceed for 3–5 days. Readouts specific to the assay used (luciferase assays and visualization of endogenous STAT3 are illustrated in this example) are then undertaken. Data is captured by luminometer/automated microscopy. Data is subsequently analyzed and potential interacting loci identified prior to further downstream validation.

Table 1
Comparison of conditions used in three JAK/STAT genome-wide reporter screens. Adapted from ref. 47

Condition	Müller screen	Baeg screen	Kallio screen
Cell line	Kc_{167}	S2-NP	S2
RNAi library	HD1	DSRC	MRC
Pathway stimulation	Transfection of *upd-GFP*	Endogenous Upd2	Transfection of Hop^{Tuml}
Pathway reporter	6×2×DrafFLuc	10×STATluc	TotM-Luc
dsRNA concentration	250 ng/well	80 ng/well	Not reported
Cells seeded	15,000/well	40,000/well	Not reported

core pathway components, namely Dome, Hop, Stat92E, Socs36E, and the previously unidentified pathway phosphatase Ptp61F, were identified in both screens (41, 42). However, technical and biological differences between the two screens (summarized in Table 1) also led to the identification of a considerable number of additional pathway regulators, only a small proportion of which were common to both screens (reviewed in ref. 47). Although not conclusively proven, the majority of differences between the screens are likely to have arisen from the alternative methods of pathway stimulation. In the Müller screen, relatively high levels of stimulation by co-transfected ligand-expressing plasmid made the assay system sensitive to the knockdown of genes that reduced pathway activity (41). By comparison, in the Baeg screen, significantly lower levels of pathway activity, induced by endogenously expressed pathway ligand, favored the identification of negative pathway regulators (42). More recently, a repeat of the Müller screen has been carried out using an improved library and analysis techniques (39). This study went on to validate hits found in both screens to rule out hits due to off-targets. A further genome-wide screen has also been undertaken to identify JAK/STAT pathway regulators functioning downstream of the constitutive gain-of-function mutation Hop^{Tuml} (Table 1) (48). This constitutively active mutation in the *Drosophila* JAK homologue results in the over-proliferation of haematopoietic cells and the development of melanized blood cell tumors (49), a phenotype with similarities to the gain-of-function mutations in human JAK2 that underlies the development of human myeloproliferative neoplasias (50). This group used a further STAT92E responsive luciferase reporter based on sequences within the promoter of the immune response gene TotM, and identified five novel regulators, including a short co-receptor, named Eye transformer, which represses pathway activity.

While genome-wide screening methods have been successful in identifying many JAK/STAT regulators, smaller scale focused screens offer a cheaper alternative to investigate specific processes or stages of pathway regulation. Furthermore, the smaller library size has the potential to allow more replicates as well as more focused assay development and optimization. For example, a focused, high-resolution screen of the cellular endocytic machinery has recently addressed the role of endocytic trafficking as a key mechanism by which the JAK/STAT pathway is regulated, via the degradation of the ligand–receptor complex (51). While this screen reidentified a number of effectors including Rab5 and TSG101 that had previously been discovered in genome scale screens (41), its greater sensitivity also identified a number of weaker effectors, which may not have been recognized otherwise.

Although *Drosophila* JAK/STAT RNAi screens have successfully identified multiple regulators, similar genome-wide efforts to identify human pathway interactors have not yet been undertaken. However, signaling mediated by the four JAK-like kinases and seven STAT transcription factors has benefited from an extensive history of molecular tool development. This includes *Firefly* luciferase-based transcriptional reporters, which report STAT activity in a wide range of tissues and stimulation conditions (52, 53). Similarly, both genome-wide and subset libraries of siRNAs/shRNAs are available from a number of commercial companies including Dharmacon, Invitrogen, and Sigma and have proven their worth in a range of other screens (54, 55).

One publication has recently bridged the gap between the completed *Drosophila*-based genome-wide screens and the higher complexity of the vertebrate system (56). Using assays for STAT1 and STAT3 activity, the authors investigated human homologues of JAK/STAT regulators originally identified in *Drosophila*, by siRNA. While the applicability of this study in other cell types and in vivo systems remains to be established, no less than 95% of proteins with sequence conservation between *Drosophila* and humans also displayed functional conservation and had a significant effect on JAK/STAT signaling in HeLa cells.

3.3. Non-transcriptional Assays

Although no RNAi screens for JAK/STAT pathway regulators based on non-transcriptional assays have been published, a number of approaches do suggest themselves as potentially worthwhile. Some possibilities are discussed below in more detail.

Functional assays. A number of cell lines require JAK/STAT pathway activity for their survival in culture. These include mouse Ba/F3 cells, which require interleukin (IL)-3 stimulation for their proliferation and growth, and mouse embryonic stem cells (ESCs) which require the JAK/STAT pathway ligand leukemia inhibitory factor (LIF) for their continued proliferation and pluripotency (57). Thus, both cell systems could potentially represent functional models to identify loci whose knockdown is sufficient to either induce signaling or bypass the need for the pathway altogether. Given that such a screen would have a very simple cell death/cell survival readout and could identify genes who's loss-of-function phenotype is to allow proliferation, such a screen clearly has the possibility to identify pathway regulators within a disease relevant cellular context.

STAT subcellular translocation. One fundamental aspect of JAK/STAT signaling is the requirement for STATs to be activated by JAKs at the plasma membrane, before translocating to the nucleus to bind to the promoters of their target genes (58, 59). Thus, it is not surprising that this cytoplasmic/nuclear transport is a potential mechanism of pathway regulation. Indeed, given the recent insights into the potential roles of STATs in heterochromatin (60–62), the regulation of STAT shuttling may well represent an important epigenetic factor even in the absence of canonical pathway stimulation. By assessing this translocation using a cell line expressing a GFP-tagged STAT or immunohistochemical staining of endogenously expressed STATs, a screen could relatively easily be developed to identify these regulators of this aspect of pathway signaling using high-content imaging approaches (Fig. 3).

Dimerization of pathway components. The cell biology that regulates the physical interactions between JAK/STAT pathway components is a key requirement for signaling and an obvious regulatory mechanism. As such, an assay could readily be developed to investigate these interactions using the well-established bimolecular complementation techniques, which are now available (63, 64). Previously, one study used a β-galactosidase complementation system to interrogate dimerization of the *Drosophila* receptor, Dome, in vivo (65). This work demonstrated that Dome is only dimerized in selected embryonic tissues, and does not require pathway ligand for this process. Given that the factors conferring tissue specificity of dimerization in vivo are yet to be established, a cell-based screen for factors that can either promote or hinder the dimerization of either Dome:Dome complexes or even the recently demonstrated Dome:Latran complex (66) have the potential to cast a light on this previously unexplored aspect of JAK/STAT pathway regulation. Ultimately, similar bimolecular complementation approaches could also be applied to study other protein:protein interactions central to JAK/STAT pathway activity. For example, interaction between receptors and JAKs, JAKs and STATs, or even STATs and HP1 (67) would have the potential to dissect the cell biology of pathway signaling at a previously unattainable resolution.

3.4. Future Directions

Although the field of JAK/STAT-related RNAi screening is still young, a number of questions arising in the wider field of JAK/ STAT research could be addressed using this technology. Such potential screens could include as follows.

Differential regulators of ligand stimulated JAK2 vs. JAK2 V617F-mediated signaling—differences that are potentially key to the development of drugs specific to the mutant gain-of-function JAK2 but would still allow normal signaling by the wild-type protein.

Assays to identify the factors required to mediate the noncanonical heterochromatin-related aspects of STAT—a field that is likely to become increasingly important as our understanding of the noncanonical roles of the JAK/STAT pathway grows.

Screens to identify the factors that provide pathway specificity—an important and largely unanswered question for a pathway in which many ligands activate many STATs all of which have very similar DNA binding sequences yet still manage to produce tissue and ligand appropriate responses.

3.5. Conclusions

One of the key features to emerge from RNAi screens investigating JAK/STAT signaling, is that while the number of "core" pathway components is modest, the landscape of genes that can be identified as regulators or modifiers of pathway signaling is surprisingly broad. Indeed, multiple cellular processes such as the modification, secretion and trafficking of ligands, the dimerization and endocytic processing of receptor complexes and the stability, modification, and translocation of protein complexes all represent cell biological mechanisms likely to impact on the JAK/STAT signaling cascade. Future, more sophisticated, screens will identify genes with specific roles in these particular aspects of pathway signaling and will thus allow the classification of regulators specific for particular aspects of JAK/STAT signaling. In this way, genes already identified will be validated and their signaling contexts established, progress that will no doubt be furthered by parallel efforts to directly examine both their in vivo and *in vitro* roles.

Crosstalk between the JAK/STAT and other intracellular signaling pathways is a critical requirement for development, homeostasis, and disease (68). Indeed, a significant proportion of the "non-core" interacting loci identified by JAK/STAT screens are also likely to play roles in other signaling pathways. The interrogation of RNAi screen databases using bioinformatics tools will increasingly identify such promiscuous pathway regulators. The biological elucidation of their functions and mechanisms will no doubt represent an important aspect of signaling research in future years. As such, the establishment of assays that can investigate crosstalk will ultimately be of great benefit.

Perhaps the biggest future challenge, however, rests in the translation of information gained in low-complexity *Drosophila* screening systems to vertebrate signaling. Results from modern, second-generation siRNA libraries combined with both transcriptional and non-transcriptional assays designed to interrogate the vertebrate JAK/STAT pathway have the potential to revolutionize the field. Furthermore, disease-derived cell lines, such as JAK2 V617F positive myeloid leukemia cells, will no doubt provide cellular models well suited for the identification of future drug targets. It will be fascinating to experience the progress that will undoubtedly be made in the coming years.

4. Notes

A number of considerations that should be made when planning an RNAi screen are listed below.

1. Batch transfect cells or generate stable reporters that are mixed before addition to the dsRNA library, so as to ensure that all wells have a similar population of cells in which to perform the assay.
2. If possible use a single batch of media, serum, antibiotics, and transfection reagents for the whole screen.
3. Stagger start times to ensure that all plates are read at a similar time. Purchase enough assay reagents (e.g., luciferin or antibodies) of a single batch to complete the entire screen.
4. Consider the logistics of handling the number of plates required to undertake a genome scale screen, sometimes in multiple replicates. Think carefully about the steps required and the time they will take within a practical workflow that is also compatible with the assay you have developed.
5. Choose the number and position of controls carefully. The choice of controls will largely depend on the assay, but broadly speaking should include, at least one negative RNAi control, which has no effect on the cells (LacZ, GFP, and Gal4 are common choices), plus positive controls, which are known to increase or decrease the phenotype of interest.
6. After primary screening, rescreen to identify initial false positives. Recent examination, by the Harvard *Drosophila* RNAi Screening Center, has established that robust assays identify fewer false positives; however, no screen is exempt from false positives (69).
7. Consider designing alternative RNAi probes to genes that were identified as hits to ensure against off-target effects. Secondary assays often begin with a repeat of the original screen, but

alternatives may also be appropriate, using different reporter constructs and the use of different RNAi probes.

8. Data analysis, hit validation, and the eventual biological confirmation of the results arising are likely to represent the most challenging aspect of any RNAi screen. Screening is ultimately an enrichment process and while false negatives will mean some loci are not identified, due to factors such as inefficient knockdown or protein perdurance, false positives will also occur, in particular due to off-target effects of dsRNA probes. As such, even the best-planned assays will be prone to artifacts that need to be identified during subsequent validation approaches (69, 70).
9. CellHTS2, which is implemented in R/Bioconductor, has been written to analyze cell-based high-throughput RNAi screens (71). This software is ideal for luciferase-based data analysis, as it offers a number of normalization and hit scoring methods, as well as allowing for the removal of contaminated wells and plates. Outputs to assess data quality can be easily created, allowing for the removal of data consisting of edge effects or from bulk liquid handling errors. Although dual luciferase reporter screens create two sets of data, which therefore provide a ratio-based output (i.e., FL/RL), we also recommend that the data sets be treated separately during the quality control process and prior to hit selection.
10. High-content microscopy-based screens are usually analyzed through microscope-specific software, collecting and analyzing large image-based datasets from each well. However, when compared with plate reader screens, assay development can take longer, the experiment can be more involved and it is more time consuming to gather and process screen data. The long-term storage of images and the accompanying meta-data can also be a significant challenge. The primary route to a successful image-based screen is to establish a cell density at plating that allows your microscope to take focused images of enough cells that can be segmented without confusion from clumps or overly confluent populations, while maintaining a large enough statistical sample. The result of this is partly based upon cell growth and processing of the sample (aspiration and dispensing) before data acquisition.
11. During the development of the experimentation, analysis algorithms need to be developed and may even require fine tuning after screening. Furthermore, with increases in processor and algorithm speed, multiple analyses are becoming possible and will undoubtedly allow additional multi-parametric data to be extracted from screens in the future (72).

Acknowledgments

The authors would like to thank Amy Taylor for technical assistance with RNAi screens undertaken in the Zeidler lab as well as Patrick Müller and Sabina Haeder for their initial work on developing cell-based JAK/STAT assays. Michael Boutros kindly provided the HD2 library used at the SRSF, which is supported by a Wellcome Trust equipment award. MZ is a Cancer Research UK Senior Cancer Research Fellow and member of the MRC Centre for Biomedical and Developmental Biology. KF was supported by the EU Framework 7 "Cancer Pathways" project.

References

1. Altomare DA, You H, Xiao GH, Ramos-Nino ME, Skele KL, De Rienzo A, Jhanwar SC, Mossman BT, Kane AB, Testa JR (2005) Human and mouse mesotheliomas exhibit elevated AKT/PKB activity, which can be targeted pharmacologically to inhibit tumor cell growth. Oncogene 24:6080–6089
2. Bienz M, Clevers H (2000) Linking colorectal cancer to Wnt signaling. Cell 103:311–320
3. Taipale J, Beachy PA (2001) The Hedgehog and Wnt signalling pathways in cancer. Nature 411:349–354
4. Fire A, Xu S, Montgomery MK, Kostas SA, Driver SE, Mello CC (1998) Potent and specific genetic interference by double-stranded RNA in Caenorhabditis elegans. Nature 391:806–811
5. Tabara H, Grishok A, Mello CC (1998) RNAi in C. elegans: soaking in the genome sequence. Science 282:430–431
6. Timmons L, Fire A (1998) Specific interference by ingested dsRNA. Nature 395:854
7. Fraser AG, Kamath RS, Zipperlen P, Martinez-Campos M, Sohrmann M, Ahringer J (2000) Functional genomic analysis of C. elegans chromosome I by systematic RNA interference. Nature 408:325–330
8. Kamath RS, Martinez-Campos M, Zipperlen P, Fraser AG, Ahringer J (2001) Effectiveness of specific RNA-mediated interference through ingested double-stranded RNA in Caenorhabditis elegans. Genome Biol 2(1):RESEARCH0002
9. Kennerdell JR, Carthew RW (1998) Use of dsRNA-mediated genetic interference to demonstrate that frizzled and frizzled 2 act in the wingless pathway. Cell 95:1017–1026
10. Clemens JC, Worby CA, Simonson-Leff N, Muda M, Maehama T, Hemmings BA, Dixon JE (2000) Use of double-stranded RNA interference in *Drosophila* cell lines to dissect signal transduction pathways. Proc Natl Acad Sci U S A 97:6499–6503
11. Boutros M, Kiger AA, Armknecht S, Kerr K, Hild M, Koch B, Haas SA, Paro R, Perrimon N (2004) Genome-wide RNAi analysis of growth and viability in *Drosophila* cells. Science 303:832–835
12. Goshima G, Wollman R, Goodwin SS, Zhang N, Scholey JM, Vale RD, Stuurman N (2007) Genes required for mitotic spindle assembly in *Drosophila* S2 cells. Science 316:417–421
13. Saj A, Arziman Z, Stempfle D, van Belle W, Sauder U, Horn T, Durrenberger M, Paro R, Boutros M, Merdes G (2010) A combined ex vivo and in vivo RNAi screen for notch regulators in *Drosophila* reveals an extensive notch interaction network. Dev Cell 18:862–876
14. Macrae IJ, Zhou K, Li F, Repic A, Brooks AN, Cande WZ, Adams PD, Doudna JA (2006) Structural basis for double-stranded RNA processing by Dicer. Science 311:195–198
15. Bernstein E, Caudy AA, Hammond SM, Hannon GJ (2001) Role for a bidentate ribonuclease in the initiation step of RNA interference. Nature 409:363–366
16. Hammond SM, Bernstein E, Beach D, Hannon GJ (2000) An RNA-directed nuclease mediates post-transcriptional gene silencing in *Drosophila* cells. Nature 404:293–296
17. Hannon GJ (2002) RNA interference. Nature 418:244–251
18. Reynolds A, Anderson EM, Vermeulen A, Fedorov Y, Robinson K, Leake D, Karpilow J, Marshall WS, Khvorova A (2006) Induction of the interferon response by siRNA is cell type- and duplex length-dependent. RNA 12: 988–993

19. Manche L, Green SR, Schmedt C, Mathews MB (1992) Interactions between double-stranded RNA regulators and the protein kinase DAI. Mol Cell Biol 12:5238–5248
20. Elbashir SM, Harborth J, Lendeckel W, Yalcin A, Weber K, Tuschl T (2001) Duplexes of 21-nucleotide RNAs mediate RNA interference in cultured mammalian cells. Nature 411:494–498
21. Chang K, Elledge SJ, Hannon GJ (2006) Lessons from nature: microRNA-based shRNA libraries. Nat Methods 3:707–714
22. Dietzl G, Chen D, Schnorrer F, Su KC, Barinova Y, Fellner M, Gasser B, Kinsey K, Oppel S, Scheiblauer S, Couto A, Marra V, Keleman K, Dickson BJ (2007) A genome-wide transgenic RNAi library for conditional gene inactivation in *Drosophila*. Nature 448:151–156
23. Caplen NJ, Parrish S, Imani F, Fire A, Morgan RA (2001) Specific inhibition of gene expression by small double-stranded RNAs in invertebrate and vertebrate systems. Proc Natl Acad Sci U S A 98:9742–9747
24. Amarzguioui M, Prydz H (2004) An algorithm for selection of functional siRNA sequences. Biochem Biophys Res Commun 316: 1050–1058
25. Reynolds A, Leake D, Boese Q, Scaringe S, Marshall WS, Khvorova A (2004) Rational siRNA design for RNA interference. Nat Biotechnol 22:326–330
26. McCaffrey AP, Meuse L, Pham TT, Conklin DS, Hannon GJ, Kay MA (2002) RNA interference in adult mice. Nature 418:38–39
27. Steinbrink S, Boutros M (2008) RNAi screening in cultured *Drosophila* cells. Methods Mol Biol 420:139–153
28. Horn T, Sandmann T, Boutros M (2010) Design and evaluation of genome-wide libraries for RNA interference screens. Genome Biol 11:R61
29. Horn T, Boutros M (2010) E-RNAi: a web application for the multi-species design of RNAi reagents–2010 update. Nucleic Acids Res 38:W332–W339
30. Ragab A, Buechling T, Gesellchen V, Spirohn K, Boettcher AL, Boutros M (2011) *Drosophila* Ras/MAPK signalling regulates innate immune responses in immune and intestinal stem cells. EMBO J 30:1123–1136
31. Yi CH, Sogah DK, Boyce M, Degterev A, Christofferson DE, Yuan J (2007) A genome-wide RNAi screen reveals multiple regulators of caspase activation. J Cell Biol 179:619–626
32. Arbouzova NI, Zeidler MP (2006) JAK/STAT signalling in *Drosophila*: insights into conserved regulatory and cellular functions. Development 133:2605–2616
33. Darnell JEJ (1997) STATs and gene regulation. Science 277:1630–1635
34. O'Sullivan LA, Liongue C, Lewis RS, Stephenson SE, Ward AC (2007) Cytokine receptor signaling through the Jak-Stat-Socs pathway in disease. Mol Immunol 44:2497–2506
35. Kisseleva T, Bhattacharya S, Braunstein J, Schindler CW (2002) Signaling through the JAK/STAT pathway, recent advances and future challenges. Gene 285:1–24
36. Endo TA, Masuhara M, Yokouchi M, Suzuki R, Sakamoto H, Mitsui K, Matsumoto A, Tanimura S, Ohtsubo M, Misawa H, Miyazaki T, Leonor N, Taniguchi T, Fujita T, Kanakura Y, Komiya S, Yoshimura A (1997) A new protein containing an SH2 domain that inhibits JAK kinases. Nature 387:921–924
37. Naka T, Narazaki M, Hirata M, Matsumoto T, Minamoto S, Aono A, Nishimoto N, Kajita T, Taga T, Yoshizaki K, Akira S, Kishimoto T (1997) Structure and function of a new STAT-induced STAT inhibitor. Nature 387:924–929
38. Starr R, Willson TA, Viney EM, Murray LJ, Rayner JR, Jenkins BJ, Gonda TJ, Alexander WS, Metcalf D, Nicola NA, Hilton DJ (1997) A family of cytokine-inducible inhibitors of signalling. Nature 387:917–921
39. Fisher KH, Wright VM, Taylor A, Zeidler MP, Brown S (2012) Advances in genome-wide RNAi cellular screens: a case study using the Drosophila JAK/STAT pathway. BMC Genomics 13:506
40. Armknecht S, Boutros M, Kiger A, Nybakken K, Mathey-Prevot B, Perrimon N (2005) High-throughput RNA interference screens in *Drosophila* tissue culture cells. Methods Enzymol 392:55–73
41. Müller P, Kuttenkeuler D, Gesellchen V, Zeidler MP, Boutros M (2005) Identification of JAK/STAT signalling components by genome-wide RNA interference. Nature 436:871–875
42. Baeg GH, Zhou R, Perrimon N (2005) Genome-wide RNAi analysis of JAK/STAT signaling components in *Drosophila*. Genes Dev 19:1861–1870
43. Kwon EJ, Park HS, Kim YS, Oh EJ, Nishida Y, Matsukage A, Yoo MA, Yamaguchi M (2000) Transcriptional regulation of the *Drosophila* raf proto-oncogene by *Drosophila* STAT during development and in immune response. J Biol Chem 275:19824–19830
44. Karsten P, Hader S, Zeidler MP (2002) Cloning and expression of *Drosophila* SOCS36E and its potential regulation by the JAK/STAT pathway. Mech Dev 117:343–346
45. Callus BA, Mathey-Prevot B (2002) SOCS36E, a novel *Drosophila* SOCS protein, suppresses

JAK/STAT and EGF-R signalling in the imaginal wing disc. Oncogene 21:4812–4821

46. Gould SJ, Subramani S (1988) Firefly luciferase as a tool in molecular and cell biology. Anal Biochem 175:5–13
47. Müller P, Boutros M, Zeidler MP (2008) Identification of JAK/STAT pathway regulators–insights from RNAi screens. Semin Cell Dev Biol 19:360–369
48. Kallio J, Myllymaki H, Gronholm J, Armstrong M, Vanha-aho LM, Makinen L, Silvennoinen O, Valanne S, Ramet M (2010) Eye transformer is a negative regulator of *Drosophila* JAK/STAT signaling. FASEB J 24:4467–4479
49. Luo H, Hanratty WP, Dearolf CR (1995) An amino acid substitution in the *Drosophila* hop-Tum-l Jak kinase causes leukemia-like hematopoietic defects. EMBO J 14:1412–1420
50. Kilpivaara O, Levine RL (2008) JAK2 and MPL mutations in myeloproliferative neoplasms: discovery and science. Leukemia 22: 1813–1817
51. Vidal OM, Stec W, Bausek N, Smythe E, Zeidler MP (2010) Negative regulation of *Drosophila* JAK-STAT signalling by endocytic trafficking. J Cell Sci 123:3457–3466
52. Chau MN, Banerjee PP (2008) Development of a STAT3 reporter prostate cancer cell line for high throughput screening of STAT3 activators and inhibitors. Biochem Biophys Res Commun 377:627–631
53. Sillaber C, Gesbert F, Frank DA, Sattler M, Griffin JD (2000) STAT5 activation contributes to growth and viability in Bcr/Abl-transformed cells. Blood 95:2118–2125
54. Westbrook TF, Martin ES, Schlabach MR, Leng Y, Liang AC, Feng B, Zhao JJ, Roberts TM, Mandel G, Hannon GJ, Depinho RA, Chin L, Elledge SJ (2005) A genetic screen for candidate tumor suppressors identifies REST. Cell 121:837–848
55. Slabicki M, Theis M, Krastev DB, Samsonov S, Mundwiller E, Junqueira M, Paszkowski-Rogacz M, Teyra J, Heninger AK, Poser I, Prieur F, Truchetto J, Confavreux C, Marelli C, Durr A, Camdessanche JP, Brice A, Shevchenko A, Pisabarro MT, Stevanin G, Buchholz F (2010) A genome-scale DNA repair RNAi screen identifies SPG48 as a novel gene associated with hereditary spastic paraplegia. PLoS Biol 8:e1000408
56. Müller P, Pugazhendhi D, Zeidler MP (2012) Modulation of human JAK/STAT pathway signalling by functionally conserved regulators. JAK-STAT 1(1):1–11
57. Williams RL, Hilton DJ, Pease S, Willson TA, Stewart CL, Gearing DP, Wagner EF, Metcalf D, Nicola NA, Gough NM (1988) Myeloid leukaemia inhibitory factor maintains the developmental potential of embryonic stem cells. Nature 336:684–687
58. Marg A, Shan Y, Meyer T, Meissner T, Brandenburg M, Vinkemeier U (2004) Nucleocytoplasmic shuttling by nucleoporins Nup153 and Nup214 and CRM1-dependent nuclear export control the subcellular distribution of latent Stat1. J Cell Biol 165: 823–833
59. Liu L, McBride KM, Reich NC (2005) STAT3 nuclear import is independent of tyrosine phosphorylation and mediated by importin-alpha3. Proc Natl Acad Sci U S A 102:8150–8155
60. Brown S, Zeidler MP (2008) Unphosphorylated STATs go nuclear. Curr Opin Genet Dev 18: 455–460
61. Griffiths DS, Li J, Dawson MA, Trotter MW, Cheng YH, Smith AM, Mansfield W, Liu P, Kouzarides T, Nichols J, Bannister AJ, Green AR, Gottgens B (2011) LIF-independent JAK signalling to chromatin in embryonic stem cells uncovered from an adult stem cell disease. Nat Cell Biol 13:13–21
62. Shi S, Calhoun HC, Xia F, Li J, Le L, Li WX (2006) JAK signaling globally counteracts heterochromatic gene silencing. Nat Genet 38:1071–1076
63. Rossi FM, Blakely BT, Charlton CA, Blau HM (2000) Monitoring protein-protein interactions in live mammalian cells by beta-galactosidase complementation. Methods Enzymol 328:231–251
64. Kerppola TK (2006) Design and implementation of bimolecular fluorescence complementation (BiFC) assays for the visualization of protein interactions in living cells. Nat Protoc 1:1278–1286
65. Brown S, Hu N, Hombria JC (2003) Novel level of signalling control in the JAK/STAT pathway revealed by in situ visualisation of protein-protein interaction during *Drosophila* development. Development 130: 3077–3084
66. Makki R, Meister M, Pennetier D, Ubeda JM, Braun A, Daburon V, Krzemien J, Bourbon HM, Zhou R, Vincent A, Crozatier M (2010) A short receptor downregulates JAK/STAT signalling to control the *Drosophila* cellular immune response. PLoS Biol 8:e1000441
67. Shi S, Larson K, Guo D, Lim SJ, Dutta P, Yan SJ, Li WX (2008) *Drosophila* STAT is required for directly maintaining HP1 localization and heterochromatin stability. Nat Cell Biol 10: 489–496

68. Ugo V, Marzac C, Teyssandier I, Larbret F, Lecluse Y, Debili N, Vainchenker W, Casadevall N (2004) Multiple signaling pathways are involved in erythropoietin-independent differentiation of erythroid progenitors in polycythemia vera. Exp Hematol 32:179–187
69. Booker M, Samsonova AA, Kwon Y, Flockhart I, Mohr SE, Perrimon N (2011) False negative rates in *Drosophila* cell-based RNAi screens: a case study. BMC Genomics 12:50
70. Mohr S, Bakal C, Perrimon N (2010) Genomic screening with RNAi: results and challenges. Annu Rev Biochem 79:37–64
71. Boutros M, Bras LP, Huber W (2006) Analysis of cell-based RNAi screens. Genome Biol 7:R66
72. Axelsson E, Sandmann T, Horn T, Boutros M, Huber W, Fischer B (2011) Extracting quantitative genetic interaction phenotypes from matrix combinatorial RNAi. BMC Bioinformatics 12:342

Chapter 7

The Use of JAK-Specific Inhibitors as Chemical Biology Tools

Christopher J. Burns, David Segal, and Andrew F. Wilks

Abstract

The JAK family of protein tyrosine kinases are now recognized as important participants in a wide range of pathologies, from cancer to inflammatory diseases. In the last decade, the drive to develop drugs targeting members of this family has begun to deliver a panel of small molecule inhibitors of JAK family members, with a range of potencies and specificities. A number of these compounds have already found widespread use as biochemical tools in the elucidation of JAK activity in specific signaling and disease processes; however, many of the first generation compounds are poorly characterized with suboptimal potencies and selectivities.

Herein, we present the data for those small molecule JAK inhibitors that have been described in the peer-reviewed literature and the benefits and potential issues that may be associated with the use of these tool compounds.

Key words: JAK1, JAK2, JAK3, TYK2, Kinase inhibitor, JAK inhibitor, Chemical probe

1. Introduction

The JAK family of protein tyrosine kinases is an evolutionarily "recent" addition to the broader kinase family, coming into being some time after the Cambrian explosion ~600 million years ago, when insects (which have a member of the JAK family, viz. *Drosophila melanogaster Hopscotch*) and nematodes (*Caenorhabditis elegans* has no JAK family member) last shared a common ancestor. In the human kinome, there are four JAK family members: JAK1, JAK2, JAK3, and TYK2. Each are cytoplasmic kinases that are critical components of a number of intracellular pathways, but the family is best known as transducers of signals that flow from the binding of a range of cytokines to their cognate receptors. In this capacity, one of their key roles is the activation of members of the STAT family of transcription factors from latency to potent and specific activators of genetic programs that define cellular fate, proliferation, and purpose. In particular, members of the JAK family

Sandra E. Nicholson and Nicos A. Nicola (eds.), *JAK-STAT Signalling: Methods and Protocols*, Methods in Molecular Biology, vol. 967, DOI 10.1007/978-1-62703-242-1_7,

play critical roles in a wide range of physiological settings involving cytokines, including hematopoiesis and humoral and cellular immune responses. In addition, (or perhaps because of this) the JAKs are integral to the pathophysiology of a wide range of disease states such as immunodeficiency, autoimmunity, and cancer.

A number of mutations and rearrangements in the genes of members of the JAK family have been associated with leukemias and myeloproliferative disorders. The JAK2 gene, for example, has been shown to recombine with several other genes, a process that leads to the formation of a fused oncogenic JAK2 allele, and the expression of hyperactive JAK2 fusion proteins that drive the oncogenic process. The JAK2 gene has been shown to be recombined with the transcription factor TEL (childhood T cell and B cell acute lymphoblastic leukemia) (1), PCM1 (T cell lymphoma (2), atypical chronic myeloid leukemia (CML) (3), and acute erythroid leukemia (4)), BCR (atypical chronic myeloid leukemia (CML)) (5), and RPN1 (chronic idiopathic myelofibrosis) (6). In all cases it is believed that the kinase domain contributed by the JAK2 gene is responsible for the oncogenesis of these cells. More recently, point mutations in JAK1 and JAK2 have been identified at a high frequency in children with BCR-ABL1-negative ALL characterized by deletion of IKZF1 and poor prognosis (7), but the major driver of the recent burst of activity in JAK drug development has arisen because point mutations in JAK2 are also very common in myeloproliferative disorders, with an activating mutation of the kinase JAK2 (JAK2V617F) occurring in most Polycythemia Vera cases as well as in ~50% of patients with Essential Thrombocythemia or Primary Myelofibrosis.

The use of small molecule chemical probes to support data from genetic and proteomic studies has been widely adopted in biomedical research. Chemical probe compounds are defined as discrete molecular entities with well-defined activity (potency and selectivity) that can be used in in vitro, and ideally in vivo, studies (8). The advantages of small molecule chemical probes are manifold and include the ability to titrate effects in a dose- and time-controlled manner, the ability to rapidly add and remove the probe from the system, and the ability to use the probe as a starting point in the development of novel therapeutics. Nonetheless chemical probes do have drawbacks particularly with regard to off-target activities, an issue that is particularly acute when dealing with ATP competitive kinase inhibitors where the binding site of these chemical probes is invariably the ATP binding site, a site each kinase shares in high degree with more than 500 other cellular proteins.

Research in the field of the JAK kinases has been aided considerably by the ready availability of small molecule JAK inhibitors and with a number of JAK inhibitors undergoing clinical trials, compounds with well-defined activity and pharmacokinetic profiles can now be used to confidently inhibit JAK activity for in vitro and in vivo studies. At the time of writing, specific inhibitors of JAK2 and JAK3 have been described in the literature, along with dual inhibitors of JAK1 and JAK2 and pan-JAK inhibitors, however no selective inhibitors of JAK1 or TYK2 have been reported.

2. Materials

First generation JAK inhibitors are widely available from many international providers of biochemical reagents (e.g., Calbiochem, Tocris Bioscience, Santa Cruz Biotechnology, Biomol *inter alia*). Second generation JAK inhibitors that have been developed by biotechnology and pharmaceutical companies for clinical use may be available directly from the innovator company through a collaborative research agreement. In many cases, they are also available in small quantities for research purposes only from specialist chemical providers sourced through the internet (e.g., SynKinase, Selleck Chemicals, LC Laboratories, Symansis *inter alia*).

3. Methods

3.1. Introduction

Purported JAK inhibitors have been available to the research community for more than a decade. The initial set of JAK inhibitors are rather "rough and ready" and leave much to be desired in terms of potency, specificity, and pharmacology. These first generation JAK inhibitors, then, are those compounds that have been used as chemical probes in JAK research over the past decade and represent compounds typically with moderate JAK inhibitory activity and largely uncharacterized selectivity. More recent JAK inhibitors that have been developed are more polished, and therefore superior to their first generation counterparts. Second generation compounds are defined as those that have been rationally designed using structural biology input and structure–activity relationships, and possess optimized activity and pharmacokinetics, and minimal toxicological and off-target side effects. Compounds that fall into this category have been designed for use in vivo and, indeed, some have entered clinical trials. Such second generation compounds generally fulfill the guidelines for a useful kinase chemical probe as defined by Cohen and coworkers (9). We have therefore divided this discussion into two distinct categories based upon the first generation/second generation divide.

For clarity we have only focused on compounds that inhibit JAK activity specifically, rather than compounds with activity against STAT proteins or reported activity on the JAK-STAT signaling cascade.

The vast majority of JAK inhibitors reported to date bind to the ATP binding site of the kinase and are ATP competitive. The enzyme inhibitory activity of the compounds, determined using a variety of assay methodologies, is generally determined at low ATP concentrations (typically at Km of enzyme, 10–100 μM). The cellular activity of the compounds is typically determined in proliferation assays or through assessment of the effects on downstream phosphorylation (e.g., of STAT proteins) and given the intracellular ATP concentration is 1–5 mM, biochemical activity of the compounds often overestimates the cellular activity by 10- to 50-fold.

3.2. First Generation Inhibitors

"JAK2" Inhibitors. The most widely used (purported) JAK2 inhibitor is the tyrphostin AG490. This compound was originally described as a selective JAK2 inhibitor (10), though data published subsequently by independent researchers indicates the compound has minimal inhibitory activity on JAK2 and JAK3 (IC_{50} > 10 μM) and weak JAK1 activity (IC_{50} ~3.4 μM) (11). Furthermore the compound has been shown to inhibit activation of cyclin-dependent kinases (12) and has significant activity against c-SRC (13) and potently inhibits EGFR kinases (14). This compound has been used in vivo though we have found the compound is both chemically and metabolically unstable leading to a short-half life and minimal exposure (unpublished data). A structural analogue of AG490, LS-104 (15), has also been reported as a JAK2 inhibitor (IC_{50} 0.1–0.6 μM) however given the poor activity and stability of AG490, it is unlikely this compound possesses optimal properties as a JAK inhibitor probe. Considering the fact that AG490 is neither a potent nor specific JAK2 inhibitor, it will be important to critically review all of the data generated using this molecule. We suspect that much will have to be revised in the light of improved data collected from experiments that make use of the improved second generation inhibitors (Fig. 1).

A number of JAK2 inhibitor probes have been identified through a virtual screening process combined with cellular studies. Thus, G6 (16), SD1008 (17), Z3 (18), and 1,2,3,4,5,6-hexabromocyclohexane (19) have all been reported to inhibit JAK2 phosphorylation in cells or to inhibit cellular proliferation of cells known to be dependent on JAK2 activity. Furthermore, G6 was shown to inhibit JAK2 activity in a cell-free system (IC_{50} 60 nM). These compounds are structurally distinct and considerably different to other known JAK2 ATP competitive inhibitors and may therefore represent chemical probes with unique characteristics. Further studies will be needed to confirm the preliminary data, though the risk that 1,2,3,4,5,6-hexabromocyclohexane would nonspecifically alkylate proteins (through attack at bromine by nucleophilic aminoacids such as cysteine) is high and renders this compound the least interesting of these novel probes.

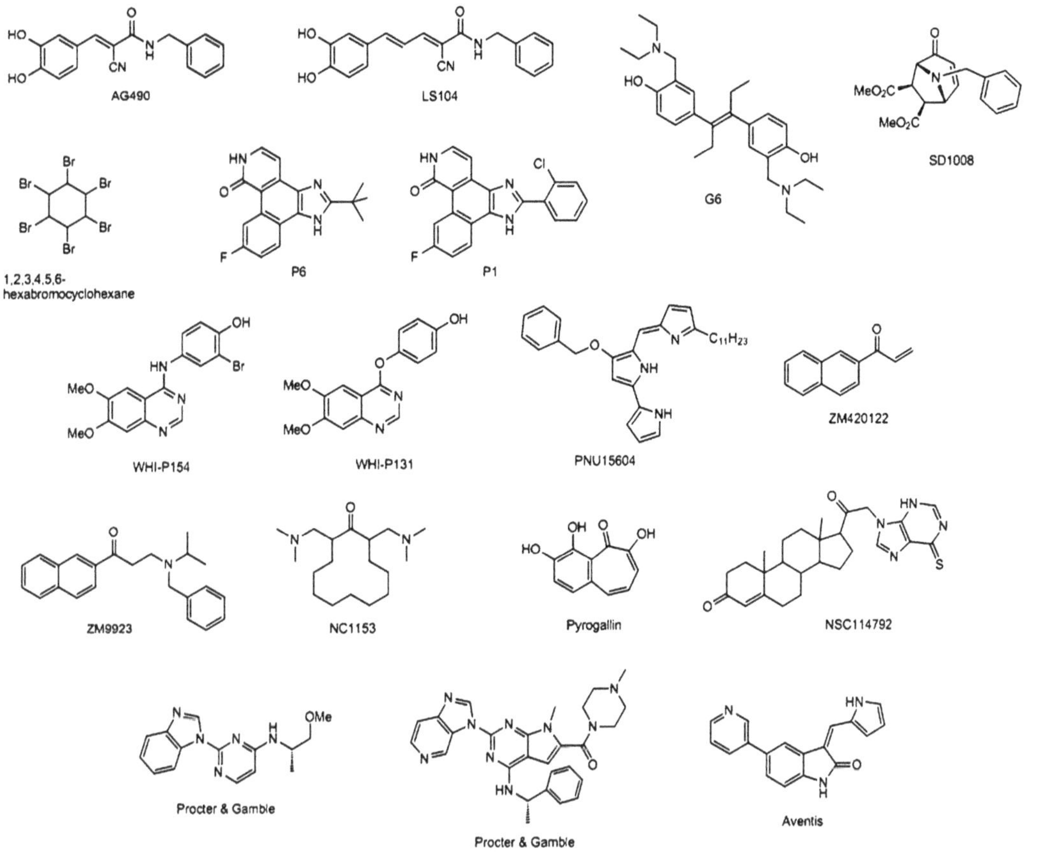

Fig. 1. Chemical structures of first generation inhibitors.

"JAK3" Inhibitors. The JAK3 probes WHI-P131 (also known as JANEX) and WHI-P154 were the first reported JAK3 inhibitors (20). These compounds have limited utility as JAK3 chemical probes as they have been shown to possess minimal JAK3 potency (11) and are potent inhibitors of EGFR (21). The naphthyl vinyl ketone ZM420122 and its chemical progenitor, the aminoethyl ketone ZM9923, have been reported as possessing moderate and selective inhibitory activity of JAK3 (IC_{50} ~100 nM) (22). The compound ZM9923 degrades and converts to ZM420122, which is the active inhibitor. As the active compound (ZM420122) possesses a chemically reactive vinyl ketone moiety, it is highly likely that this compound would react nonspecifically with other proteins, limiting the compound's utility. Another aminoethyl ketone, NC1153 has also been reported as an inhibitor of JAK3 (23). Whilst degradation to a vinyl ketone was not reported for this compound, it is possible NC1153 also inhibits JAK3 through a similar mechanism. Despite this, the compound has been used in vivo and has shown activity when dosed orally in a rat kidney transplantation model (23).

An analogue of the antibiotic undecylprodigiosin, PNU156804, has been reported to have modest activity against JAK3 in a cellular assay. Thus, the compound was shown to block JAK3-dependent T-cell proliferation with an IC_{50} of 7.5 μM with twofold greater specificity versus JAK2-dependent cell proliferation (24). The natural product pyrrogallin has recently been reported to be a weak ATP competitive JAK3 inhibitor (IC_{50} 6.4 μM) (25). A virtual screening method was used to identify the steroid derivative NSC114792 as a JAK3 inhibitor, though the characterization of the compound is limited and the cellular activity is weak (26).

3.3. Improved First Generation Compounds

The tetracyclic pyridone P6 potently inhibits all JAKs (IC_{50} 1–15 nM) and has become one of the most widely used JAK2 chemical probes in recent years. Its availability has represented an important breakthrough for the field, since it was the first molecule to exhibit a useful activity on these enzymes, with a satisfactory degree of JAK specificity (27). The compound was first reported by researchers at Merck who also patented a series of closely related compounds as PDKl inhibitors, and pyridone P6 does indeed inhibit PDKl also (IC_{50} 200 nM) (28). A recent study using a panel of cellular readouts indicates that the compound is reasonably selective (29). We are not aware of the compound being used in in vivo models, presumably indicating that the compound has poor pharmacokinetics, however a closely related pyridone derivative, pyridone P1, has demonstrated activity in a murine model of polycythemia vera, presumably through inhibition of JAK2 signaling in vivo (30). This compound possesses good potency against the JAK family of enzymes (IC_{50} 1–11 nM) and promising selectivity against a panel of 175 kinases, though potent activity against PKC (IC_{50} 3 nM) was also reported.

Two structurally related chemical series with JAK3 activity have been reported by scientists at Procter and Gamble (31, 32). Both series possess good activity against JAK3 in enzyme and cellular assays but have limited selectivity over JAK2. Researchers from Aventis Pharmaceuticals have reported inhibitors of JAK3 that are structurally related to the clinically approved multi-kinase inhibitor sunitinib (33). The optimized compounds possess good JAK3 inhibitory activity in both enzyme and cellular assays, though selectivity against other JAKs and across the kinome is not disclosed.

3.4. Second Generation Inhibitors

"JAK2" Inhibitors. The identification of an activating mutation in JAK2 (JAK2V617F) in the majority of patients with Philadelphia chromosome negative myeloproliferative neoplasms in 2005 (34–37) has led to the development of a number of compounds with highly potent JAK2 inhibitory activity. These compounds are generally well-characterized biochemically, possess pharmacokinetics allowing for in vivo use, and in certain

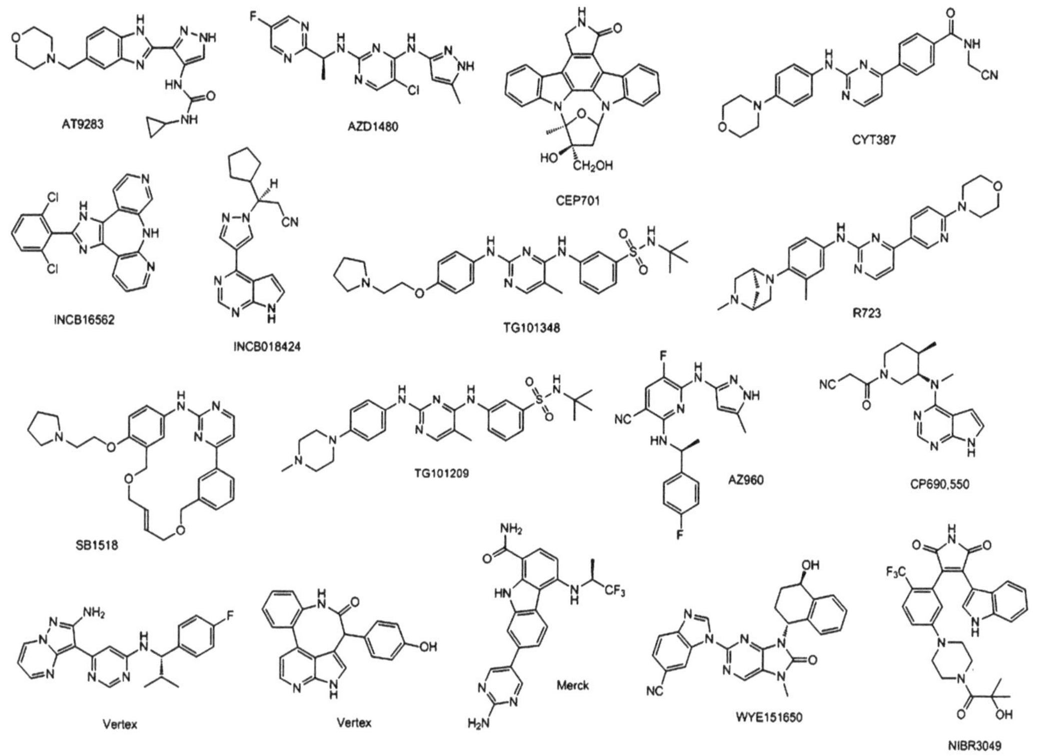

Fig. 2. Chemical structures of second generation inhibitors.

cases have been co-crystallized with the JAK2 kinase domain. Furthermore, a number of these have now entered clinical trials indicating that they possess an acceptable toxicological profile for human use (38) (Fig. 2).

Table 1 summarizes the biochemical potencies of second generation JAK2 inhibitors that have been described in the literature with demonstrated activity in vitro and in vivo. Caution should be exercised in comparing compound potencies as the IC_{50}s reported refer to compound activities determined under different assay formats and conditions.

Other rationally designed JAK2 inhibitors have been reported in the peer-reviewed literature and include TG101209 (IC_{50} 6 nM) (39), AZ960 (IC_{50} <3 nM) (40) and compounds from Vertex Pharmaceuticals (IC_{50} ~1 nM) (41, 42) and Merck pharmaceuticals (IC_{50} 0.8 nM) (43).

"JAK3" Inhibitors. The expression of JAK3 is restricted to lymphoid cells and neurological cells. The role played by JAK3 in signaling downstream of cytokine receptors that use the common γ-chain, viz. IL-2, IL-4, IL-7, IL-9, IL-15, and IL-21, suggests that inhibition of this kinase may help modulate the growth, proliferation, and end cell function of the lymphoid cells that express these receptors. An added bonus is the viability (apart from a massive

Table 1
Reported activity and selectivity profile of second generation JAK2 inhibitors

	JAK inhibition (IC_{50} nM)					
Compound	**JAK1**	**JAK2**	**JAK3**	**TYK2**	**Reported off-targets**	**Reference**
AT9283	–	1.1	1.2	<10	AurA, AurB, Abl, GSK3b, FGFR1/2/3, FLT1/3/4, MER, RET, RSK2/3/4, YES, PDK1, DRAK1, PKCμ, KDR	(49)
AZD1480	1.3	<0.4	3.9	–	TrkA, AurA, FLT4, FGFR1, Axl, FAK, ALK4, RET	(50)
CEP701	–	0.9	–	–	FLT3 (For broad kinome screen see ref. 51)	(52)
CYT387	11	18	155	–	CDK2, TBK1	(53)
INCB16562	2.2	0.25	10.1	2.7	LCK, AurA, ALK	(54)
INCB18424	3.3	2.8	428	19	"Clean" against limited panel (23 kinases)	(55)
TG101348	105	3	1002	405	FLT3, RET	(56)
R723	740	2	26	3950	PRKD2, TBK1, SYK, PAK4, DAPK3, PRKD1	(57)
SB1518	1280	23	520	50	FLT3	(58)

hole in their adaptive immune system…) of humans that carry inactivating mutations in JAK3 (44). As a consequence, a number of inhibitors of JAK3 have been developed by pharmaceutical and biotechnology companies as potential treatments for autoimmune diseases, such as rheumatoid arthritis, and to prevent allograft rejection. The most widely characterized is CP-690,550, originally reported as a selective JAK3 inhibitor (IC_{50} 1 nM) (11) but subsequently shown to be a pan-JAK inhibitor (see Table 2) with minimal off-target activity (45). Other purported selective JAK3 inhibitors currently undergoing clinical trials include VX-509 (Vertex Pharmaceuticals) and R348 (Rigel Pharmaceuticals); however the chemical structure and preclinical data for the compounds have not been published in the peer-reviewed literature. A potent inhibitor of JAK3 (WYE151650) has been reported by researchers from Wyeth Pharmaceuticals (now Pfizer) that shows good activity in cells and efficacy in an in vivo model when dosed orally. The reported selectivity of the compound across the JAK kinases is shown in Table 2; however the JAK3 selectivity of this compound has been questioned (46).

Recently, researchers at Novartis have reported a compound (NIBR3049) with good JAK3 activity (IC_{50} 8 nM) and selectivity within the JAK family; however the compound possesses significant activity against PKC and GSK3β (46).

Table 2
Reported activity and selectivity profile of second generation JAK3 Inhibitors

Compound	JAK inhibition (IC_{50} nM)				Reported off-targets	References
	JAK1	JAK2	JAK3	TYK2		
CP690,550	6.1	12	8.0	176	DCamKL3	(45, 59)
WYE-151650	32	13	0.9	31	"Clean" against limited panel (27 kinases)	(60)

3.5. Which Inhibitor to Use?

There is no correct answer to this question. Clearly, each of the aforementioned JAK inhibitors exhibit inhibitory activities against a range of off-target kinases. It is therefore not reasonable to select only one of their number and expect to obtain a definitive result regarding the involvement or otherwise of a particular JAK family member in your experimental system. We recommend a "Portfolio" approach, wherein, mindful of the particular JAK family member one wishes to interrogate, the selectivity of a number of JAK inhibitors is considered so that the consequences of inhibiting one or other of the off-targets of a particular JAK inhibitor is neutralized by dint of the use of another JAK inhibitor that has a completely different spectrum of off-target activities. For example, if one were to target JAK2 in dendritic cell differentiation, for example, might include in the inhibitor panel: TG101348 (where the key off-target kinases are FLT3 and RET) (39); R723 (where the key off-target kinases are PRKD2, TBK1, SYK, PAK4, DAPK3, and PRKD1) (56); and CYT387 (where the key off-target kinases are JAK1, CDK2, andTBK1) (52). If the data are consistent amongst all three of these inhibitors at concentrations around their cellular IC_{50}, then one may draw the conclusion that the role of JAK2 in the experimental process is "real." Inconsistencies may well point to the involvement of other off target kinases in the phenomenon under study, and suggest a more cautious interpretation of one's data.

4. Notes

Handling small molecular weight kinase inhibitors:

1. Small molecule JAK inhibitors are generally stable as solids and as solutions in DMSO when stored at 4 °C, though certain compounds may degrade over time under these conditions—(refer to suppliers' directions). As many of the compounds described below are poorly soluble in aqueous solutions, dilution of a DMSO stock solution (1–20 mM) of the compound

in assay buffer (for example 5–10% fetal calf serum for cellular assays) is advised. For in vivo studies, use of detergents and/or solubilizing excipients (e.g., cyclodextrins or polyethylene glycols—PEGs) may be necessary.

General principles:

2. Do not waste your time with first generation JAK inhibitors. With the exception of the tetracyclic pyridone P6, these compounds have very little value as chemical biology tools, and are more likely to muddy the conclusions one might wish to draw, than shed any useful light on your data.
3. Buy only from reliable providers. It is in the nature of chemical synthesis that by-products are frequent impurities in many small molecule preparations. Seek out a trustworthy provider and stick to them for as many tool compound purchases as you need.
4. It pays to run duplicates, perhaps triplicates.

For in vitro experiments:

5. As with all small molecule chemicals it is unwise to perform multiple freeze/thaw cycles; create a number of vials of small amounts of the stock solution to freeze down, and discard the tube after use.
6. Before preparing the dilutions required warm the diluent to the temperature of the experiment (4, 15, 23, and 37 °C).
7. Beware of aggregation. Aggregation and/or precipitation is the greatest cause of spurious activities (47). Simple observation of the stock solution, sample well or petri dish should alert you to the likelihood of aggregation, and suggest caution in the interpretation of data from such experiments. Addition of 0.01% Triton X-100 has been shown to limit aggregation in enzyme assays (48).
8. Select your JAK inhibitor wisely: many will have off-target activities that will affect your interpretation of the data. A portfolio approach is best.

For in vivo experiments:

9. If available, follow the published experimental conditions for the use of the selected JAK inhibitor. These are usually arrived at after much investigation by the authors of the paper. If particular excipients or detergents are used, there is usually a good reason.
10. Always run an escalating dose toxicity study before taking a compound in vivo (even one that is already in the clinic); this will save time and money (these compounds are expensive!) in the long run, and will provide a measure of support that the data are the result of a physiological process, not a pathological one.

5. Conclusion

Over the past decade a number of highly potent and well-characterized inhibitors of the JAKs have been reported. These compounds can be used in in vitro and in vivo studies to help unravel the involvement of particular JAK kinases in disease and signaling processes. Many of the first generation JAK inhibitors however do not measure up to the standards required for being useful chemical probes, being weak, non-selective or poorly characterized. It may be appropriate to revisit many of the conclusions derived from the use of such inhibitors as AG490. Whilst some of the ideas formulated and "proven" through the use of these poorly specific compounds appear to be sound, we cannot be confident that all of these observations are correct. It seems more plausible that some of the data generated with compounds are a direct consequence of off-target inhibition, or perhaps more critically, a consequence of the polypharmacology caused by the combination of multiple targets being hit.

The second generation JAK inhibitors offer a palate of useful reagents to investigate the role of JAK family members in vitro and in vivo. They remain imperfect tools, however, with differing extents of off-target activities, both within and without the JAK family. Nonetheless JAK2 and JAK3 are well covered with a number of relatively clean inhibitors available. At the time of writing however, no selective inhibitor of JAK1 or TYK2 has been disclosed and therefore judicious selection of JAK inhibitors with varying activity against these kinases will need to be made to identify the role of these specific kinases in cellular processes.

References

1. Lacronique V, Boureux A, Valle VD, Poirel H, Quang CT, Mauchauffe M, Berthou C, Lessard M, Berger R, Ghysdael J, Bernard OA (1997) A TEL-JAK2 fusion protein with constitutive kinase activity in human leukemia. Science 278:1309–1312
2. Reiter A, Walz C, Watmore A, Schoch C, Blau I, Schlegelberger B, Berger U, Telford N, Aruliah S, Yin JA, Vanstraelen D, Barker HF, Taylor PC, O'Driscoll A, Benedetti F, Rudolph C, Kolb HJ, Hochhaus A, Hehlmann R, Chase A, Cross NC (2005) The t(8;9)(p22;p24) is a recurrent abnormality in chronic and acute leukemia that fuses PCM1 to JAK2. Cancer Res 65:2662–2667
3. Bousquet M, Quelen C, De Mas V, Duchayne E, Roquefeuil B, Delsol G, Laurent G, Dastugue N, Brousset P (2005) The t(8;9)(p22;p24) translocation in atypical chronic myeloid leukaemia yields a new PCM1-JAK2 fusion gene. Oncogene 24:7248–7252
4. Murati A, Gelsi-Boyer V, Adelaide J, Perot C, Talmant P, Giraudier S, Lode L, Letessier A, Delaval B, Brunel V, Imbert M, Garand R, Xerri L, Birnbaum D, Mozziconacci MJ, Chaffanet M (2005) PCM1-JAK2 fusion in myeloproliferative disorders and acute erythroid leukemia with t(8;9) translocation. Leukemia 19:1692–1696
5. Griesinger F, Hennig H, Hillmer F, Podleschny M, Steffens R, Pies A, Wormann B, Haase D, Bohlander SK (2005) A BCR-JAK2 fusion gene as the result of a t(9;22)(p24;q11.2) translocation in a patient with a clinically typical chronic myeloid leukemia. Genes Chromosomes Cancer 44:329–333
6. Mark HF, Sotomayor EA, Nelson M, Chaves F, Sanger WG, Kaleem Z, Caughron SK (2006)

Chronic idiopathic myelofibrosis (CIMF) resulting from a unique 3;9 translocation disrupting the Janus kinase 2 (JAK2) gene. Exp Mol Pathol 81:217–223

7. Mullighan CG, Zhang J, Harvey RC, Collins-Underwood JR, Schulman BA, Phillips LA, Tasian SK, Loh ML, Su X, Liu W, Devidas M, Atlas SR, Chen I-M, Clifford RJ, Gerhard DS, Carroll WL, Reaman GH, Smith M, Downing JR, Hunger SP, Willman CL (2009) JAK mutations in high-risk childhood acute lymphoblastic leukemia. Proc Natl Acad Sci 106:9414–9418
8. Frye SV (2010) The art of the chemical probe. Nat Chem Biol 6:159–161
9. Cohen P (2009) Guidelines for the effective use of chemical inhibitors of protein function to understand their roles in cell regulation. Biochem J 425:53–54
10. Meydan N, Grunberger T, Dadi H, Shahar M, Arpaia E, Lapidot Z, Leeder JS, Freedman M, Cohen A, Gazit A, Levitzki A, Roifman CM (1996) Inhibition of acute lymphoblastic leukaemia by a Jak-2 inhibitor. Nature 379: 645–648
11. Changelian PS, Flanagan ME, Ball DJ, Kent CR, Magnuson KS, Martin WH, Rizzuti BJ, Sawyer PS, Perry BD, Brissette WH, McCurdy SP, Kudlacz EM, Conklyn MJ, Elliott EA, Koslov ER, Fisher MB, Strelevitz TJ, Yoon K, Whipple DA, Sun J, Munchhof MJ, Doty JL, Casavant JM, Blumenkopf TA, Hines M, Brown MF, Lillie BM, Subramanyam C, Shang-Poa C, Milici AJ, Beckius GE, Moyer JD, Su C, Woodworth TG, Gaweco AS, Beals CR, Littman BH, Fisher DA, Smith JF, Zagouras P, Magna HA, Saltarelli MJ, Johnson KS, Nelms LF, Des Etages SG, Hayes LS, Kawabata TT, Finco-Kent D, Baker DL, Larson M, Si MS, Paniagua R, Higgins J, Holm B, Reitz B, Zhou YJ, Morris RE, O'Shea JJ, Borie DC (2003) Prevention of organ allograft rejection by a specific Janus kinase 3 inhibitor. Science 302: 875–878
12. Kleinberger-Doron N, Shelah N, Capone R, Gazit A, Levitzki A (1998) Inhibition of Cdk2 activation by selected tyrphostins causes cell cycle arrest at late G1 and S phase. Exp Cell Res 241:340–351
13. Oda Y, Renaux B, Bjorge J, Saifeddine M, Fujita DJ, Hollenberg MD (1999) cSrc is a major cytosolic tyrosine kinase in vascular tissue. Can J Physiol Pharmacol 77:606–617
14. Osherov N, Gazit A, Gilon C, Levitzki A (1993) Selective inhibition of the epidermal growth factor and HER2/neu receptors by tyrphostins. J Biol Chem 268:11134–11142
15. Grunberger T, Demin P, Rounova O, Sharfe N, Cimpean L, Dadi H, Freywald A, Estrov Z, Roifman CM (2003) Inhibition of acute lymphoblastic and myeloid leukemias by a novel kinase inhibitor. Blood 102:4153–4158
16. Kiss R, Polgar T, Kirabo A, Sayyah J, Figueroa NC, List AF, Sokol L, Zuckerman KS, Gali M, Bisht KS, Sayeski PP, Keseru GM (2009) Identification of a novel inhibitor of JAK2 tyrosine kinase by structure-based virtual screening. Bioorg Med Chem Lett 19:3598–3601
17. Duan Z, Bradner J, Greenberg E, Mazitschek R, Foster R, Mahoney J, Seiden MV (2007) 8-benzyl-4-oxo-8-azabicyclo(3.2.1)oct-2-ene-6,7-dicarboxylic acid (SD-1008), a novel Janus kinase 2 inhibitor, increases chemotherapy sensitivity in human ovarian cancer cells. Mol Pharmacol 72:1137–1145
18. Sayyah J, Magis A, Ostrov DA, Allan RW, Braylan RC, Sayeski PP (2008) Z3, a novel Jak2 tyrosine kinase small-molecule inhibitor that suppresses Jak2-mediated pathologic cell growth. Mol Cancer Ther 7:2308–2318
19. Sandberg EM, Ma X, He K, Frank SJ, Ostrov DA, Sayeski PP (2005) Identification of 1,2,3,4,5,6-hexabromocyclohexane as a small molecule inhibitor of jak2 tyrosine kinase autophosphorylation (correction of autophophorylation). J Med Chem 48:2526–2533
20. Sudbeck EA, Liu X-P, Narla RK, Mahajan S, Ghosh S, Mao C, Uckun FM (1999) Structure-based design of specific inhibitors of Janus kinase 3 as apoptosis-inducing antileukemic agents. Clin Cancer Res 5:1569–1582
21. Changelian PS, Moshinsky D, Kuhn CF, Flanagan ME, Munchhof MJ, Harris TM, Whipple DA, Doty JL, Sun J, Kent CR, Magnuson KS, Perregaux DG, Sawyer PS, Kudlacz EM (2008) The specificity of JAK3 kinase inhibitors. Blood 111:2155–2157
22. Torr V, Culbert EJ (2000) Naphthyl ketones: a new class of Janus kinase 3 inhibitors. Bioorg Med Chem Lett 10:575–579
23. Stepkowski SM, Kao J, Wang ME, Tejpal N, Podder H, Furian L, Dimmock J, Jha A, Das U, Kahan BD, Kirken RA (2005) The Mannich base NC1153 promotes long-term allograft survival and spares the recipient from multiple toxicities. J Immunol 175:4236–4246
24. Stepkowski SM, Erwin-Cohen RA, Behbod F, Wang ME, Qu X, Tejpal N, Nagy ZS, Kahan BD, Kirken RA (2002) Selective inhibitor of Janus tyrosine kinase 3, PNU156804, prolongs allograft survival and acts synergistically with cyclosporine but additively with rapamycin. Blood 99:680–689
25. Lee BI, Ahn HJ, Han KC, Ahn DR, Shin D (2011) Pyrogallin, an ATP-competitive inhibitor of JAK3. Bull Korean Chem Soc 32: 1077–1079
26. Kim BH, Jee JG, Yin CH, Sandoval C, Jayabose S, Kitamura D, Bach EA, Baeg GH (2010)

NSC114792, a novel small molecule identified through structure-based computational database screening, selectively inhibits JAK3. Mol Cancer 9:36

27. Thompson JE, Cubbon RM, Cummings RT, Wicker LS, Frankshun R, Cunningham BR, Cameron PM, Meinke PT, Liverton N, Weng Y, DeMartino JA (2002) Photochemical preparation of a pyridone containing tetracycle: a Jak protein kinase inhibitor. Bioorg Med Chem Lett 12:1219–1223
28. Peifer C, Alessi DR (2008) Small-molecule inhibitors of PDK1. ChemMedChem 3: 1810–1838
29. Hancock MK, Lebakken CS, Wang J, Bi K (2010) Multi-pathway cellular analysis of compound selectivity. Mol Biosyst 6:1834–1843
30. Mathur A, Mo JR, Kraus M, O'Hare E, Sinclair P, Young J, Zhao S, Wang Y, Kopinja J, Qu X, Reilly J, Walker D, Xu L, Aleksandrowicz D, Marshall G, Scott ML, Kohl NE, Bachman E (2009) An inhibitor of Janus kinase 2 prevents polycythemia in mice. Biochem Pharmacol 78: 382–389
31. Chen JJ, Thakur KD, Clark MP, Laughlin SK, George KM, Bookland RG, Davis JR, Cabrera EJ, Easwaran V, De B, George Zhang Y (2006) Development of pyrimidine-based inhibitors of Janus tyrosine kinase 3. Bioorg Med Chem Lett 16:5633–5638
32. Clark MP, George KM, Bookland RG, Chen J, Laughlin SK, Thakur KD, Lee W, Davis JR, Cabrera EJ, Brugel TA, VanRens JC, Laufersweiler MJ, Maier JA, Sabat MP, Golebiowski A, Easwaran V, Webster ME, De B, Zhang G (2007) Development of new pyrrolopyrimidine-based inhibitors of Janus kinase 3 (JAK3). Bioorg Med Chem Lett 17: 1250–1253
33. Adams C, Aldous DJ, Amendola S, Bamborough P, Bright C, Crowe S, Eastwood P, Fenton G, Foster M, Harrison TK, King S, Lai J, Lawrence C, Letallec JP, McCarthy C, Moorcroft N, Page K, Rao S, Redford J, Sadiq S, Smith K, Souness JE, Thurairatnam S, Vine M, Wyman B (2003) Mapping the kinase domain of Janus Kinase 3. Bioorg Med Chem Lett 13: 3105–3110
34. Kralovics R, Passamonti F, Buser AS, Teo SS, Tiedt R, Passweg JR, Tichelli A, Cazzola M, Skoda RC (2005) A gain-of-function mutation of JAK2 in myeloproliferative disorders. N Engl J Med 352:1779–1790
35. Levine RL, Wadleigh M, Cools J, Ebert BL, Wernig G, Huntly BJ, Boggon TJ, Wlodarska I, Clark JJ, Moore S, Adelsperger J, Koo S, Lee JC, Gabriel S, Mercher T, D'Andrea A, Frohling S, Dohner K, Marynen P, Vandenberghe P, Mesa RA, Tefferi A, Griffin JD, Eck MJ, Sellers WR, Meyerson M, Golub TR, Lee SJ, Gilliland DG (2005) Activating mutation in the tyrosine kinase JAK2 in polycythemia vera, essential thrombocythemia, and myeloid metaplasia with myelofibrosis. Cancer Cell 7:387–397
36. James C, Ugo V, Le Couedic JP, Staerk J, Delhommeau F, Lacout C, Garcon L, Raslova H, Berger R, Bennaceur-Griscelli A, Villeval JL, Constantinescu SN, Casadevall N, Vainchenker W (2005) A unique clonal JAK2 mutation leading to constitutive signalling causes polycythaemia vera. Nature 434:1144–1148
37. Baxter EJ, Scott LM, Campbell PJ, East C, Fourouclas N, Swanton S, Vassiliou GS, Bench AJ, Boyd EM, Curtin N, Scott MA, Erber WN, Green AR (2005) Acquired mutation of the tyrosine kinase JAK2 in human myeloproliferative disorders. Lancet 365: 1054–1061
38. Verstovsek S (2009) Therapeutic potential of JAK2 inhibitors. Hematology Am Soc Hematol Educ Program 636–642. Volume: 2009
39. Pardanani A, Hood J, Lasho T, Levine RL, Martin MB, Noronha G, Finke C, Mak CC, Mesa R, Zhu H, Soll R, Gilliland DG, Tefferi A (2007) TG101209, a small molecule JAK2-selective kinase inhibitor potently inhibits myeloproliferative disorder-associated JAK2V617F and MPLW515L/K mutations. Leukemia 21:1658–1668
40. Gozgit JM, Bebernitz G, Patil P, Ye M, Parmentier J, Wu J, Su N, Wang T, Ioannidis S, Davies A, Huszar D, Zinda M (2008) Effects of the JAK2 inhibitor, AZ960, on Pim/BAD/BCL-xL survival signaling in the human JAK2 V617F cell line SET-2. J Biol Chem 283: 32334–32343
41. Ledeboer MW, Pierce AC, Duffy JP, Gao H, Messersmith D, Salituro FG, Nanthakumar S, Come J, Zuccola HJ, Swenson L, Shlyakter D, Mahajan S, Hoock T, Fan B, Tsai WJ, Kolaczkowski E, Carrier S, Hogan JK, Zessis R, Pazhanisamy S, Bennani YL (2009) 2-Aminopyrazolo(1,5-a)pyrimidines as potent and selective inhibitors of JAK2. Bioorg Med Chem Lett 19:6529–6533
42. Wang T, Duffy JP, Wang J, Halas S, Salituro FG, Pierce AC, Zuccola HJ, Black JR, Hogan JK, Jepson S, Shlyakter D, Mahajan S, Gu Y, Hoock T, Wood M, Furey BF, Frantz JD, Dauffenbach LM, Germann UA, Fan B, Namchuk M, Bennani YL, Ledeboer MW (2009) Janus kinase 2 inhibitors. Synthesis and characterization of a novel polycyclic azaindole. J Med Chem 52:7938–7941
43. Lim J, Taoka B, Otte RD, Spencer K, Dinsmore CJ, Altman MD, Chan G,

Rosenstein C, Sharma S, Su HP, Szewczak AA, Xu L, Yin H, Zugay-Murphy J, Marshall CG, Young JR (2011) Discovery of 1-amino-5 h-pyrido(4,3-b)indol-4-carboxamide inhibitors of Janus kinase 2 (JAK2) for the treatment of myeloproliferative disorders. J Med Chem. 54:7334–7349

44. Russell SM, Tayebi N, Nakajima H, Riedy MC, Roberts JL, Aman MJ, Migone TS, Noguchi M, Markert ML, Buckley RH, O'Shea JJ, Leonard WJ (1995) Mutation of Jak3 in a patient with SCID: essential role of Jak3 in lymphoid development. Science 270:797–800
45. Jiang JK, Ghoreschi K, Deflorian F, Chen Z, Perreira M, Pesu M, Smith J, Nguyen DT, Liu EH, Leister W, Costanzi S, O'Shea JJ, Thomas CJ (2008) Examining the chirality, conformation and selective kinase inhibition of 3-((3R,4R)-4-methyl-3-(methyl(7 H-pyrrolo(2,3-d)pyrimidin-4-yl) amino)piperi din-1-yl)-3-oxopropanenitrile (CP-690,550). J Med Chem 51:8012–8018
46. Thoma G, Nuninger F, Falchetto R, Hermes E, Tavares GA, Vangrevelinghe E, Zerwes HG (2011) Identification of a potent Janus kinase 3 inhibitor with high selectivity within the Janus kinase family. J Med Chem 54:284–288
47. McGovern SL, Caselli E, Grigorieff N, Shoichet BK (2002) A common mechanism underlying promiscuous inhibitors from virtual and high-throughput screening. J Med Chem 45:1712–1722
48. McGovern SL, Helfand BT, Feng B, Shoichet BK (2003) A specific mechanism of nonspecific inhibition. J Med Chem 46:4265–4272
49. Howard S, Berdini V, Boulstridge JA, Carr MG, Cross DM, Curry J, Devine LA, Early TR, Fazal L, Gill AL, Heathcote M, Maman S, Matthews JE, McMenamin RL, Navarro EF, O'Brien MA, O'Reilly M, Rees DC, Reule M, Tisi D, Williams G, Vinkovic M, Wyatt PG (2009) Fragment-based discovery of the pyrazol-4-yl urea (AT9283), a multitargeted kinase inhibitor with potent aurora kinase activity. J Med Chem 52:379–388
50. Hedvat M, Huszar D, Herrmann A, Gozgit JM, Schroeder A, Sheehy A, Buettner R, Proia D, Kowolik CM, Xin H, Armstrong B, Bebernitz G, Weng S, Wang L, Ye M, McEachern K, Chen H, Morosini D, Bell K, Alimzhanov M, Ioannidis S, McCoon P, Cao ZA, Yu H, Jove R, Zinda M (2009) The JAK2 inhibitor AZD1480 potently blocks Stat3 signaling and oncogenesis in solid tumors. Cancer Cell 16:487–497
51. Zarrinkar PP, Gunawardane RN, Cramer MD, Gardner MF, Brigham D, Belli B, Karaman MW, Pratz KW, Pallares G, Chao Q, Sprankle KG, Patel HK, Levis M, Armstrong RC, James J, Bhagwat SS (2009) AC220 is a uniquely potent and selective inhibitor of FLT3 for the treatment of acute myeloid leukemia (AML). Blood 114:2984–2992
52. Hexner EO, Serdikoff C, Jan M, Swider CR, Robinson C, Yang S, Angeles T, Emerson SG, Carroll M, Ruggeri B, Dobrzanski P (2008) Lestaurtinib (CEP701) is a JAK2 inhibitor that suppresses JAK2/STAT5 signaling and the proliferation of primary erythroid cells from patients with myeloproliferative disorders. Blood 111:5663–5671
53. Pardanani A, Lasho T, Smith G, Burns CJ, Fantino E, Tefferi A (2009) CYT387, a selective JAK1/JAK2 inhibitor: in vitro assessment of kinase selectivity and preclinical studies using cell lines and primary cells from polycythemia vera patients. Leukemia 23:1441–1445
54. Li J, Favata M, Kelley JA, Caulder E, Thomas B, Wen X, Sparks RB, Arvanitis A, Rogers JD, Combs AP, Vaddi K, Solomon KA, Scherle PA, Newton R, Fridman JS (2010) INCB16562, a JAK1/2 selective inhibitor, is efficacious against multiple myeloma cells and reverses the protective effects of cytokine and stromal cell support. Neoplasia 12:28–38
55. Quintas-Cardama A, Vaddi K, Liu P, Manshouri T, Li J, Scherle PA, Caulder E, Wen X, Li Y, Waeltz P, Rupar M, Burn T, Lo Y, Kelley J, Covington M, Shepard S, Rodgers JD, Haley P, Kantarjian H, Fridman JS, Verstovsek S (2010) Preclinical characterization of the selective JAK1/2 inhibitor INCB018424: therapeutic implications for the treatment of myeloproliferative neoplasms. Blood 115:3109–3117
56. Wernig G, Kharas MG, Okabe R, Moore SA, Leeman DS, Cullen DE, Gozo M, McDowell EP, Levine RL, Doukas J, Mak CC, Noronha G, Martin M, Ko YD, Lee BH, Soll RM, Tefferi A, Hood JD, Gilliland DG (2008) Efficacy of TG101348, a selective JAK2 inhibitor, in treatment of a murine model of JAK2V617F-induced polycythemia vera. Cancer Cell 13:311–320
57. Shide K, Kameda T, Markovtsov V, Shimoda HK, Tonkin E, Fang S, Liu C, Gelman M, Lang W, Romero J, McLaughlin J, Bhamidipati S, Clough J, Low C, Reitsma A, Siu S, Pine P, Park G, Torneros A, Duan M, Singh R, Payan DG, Matsunaga T, Hitoshi Y, Shimoda K (2011) R723, a selective JAK2 inhibitor, effectively treats JAK2V617F-induced murine myeloproliferative neoplasm. Blood 117:6866–6875
58. William AD, Lee AC, Blanchard S, Poulsen A, Teo EL, Nagaraj H, Tan E, Chen D, Williams M, Sun ET, Goh KC, Ong WC, Goh SK, Hart S, Jayaraman R, Pasha MK, Ethirajulu K,

Wood JM, Dymock BW (2011) Discovery of the macrocycle 11-(2-pyrrolidin-1-yl-ethoxy)-14,19-dioxa-5,7,26-triaza-tetracyclo[19.3.1.1(2,6).1(8,12)]heptacosa-1(25),2(26),3,5,8,10,12(27),16,21,23-decaene (SB1518), a potent Janus kinase 2/fms-like tyrosine kinase-3 (JAK2/FLT3) inhibitor for the treatment of myelofibrosis and lymphoma. J Med Chem 54:4638–4658

59. Fabian MA, Biggs WH 3rd, Treiber DK, Atteridge CE, Azimioara MD, Benedetti MG, Carter TA, Ciceri P, Edeen PT, Floyd M, Ford JM, Galvin M, Gerlach JL, Grotzfeld RM, Herrgard S, Insko DE, Insko MA, Lai AG, Lelias JM, Mehta SA, Milanov ZV, Velasco AM, Wodicka LM, Patel HK, Zarrinkar PP, Lockhart DJ (2005) A small molecule-kinase interaction map for clinical kinase inhibitors. Nat Biotechnol 23:329–336

60. Lin TH, Hegen M, Quadros E, Nickerson-Nutter CL, Appell KC, Cole AG, Shao Y, Tam S, Ohlmeyer M, Wang B, Goodwin DG, Kimble EF, Quintero J, Gao M, Symanowicz P, Wrocklage C, Lussier J, Schelling SH, Hewet AG, Xuan D, Krykbaev R, Togias J, Xu X, Harrison R, Mansour T, Collins M, Clark JD, Webb ML, Seidl KJ (2010) Selective functional inhibition of JAK-3 is sufficient for efficacy in collagen-induced arthritis in mice. Arthritis Rheum 62:2283–2293

Chapter 8

Methods for Detecting Mutations in the Human JAK2 Gene

Anthony J. Bench, E. Joanna Baxter, and Anthony R. Green

Abstract

Mutations in the JAK2 gene are prevalent in the human myeloid malignancies, being present in virtually all cases of polycythemia vera, and a significant proportion of patients with other myeloproliferative disorders. Various methods for the detection of acquired mutations in this gene are available depending on the need for sensitivity, quantification, or the ability to detect many different mutations. We summarize the various methods published and discuss their relative merits for each application. Two commonly used methods, quantitative real-time PCR (QPCR) for the detection of the JAK2 V617F mutation and high resolution melt-curve analysis (HRM) for the detection of multiple mutations within JAK2 exon 12, demonstrate the utility of each method and their limitations. The choice of methodology is dependent on the application; therefore there is no gold standard for detecting mutations in this gene.

Key words: JAK2, Myeloproliferative, Pseudokinase, V617F, Mutation, Quantitative PCR, High resolution melting analysis

1. Introduction

Human myeloproliferative neoplasms (MPNs) are associated at the molecular level with increased intracellular signaling as a result of mutations leading to an activated tyrosine kinase. The first example of this is the BCR-ABL tyrosine kinase formed by the t(9;22) translocation in chronic myeloid leukemia (CML) (1), and in 2005 it was found that many patients with BCR-ABL negative MPNs harbored mutations in the JAK2 tyrosine kinase gene (2–6). These mutations result in constitutive activation of the JAK2 kinase leading to activated STAT, Akt, and ERK signaling in the absence of cytokine binding (2). The first and most common mutation discovered was G1849T in exon 14, which results in the substitution V617F in the JH2 pseudokinase domain. This mutation occurs in 95% of patients with polycythemia vera (PV), and approximately 50% of essential thrombocythemia (ET) and primary myelofibrosis

Sandra E. Nicholson and Nicos A. Nicola (eds.), *JAK-STAT Signalling: Methods and Protocols*, Methods in Molecular Biology, vol. 967, DOI 10.1007/978-1-62703-242-1_8, © Springer Science+Business Media New York 2013

(PMF) (2–6). It is also found in RARS-T (7) but is much less common in atypical MPNs such as chronic myelomonocytic leukemia (CMML) or myelodysplastic syndromes (6, 8). Subsequent studies showed that among cell lines, JAK2 V617F was restricted to myeloid malignancies (9). Mutations elsewhere in the gene were identified in hematological malignancies, firstly mutations or small deletions in exon 12 associated with patients with PV with mainly red cell hyperplasia (10) and mutations or small deletions in exon 14 centered around amino acid R683, in about one-fifth of patients with Down's Syndrome-associated B-ALL (11). Mutation screening for V617F in large cohorts has identified rare variants involving the amino acids adjacent to V617 (e.g., ref. 12). Although these are only occasionally detected, they are significant because DNA changes that lie in primer or probe binding regions may destabilize interactions to reduce or abolish amplification, preventing detection of V617F.

The discovery of JAK2 mutations in nearly all PV patients and half of those with ET or PMF had immediate implications for diagnosis of these conditions and testing for V617F and exon 12 mutations is now part of the diagnostic work up for patients with raised hematocrit or red cell mass, persistent thrombocytosis, or unexplained marrow fibrosis according to the world health organization (WHO) guidelines for diagnosis of MPNs (13). The finding of a mutation in JAK2 confirms the primary and clonal nature of the disease therefore it is a key part of the diagnosis of an MPN, even in those with no other symptoms. Although the mutations can be readily detectable in the majority of PV patients, the mutation burden can be low in other PV, ET, and PMF patients. By using methods more sensitive than sequencing, the percentage of ET patients with detectable V617F increased from 12 to 57% (3), therefore there is a strong argument for use of highly sensitive methods for diagnosis and potentially for minimal residual disease monitoring post-therapy in these patients.

2. JAK2 Mutation Detection to Date

Previously, we reported two semiquantitative methods for the detection of V617F, using a mutation allele-specific PCR, and a restriction enzyme digestion assay (14). These are both more sensitive than sequencing but were difficult to quantitate accurately. Since then, a plethora of methods have been described for the detection of V617F mutation and mutations within exon 12. These utilize a range of common methods for mutation screening, summarized in Table 1. The approximate sensitivity (defined as the minimum detectable percentage of mutant over total alleles) of each is given though this will vary between laboratories.

Table 1
Summary of different methodologies used to detect mutations in the JAK2 gene

Method	Approximate sensitivity	Example reference(s)
Direct sequencing	10–20%	(3, 15)
Pyrosequencing	5%	(6, 8)
Restriction enzyme digestion	2–10%	(14, 16)
Melting curve analysis	1–5%	(16–18)
dHPLC	1–5%	(19)
ARMS/Allele-specific PCR	0.1–5%	(3, 6, 18, 20, 21)
High resolution melt analysis	0.5–1.0%	(22, 23)
Real-time allele-specific PCR	0.01–1%	(16, 24–27)
MassARRAY	0.01%	(28)

3. Considerations for JAK2 Mutation Testing

1. *Sample Type*

Peripheral blood or granulocytes are commonly used for mutation detection in the MPNs. Potentially some, but not all, granulocyte samples will be pure tumor, whereas peripheral blood will be a mixture of lineages, not all of which will be clonally derived. Few studies have compared isolated granulocytes with whole blood for V617F quantitation, and these have found no significant difference when using very sensitive methods such as allele-specific qPCR (16, 29). Arguably, with the growing numbers of patients reported to have oligo-clonal or coexistent MPNs, even the granulocytic compartment is likely to contain normal cells or clones lacking the JAK2 mutation (30). In either case, the need for a greater sensitivity of detection is clear, whether testing whole blood or granulocytes.

The other main consideration is whether to use genomic DNA or cDNA in the assay. The advantages of making genomic DNA are that it is a more readily automatable process, and the DNA quality can easily be assessed before testing, whereas cDNA is derived from less stable RNA, and its synthesis is a more lengthy protocol. ET patients often have a very low burden of V617F in their granulocytes therefore platelet cDNA could be tested instead as they are more likely to contain the malignant clone. However, in one study, the V617F mutation can be detected at a similar frequency in granulocytes and platelets of ET patients, suggesting it is not necessary to study platelets specifically (31). In addition, more sensitive tests can reliably detect very small V617F positive clones in these patients.

2. *Sensitivity Versus Specificity*
   In designing a quantitative assay regardless of the methodology to be used, a key initial consideration is the final aim of the test. It is important to decide whether the priority is sensitivity or specificity. A sensitive test may detect background amplification of the opposite allele, but still be quantitative, but a specific test will not detect background amplification of the opposite allele. This is often achieved at the expense of sensitivity, so that the lowest value detectable is only around 0.5–1%. Designing specific assays to detect single nucleotide substitutions is more difficult than for gene fusions, deletions, and duplications due to the similarity of wild type to mutant allele. The clinical significance of a V617F clone at less than 1% of the total blood or granulocyte DNA is unclear, and where known, other driver mutations may be a better marker of the clonal nature of the disease. It has been reported that JAK2 V617F was not detectable in normal individuals (18) but with the use of more sensitive detection methods for JAK2 V617F it has been possible to find low levels of the mutation in normals. Sidon et al. reported a quantitative PCR (qPCR)-based assay using a molecular beacon probe to detect V617F and a locked nucleic acid (LNA) probe to block the wild-type sequence (32). With a sensitivity limit of approximately 0.01%, a positive result was found in 5/52 normals, which was estimated to be about ten copies of the mutant DNA. It is possible that the mutation exists before a clinical disease is evident, or in a manner analogous to BCR-ABL detection in normals, that the mutation arises in a blood cell not capable of being transformed by it. Finally, it is also possible that the test is not as specific as the authors think and that the mutant probe can also detect the wild-type sequence, therefore appearing to detect mutant sequence. When deciding which test to use these reports should be kept in mind as it is difficult to understand the meaning of such results.

3. *Reproducibility of Quantitative Data*
   For quality control of the assay to assess reproducibility and consistency between experiments, it is important use standard material. Some researchers use small numbers of defined standard DNAs, to check that they amplify in the same way in each experiment. Others use dilution series of standards to give a standard curve. In qPCR the major advantage is that the standard curve equation gives a value for efficiency and a correlation coefficient for the individual replicates made, and in melt-curve analysis it also gives a limit of sensitivity. The major disadvantage of this is that the standards take up much needed

space on the plate. These DNAs must be kept under standardized conditions to prevent degradation. For standards to be comparable or to be mixed together, the DNAs must have the same copy number of the allele it represents otherwise the values will be skewed. Cell line DNAs must be checked for copy number of JAK2 before use as controls, for example UKE1 is V617F positive with a copy number of 2, whereas HEL is estimated to have between 8 and 10 copies (33). Plasmids can be used for standard curves but as they are not representative of the composition of genomic DNA, mixing curves may not reflect the PCR kinetics in a patient gDNA sample. Samples should be prepared using the same methodology so that sample results are comparable over time. Factors such as high salt concentration in the DNA can interfere with the efficiency of PCR amplification and particularly with the appearance of melt curves, which may lead to misinterpretation of data.

4. Method 1: Quantitative PCR Analysis: JAK2 V617F Detection

qPCR is classically employed in research contexts to give a more accurate quantification of gene copy number and RNA transcripts. In diagnostics it can be either used for reliable detection of a disease-associated mutation (specificity) or monitoring disease dynamics over time using the disease-associated mutation as a marker of residual disease (sensitivity). There are several methodologies for qPCR; firstly, SYBR green and related techniques, which detect all amplifiable double-stranded DNAs produced in PCR, or secondly, probe-based techniques where amplification of all amplifiable double-stranded DNAs occurs, but only those specific to the probe are detected. Secondly, qPCR quantitation can be relative or absolute—either data are expressed relative to a standard or an external control region, or in absolute terms using a conversion from plasmids of known concentration, molarity or copy number.

The most commonly used probe for qPCR is a fluorescently labeled DNA oligonucleotide which binds a sequence between the primers. This sequence can be very long (up to 40 nucleotides), though this may be detrimental to optimal binding of the probe. Probes have a fluorescent reporter, such as 6-FAM (6-carboxy fluorescein), at the 5′ end and a quencher, often a non-fluorescent molecule such as BHQ1 (black hole quencher 1), at the 3′ end. When the probe is not denatured it folds up and the quencher prevents fluorescence. When the probe is bound or denatured it is in a linear conformation and the fluorescent signal is unimpeded. The probe must always have a melting temperature significantly higher than the primers, so that it is already bound to the sequence as the DNA is amplified. If the melting temperature is too low then the probe will be in its native state, with no fluorescence, when signal should be detected. Several different designs of probes have been described, such as molecular beacons and scorpions. The advantages of these probes are their specificity, as they require sufficient melting of the probe as well as the target sequence to become fluorescently active. Their major drawback is that they can be difficult to design to work well with PCR kinetics. Finally, modifications of the DNA bases to make LNAs (locked nucleic acids), PNAs (peptide nucleic acids), or use of a minor groove binding (MGB) moiety in the probe sequence can be beneficial in increasing the melting temperature of the probe, and reducing the need for very long probe sequences. These modifications significantly raise the probe melting temperature, especially when multiple modifications are in place. Modified probes can be as small as 11 nucleotides long but still retain melting characteristics of much longer sequences. Some assays also have a non-labeled oligonucleotide that can block binding of the probe to the wrong sequence. Most commonly this will be a blocker of the wild-type sequence with a labeled mutant probe. As with all qPCR methods, the primers should be less than 200 bp apart, ideally 100–150, for optimal amplification.

qPCR assays for SNP detection, such as the Taqman assays, can be modified for the detection of mutations; however, these duplexed assays are not optimized for quantitation of very low allele burdens. The two reactions can be separated to reduce competition for the probes or primers, and prevent mis-annealing of the allele-specific primers to the wrong sequence. To design an allele-specific test, the variant nucleotide can be located either at the 3′ end of one of the primers or in the middle of the probe.

Finally, just as was needed in the original allele-specific PCR (14), a second variation can be introduced a few bases away in the primer or probe to further destabilize mismatched interactions. In our hands, qPCR with extra mismatches in the primers has low efficiency so we have not pursued this option.

4.1. Methodology: LNA Probe-Based Assay Using Amplitaq Gold

4.1.1. Materials

1. AmpliTaq Gold® (Applied Biosystems, Chester, UK).
2. 50× ROX (6-carboxy-X-rhodamine) (ThermoScientific).
3. dNTPs (GE Healthcare, Amersham, UK).

   NB. These can be combined by the user, or ready mixes can be obtained from a variety of companies, pre-optimized for qPCR, all that is needed are the primers and probe. The ready mixes may need to be adjusted for $MgCl_2$ concentration.
4. Probe sequence: 5′ 6FAM-CT(+G)TGG(+A)GAC(+G)AGA-BHQ1 3′ the (+) around a nucleotide indicates the position of LNA bases.

NB. Probes should be treated as light sensitive to preserve their activity—keep stocks and dilutions covered and on ice. Order probes with 6-FAM at the 5′ end and BHQ1 as the quencher at the 3′ end.

5. Primer sequences:

   JAK2_WT_Spec_F 5′ TTTTAAATTATGGAGTATGTC 3′.

   JAK2_MUT_Spec_F 5′ TTTTAAATTATGGAGTATGTT 3′.

   JAK2_R1 5′ CAGATGCTCTGAGAAAGGC 3′.

   NB. Primers should be ordered at HPLC purity to make sure that the allele-specific nucleotide at the 3′ end is present on all molecules, otherwise the primers may be able to amplify target independently of the allele present.

   Per reaction: 10 μl total reaction (best for 384-well plates, 20 μl can be used with 96-well plates).

6. Optical plates either 96- or 384-well size—with optical covers can be obtained from Applied Biosystems.
7. Applied Biosystems Quantitative PCR instrument—we use the 7900HT platform.

4.1.2. Method

1. Prepare on ice, a master mix for each of the following. These mixes are per reaction.

Wild-type (WT) allele detection		**Mutant (MUT) allele detection**	
1 μl	10× Amplitaq Gold PCR buffer	1 μl	10× Amplitaq Gold PCR buffer
3.2 μl	25 mM $MgCl_2$	3.2 μl	25 mM $MgCl_2$
0.32 μl	25 mM dNTPs	0.32 μl	25 mM dNTPs
0.2 μl	1 in 50 ROX (dilute in dH_2O)	0.2 μl	1 in 50 ROX (dilute in dH_2O)
0.1 μl	10 mM Probe	0.2 μl	10 mM Probe
0.02 μl	WT_SPEC_F primer 100 μM	0.02 μl	MUT_SPEC_F primer 100 μM
0.02 μl	JAK2_R1 primer 100 μM	0.04 μl	JAK2_R1 primer 100 μM
4.015 μl	dH_2O	3.895 μl	dH_2O
0.125 μl	AmpliTaq Gold enzyme	0.125 μl	AmpliTaq Gold enzyme

2. After making the master mix, briefly centrifuge the tube in a microfuge to collect the contents.
3. Aliquot on ice into the required wells of your plate, then add DNAs.
4. For each well, use 1 μl of genomic DNA at 10–20 ng/μl.

5. For 384-well experiments, prepare 96-well template plates of the genomic DNA to enable multichannel pipette transfer into the final 384-well plate. Note that the standard 96-well plate and multichannel pipettor when used in columns will be able to fill every other well on a 384-well plate. We use the intermediate wells for the duplicate sample, therefore make 2× stock of each DNA to be analyzed in the template 96-well plate. Remember to prepare enough DNA for both the wild-type and mutant PCR plates.
6. Before loading into the instrument, briefly spin the plate in a plate rotor, at 400×*g* in a Beckman Allegra 6R centrifuge or equivalent to collect the reactions into the bottom of the wells.
7. On an ABI7900HT the PCR can be performed on a standard mode using either 96- or 384-well plate format.
8. Use the following cycling parameters:
   10 min 95°C to activate the Taq enzyme.
   15 s 95°C.
   45 s 57°C for the mutant plate, 45 s 59°C for the wild-type plate.
   Continue for 40 cycles (or up to 45 cycles for greater sensitivity).
   Hold at 10°C.
9. Program into the instrument the location and concentration in nanogram of the standards, all other samples are either called unknowns or NTCs (no template controls).
10. At the end of the run the software uses the pre-programmed standards to analyze the unknown data and can export both Ct values and true quantities of the fragment of interest in the starting gDNA. At this point always check the raw data profiles to make sure that there are no abnormally shallow amplification curves in the unknowns compared to standards and that the standard curve falls within an acceptable range (gradient between −2.9 and −3.7) to ensure efficient amplification has occurred.

4.1.3. Analysis and Quality Control

We run each experiment with a set of plasmid standards to determine the standard curve for each run. As the plasmids have a defined molecular weight we can determine absolute copy number data from each experiment to determine the final percentage of V617F. This can be set up using the ABI software on absolute quantitation.

% JAK2V617F calculation: (Quantity of MUT/(Quantity of MUT + Quantity of WT)) × 100

This calculation assumes that the ability of each allele to amplify in its single reaction tube is equal, though our data suggest that the relationship is not linear when one allele is at a very low percentage. This can be adjusted by adding in a mixing series of DNAs known to contain only the wild-type or only the mutant sequence, such as plasmids or patient DNAs. For a wide coverage, we have used mixes with final concentration of 10 ng/μl and total V617F% of: 0, 0.01, 0.1, 1, 10, 50, 90, 99, 100%. A good experiment should give mutant values in the 0.01% mix, and wild-type should be detectable in the 99% mix. A background level of amplification of the wrong allele is generally present in this assay and will give a very low value in the 0% or 100% mixes. Dilutions to make mixes lower than 10% can be made by performing serial 1 in 10 dilutions of 10% mutant DNA in wild-type DNA at the same concentration.

Plasmids and genomic DNAs generate different results for this mixing series. We designed plasmids by inserting PCR-generated products from JAK2 exon 14 into the pGEM-T cloning vector (Invitrogen), then sequencing clones. Plasmid DNAs mixed together do not make good controls because they readily compete with each other for probe binding. This is apparent in the curve in Fig. 1a, where the relationship between input and observed allele frequency has a nonlinear relationship. For plasmids, the relationship is mainly linear where there is no minor allele present, i.e., below 10%. For this reason, the data in Fig. 1b are separated into

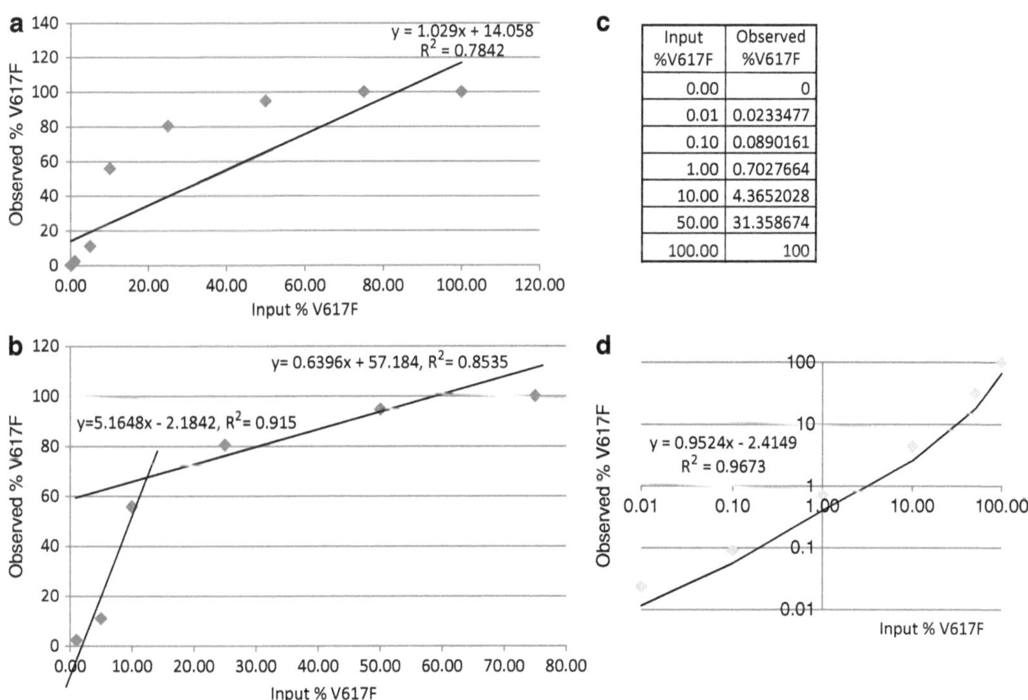

Input %V617F	Observed %V617F
0.00	0
0.01	0.0233477
0.10	0.0890161
1.00	0.7027664
10.00	4.3652028
50.00	31.358674
100.00	100

Fig. 1. (**a**) Mixing curve experiment using PCR product-derived plasmids, with a range of mutant plasmid proportions from 0.01 to 100%. (**b**) Same mixing curve data separated into two curves for minor allele frequency (i.e., <10% mutant) and where there is no minor allele (i.e., >10% mutant). Input—known percentage of DNA used, Observed—measurement obtained in the experiment. The equation gives the gradient and intercept of the trendline, R_2 is the correlation coefficient. (**c**) Input versus observed values obtained for genomic DNA mixtures. The wild-type DNA was from a hematologically normal individual's peripheral blood mononuclear cells, and the mutant DNA was from a myelofibrosis patient with no evidence of the wild-type allele or minor variants by exome sequencing. (**d**) Graphical representation of (**c**), with a moving average trendline.

points below, and those above 10% mutant, resulting in two curves with much better correlation between input and observed allele frequency. The higher gradient of the rare allele curve is probably caused by excessive competition in the assay by the wild-type DNAs for the probe. One immediate consequence of this competition is the loss of sensitivity below 1%, as the mutant allele is excluded by the excess of wild-type allele, even where the mutant plasmid copy number is greater than 10,000. In addition, curves to compensate for this competition are needed to interpret data for every single experiment. These problems are best resolved by using genomic DNAs for standards, such as cell lines of equal copy number or patient DNAs. The sequences of all standards must be confirmed before use. Figure 1c shows the observed allele frequency vs. input with genomic DNAs, which is plotted in Fig. 1d to show the greater correlation and linearity of this set of standards. For this experiment, normal individual genomic DNA was mixed with DNA from a patient shown by exome sequencing to have only the mutant allele.

5. Method 2: High Resolution Melting Analysis—JAK2 Exon 12 Mutations

Mutations within exon 12 of *JAK2* are observed in approximately 2–3% of patients with polycythemia vera and are often detected in patients previously characterized as idiopathic erythrocytosis (34). At least 17 different mutations have been described often as a result of a 6 bp deletion (10, 35). These mutations fall into three main groups—those that result in a deletion of glutamic acid at codon 543 (E543del); those that lead to a lysine to leucine substitution at codon 539 (K539L); and duplications that lead to substitution of the phenylalanine at codon 547. Due to the large number of possible mutations, techniques that target specific mutations such as allele specific PCR are of limited value for the detection of JAK2 exon 12 mutations. Direct sequencing remains an option but the often low level of disease and inherent poor sensitivity of direct sequencing makes choice of this approach far from ideal. Highly sensitive mutation scanning methods have been developed for the identification of JAK2 exon 12 mutations. The most commonly used are high resolution melting analysis (HRMA; (22, 23, 36)), melting curve assay (37), and dHPLC. The sensitivity of these assays range from 1 to 10% depending on the mutation present.

HRMA is a powerful technique applicable for the detection of any acquired mutation within a region of interest. A DNA-intercalating fluorescent dye is incorporated into the PCR product during the reaction. After PCR, the product is slowly heated until it becomes entirely denatured resulting in release of the intercalating dye. Measurement of fluorescence generates a characteristic melting profile whose features depend on fragment length and sequence. A fragment derived from a DNA sample carrying a mutation, such as a point mutation, deletion, or insertion, yields a subtly different melting profile. The melting profile can be compared to the "wild-type" profile to generate a difference plot. Routine use of HRMA has become possible due to improvements in saturating DNA dyes, instruments, and software (reviewed in ref. 38). In particular, new DNA-binding dyes such as LC Green Plus+(Idaho Technology, Inc.), EvaGreen (Biotium), and SYTO9 (Life Technologies) are less inhibitory to PCR than SYBR Green. This means that they can be used at a higher concentration allowing the dye to saturate the double-stranded DNA thereby reducing any dye redistribution during DNA melting. The use of real-time PCR instruments with HRMA capacity allows the procedure to be performed without the need for any post-PCR manipulation.

Since HRMA relies on comparison between different templates, a panel of normal controls should always be included with every run. This also enables variation in the normal melting profile, due to slight variation in salt concentration, to be observed. In addition, it is helpful to include samples from patients known to carry a JAK2 exon 12 mutation—ideally at least one each with an E543del type and a K539L type mutation. The DNA should be of good quality since carryover of salts or ethanol can reduce the efficiency of PCR resulting in aberrant melting profiles. Ideally, the DNA under investigation should be prepared and quantified in the same way as the DNA being used as the "normal" control. The actual methods themselves are not crucial, as long as similar methods of extraction and quantification are employed. The method described here is an HRMA method described by ref. 23.

5.1. HRMA Methodology: Adapted from Ref. (23)

5.1.1. Materials

1. Platinum *Taq* DNA polymerase (5 U/μl); 10× PCR Buffer, Minus Mg; 50 mM magnesium chloride (Invitrogen).
2. Deoxynucleoside triphosphate (dNTPs)—20 mM each dNTP.
3. 10×LC Green Plus+melting dye (Idaho Technology, Inc.). Once thawed, store at 4°C.
4. Oligonucleotide primers (prepare 5 μM working solution):

   F: AATGGTGTTTCTGATGTACC

   R: AGACAGTAATGAGTATCTAATGAC
5. Ultra pure water (UV irradiation is advised).
6. 0.1 ml strip tubes and caps (used here), 0.2 ml PCR tubes or Rotor-Disc (compatible with instrument).
7. Rotorgene/Rotor-gene Q Real-time PCR cycler (2-plex or 5-plex) with High resolution melt analyzer (Corbett Life Sciences/Qiagen). Also required are the appropriate aluminum loading block, rotor, locking ring, and rotor holder (Qiagen).

5.1.2. Method

Master mix for the detection of *JAK2* exon 12 mutations by HRMA: per reaction

10.3 µl	Water
2.0 µl	10× Buffer minus Mg
1.0 µl	50 mM magnesium chloride
0.2 µl	20 mM dNTPs
2.0 µl	5 µM Primer F
2.0 µl	5 µM Primer R
2.0 µl	10× LC Green Plus+
0.1 µl	Platinum Taq (5 U/µl)
0.4 µl	100 ng/µl genomic DNA

1. Prepare a mastermix as described above (excluding template DNA) with sufficient reagent for $N+1$ samples, where N is the number of tubes required, including the no template, normal, and positive controls.
2. Aliquot 19.6 µl of the mastermix into each 0.1 ml strip tube.
3. Carefully add 0.4 µl of water or DNA into the master mix within the appropriate tube.
4. After each batch of four tubes, carefully add the strip of four caps to the strip tubes. The caps may be labeled if required.
5. Once all DNA samples have been added, place the rotor onto the rotor holder and load the strip tubes into the appropriate numbered location. Ensure that all locations are filled (use a blank tube if necessary).
6. Place the locking ring onto the rotor, ensuring it is secure.
7. Place the rotor/locking ring into the Rotorgene, ensuring it is secure.
8. Open the Rotor-Gene 6000 software (or equivalent).
9. Perform the following PCR/HRM procedure:
   (a) 95°C for 10 min.
   (b) 40 cycles of: 95°C for 15 s, 58°C for 30 s, 72°C for 20 s.
   (c) 50°C for 30 s.
   (d) Pre melt conditioning at 70°C for 90 s.
   (e) High resolution melt from 70 to 95°C rising by 0.1°C each step with 2 s for each step.
10. After the PCR/HRM has finished, select the "HRM" tab under the Analysis window. Set the leading range at 72–75°C and the trailing range at 92–93°C. Open the HRM genotypes window (located within the HRM Analysis window) and define the samples that represent the normal/wild-type control.

5.1.3. Analysis and Quality Control

To correctly interpret the results generated by the HRMA software, it is imperative that the quality of the raw data is acceptable. Interrogation of the PCR amplification and raw fluorescence enables an assessment of the quality of the raw data to be made and ensures that the results generated by the HRMA software are reliable. It is usually not possible to assess all samples simultaneously. Hence, interrogate each sample, along with appropriate controls, independently.

1. Select the "Quantitation" tab and open the Quantitation Analysis window. Ensure that the samples have amplified adequately (see Notes 1–3). Samples for which the presence of PCR inhibitors causes a problem should be repeated using a dilution of the DNA. Diluting the DNA reduces the effect of any PCR inhibitors.
2. Select the "Melt" tab and open the Melt curve analysis window. The melt curve analysis is not used for interpretation of results but can give a helpful guide as to the success of PCR amplification. Successful PCR amplification should result in a single peak at approximately 80–81°C. The "no template control" should not show any amplification. DNA of poor quality/quantity will yield a smaller peak than controls (see Notes 4 and 5).
3. Select the "HRM" tab and open the HRM Analysis and HRM Normalized graph windows. The HRM Analysis window shows the raw fluorescence and is a helpful indicator of the quality of the sample (see Note 6).
4. The "Normalized Graph" and "Difference Graph" display the melting profile for selected samples allowing comparison between the sample under investigation and normal and/or positive controls. When the "Difference Graph" is viewed, select the genotype for comparison. It is usual to compare the "test" sample against a "normal" control (Note 7). Samples containing an "E543del" mutation type display greatest divergence around 77.5°C, whereas samples containing a "K539L" mutation type display greatest divergence around 79°C (Fig. 2).

5.1.4. Notes for Method 2

1. The C_T value should be less than 30. Higher C_T values imply reduced amounts of starting material, significant degradation, and/or the presence of PCR inhibitors. Any of these problems may affect the ability to generate a reliable difference plot after HRM.
2. The fluorescence at the plateau phase of the reaction should be similar for all samples. A low fluorescence at the end of the reaction suggests inhibition of the PCR.
3. The shape of the amplification plot should be similar for controls and samples. A graph that is not steep in the log phase indicates inefficient amplification, again suggesting PCR inhibition.
4. It may not be possible to generate a reliable difference plot for samples with reduced amounts of PCR product.

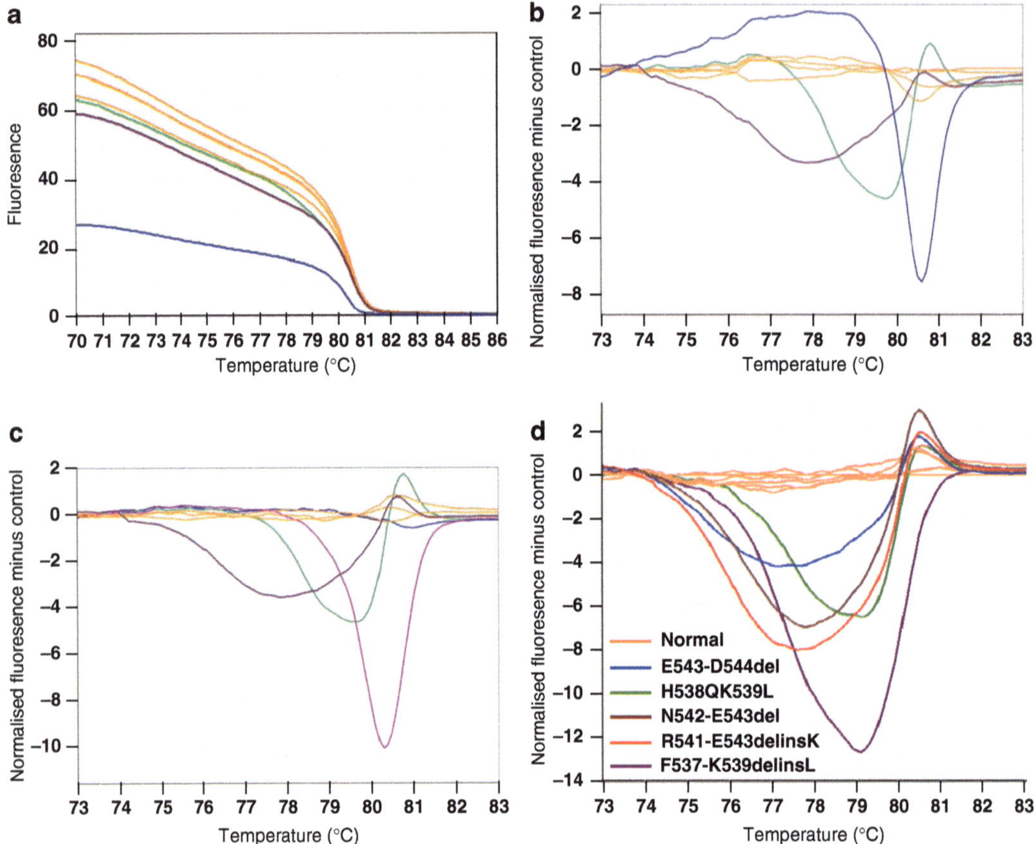

Fig. 2. High resolution melting analysis output data for the detection of JAK2 exon 12 mutations. (**a**) Use of low-quality DNA (*blue*) resulting in poor amplification, demonstrated by the low level of initial raw fluorescence. (**b**) The apparent difference between normal controls and the sample which does not indicate the presence of a mutation with low level fluorescence in (**a**). (**c**) Use of too much DNA for amplification (*pink*) generated an apparent difference at 80.5°C which does not represent the presence of a mutation. Use of less DNA (*blue*) demonstrated a normal melting profile. (**d**) Difference plot for patients carrying known mutations within JAK2 exon 12.

5. For "positive" samples, especially those carrying a high level of the "K539L" type an additional small peak may be observed around the 77–79°C region.
6. Samples with particularly high or low levels of starting fluorescence (at 70°C) compared with control samples may not yield reliable difference plots even though the software is able to normalize the fluorescence (Fig. 1a, b). Consider varying the amount of starting material.
7. An apparent difference between the sample under investigation and a normal control may be observed around 80.5°C (Fig. 1c). This does not indicate the presence of a mutation and is often generated if too much DNA is assessed. Dilute the starting DNA and repeat the assay. A "true" positive will generate the same difference plot if the DNA is diluted.

6. Conclusions

In reviewing the methods published to date we have found that there is no gold-standard method for detecting JAK2 mutations. Where the mutation may not be known, high resolution melting analysis is preferred, but for sensitive detection of a known mutation, we prefer using allele-specific real-time PCR as described here.

Whatever method is chosen, it is clearly important to use appropriate controls and it is critical to establish the specificity and sensitivity of the method in the host laboratory.

The type of methodology may change again with the advent of deep sequencing techniques. Sensitivity and detection of new mutations can be achieved in parallel if a gene such as JAK2 is sequenced many times, giving a frequency to the number of times a particular mutant allele appears in sequencing reads. Development of iron torrent, 454 and other massively parallel sequencing platforms is transforming molecular genetics in both research and diagnostics (39, 40).

References

1. Ben-Neriah Y, Daley GQ, Mes-Masson AM, Witte ON, Baltimore D (1986) The chronic myelogenous leukemia-specific P210 protein is the product of the bcr/abl hybrid gene. Science 233:212–214
2. James C, Ugo V, Le Couedic JP, Staerk J, Delhommeau F, Lacout C, Garcon L, Raslova H, Berger R, Bennaceur-Griscelli A, Villeval JL, Constantinescu SN, Casadevall N, Vainchenker W (2005) A unique clonal JAK2 mutation leading to constitutive signalling causes polycythaemia vera. Nature 434: 1144–1148
3. Baxter EJ, Scott LM, Campbell PJ, East C, Fourouclas N, Swanton S, Vassiliou GS, Bench AJ, Boyd EM, Curtin N, Scott MA, Erber WN, Green AR (2005) Acquired mutation of the tyrosine kinase JAK2 in human myeloproliferative disorders. Lancet 365:1054–1061
4. Levine RL, Wadleigh M, Cools J, Ebert BL, Wernig G, Huntly BJ, Boggon TJ, Wlodarska I, Clark JJ, Moore S, Adelsperger J, Koo S, Lee JC, Gabriel S, Mercher T, D'Andrea A, Frohling S, Dohner K, Marynen P, Vandenberghe P, Mesa RA, Tefferi A, Griffin JD, Eck MJ, Sellers WR, Meyerson M, Golub TR, Lee SJ, Gilliland DG (2005) Activating mutation in the tyrosine kinase JAK2 in polycythemia vera, essential thrombocythemia, and myeloid metaplasia with myelofibrosis. Cancer Cell 7:387–397
5. Kralovics R, Passamonti F, Buser AS, Teo SS, Tiedt R, Passweg JR, Tichelli A, Cazzola M, Skoda RC (2005) A gain-of-function mutation of JAK2 in myeloproliferative disorders. N Engl J Med 352:1779–1790
6. Jones AV, Kreil S, Zoi K, Waghorn K, Curtis C, Zhang L, Score J, Seear R, Chase AJ, Grand FH, White H, Zoi C, Loukopoulos D, Terpos E, Vervessou EC, Schultheis B, Emig M, Ernst T, Lengfelder E, Hehlmann R, Hochhaus A, Oscier D, Silver RT, Reiter A, Cross NC (2005) Widespread occurrence of the JAK2 V617F mutation in chronic myeloproliferative disorders. Blood 106:2162–2168
7. Schmitt-Graeff AH, Teo SS, Olschewski M, Schaub F, Haxelmans S, Kirn A, Reinecke P, Germing U, Skoda RC (2008) JAK2V617F mutation status identifies subtypes of refractory anemia with ringed sideroblasts associated with marked thrombocytosis. Haematologica 93: 34–40
8. Jelinek J, Oki Y, Gharibyan V, Bueso-Ramos C, Prchal JT, Verstovsek S, Beran M, Estey E, Kantarjian HM, Issa JP (2005) JAK2 mutation 1849G>T is rare in acute leukemias but can be found in CMML. Philadelphia chromosome-negative CML, and megakaryocytic leukemia. Blood 106:3370–3373
9. Scott LM, Campbell PJ, Baxter EJ, Todd T, Stephens P, Edkins S, Wooster R, Stratton MR,

Futreal PA, Green AR (2005) The V617F JAK2 mutation is uncommon in cancers and in myeloid malignancies other than the classic myeloproliferative disorders. Blood 106: 2920–2921

10. Scott LM, Tong W, Levine RL, Scott MA, Beer PA, Stratton MR, Futreal PA, Erber WN, McMullin MF, Harrison CN, Warren AJ, Gilliland DG, Lodish HF, Green AR (2007) JAK2 exon 12 mutations in polycythemia vera and idiopathic erythrocytosis. N Engl J Med 356:459–468
11. Bercovich D, Ganmore I, Scott LM, Wainreb G, Birger Y, Elimelech A, Shochat C, Cazzaniga G, Biondi A, Basso G, Cario G, Schrappe M, Stanulla M, Strehl S, Haas OA, Mann G, Binder V, Borkhardt A, Kempski H, Trka J, Bielorei B, Avigad S, Stark B, Smith O, Dastugue N, Bourquin JP, Tal NB, Green AR, Izraeli S (2008) Mutations of JAK2 in acute lymphoblastic leukaemias associated with Down's syndrome. Lancet 372:1484–1492
12. Cleyrat C, Jelinek J, Girodon F, Boissinot M, Ponge T, Harousseau JL, Issa JP, Hermouet S (2010) JAK2 mutation and disease phenotype: a double L611V/V617F in cis mutation of JAK2 is associated with isolated erythrocytosis and increased activation of AKT and ERK1/2 rather than STAT5. Leukemia 24:1069–1073
13. Swerdlow SH, Jaffe ES, International Agency for Research on Cancer., and World Health Organization (2008) WHO classification of tumours of haematopoietic and lymphoid tissues. International Agency for Research on Cancer, Lyon
14. Campbell PJ, Scott LM, Baxter EJ, Bench AJ, Green AR, Erber WN (2006) Methods for the detection of the JAK2 V617F mutation in human myeloproliferative disorders. Methods Mol Med 125:253–264
15. Lippert E, Girodon F, Hammond E, Jelinek J, Reading NS, Fehse B, Hanlon K, Hermans M, Richard C, Swierczek S, Ugo V, Carillo S, Harrivel V, Marzac C, Pietra D, Sobas M, Mounier M, Migeon M, Ellard S, Kroger N, Herrmann R, Prchal JT, Skoda RC, Hermouet S (2009) Concordance of assays designed for the quantification of JAK2V617F: a multicenter study. Haematologica 94:38–45
16. Cankovic M, Whiteley L, Hawley RC, Zarbo RJ, Chitale D (2009) Clinical performance of JAK2 V617F mutation detection assays in a molecular diagnostics laboratory: evaluation of screening and quantitation methods. Am J Clin Pathol 132:713–721
17. James C, Delhommeau F, Marzac C, Teyssandier I, Couedic JP, Giraudier S, Roy L, Saulnier P, Lacroix L, Maury S, Tulliez M, Vainchenker W, Ugo V, Casadevall N (2006) Detection of JAK2 V617F as a first intention diagnostic test for erythrocytosis. Leukemia 20:350–353
18. McClure R, Mai M, Lasho T (2006) Validation of two clinically useful assays for evaluation of JAK2 V617F mutation in chronic myeloproliferative disorders. Leukemia 20:168–171
19. Sattler M, Walz C, Crowley BJ, Lengfelder E, Janne PA, Rogers AM, Kuang Y, Distel RJ, Reiter A, Griffin JD (2006) A sensitive high-throughput method to detect activating mutations of Jak2 in peripheral-blood samples. Blood 107:1237–1238
20. Chen Q, Lu P, Jones AV, Cross NC, Silver RT, Wang YL (2007) Amplification refractory mutation system, a highly sensitive and simple polymerase chain reaction assay, for the detection of JAK2 V617F mutation in chronic myeloproliferative disorders. J Mol Diagn 9: 272–276
21. Tan AY, Westerman DA, Dobrovic A (2007) A simple, rapid, and sensitive method for the detection of the JAK2 V617F mutation. Am J Clin Pathol 127:977–981
22. Rapado I, Grande S, Albizua E, Ayala R, Hernandez JA, Gallardo M, Gilsanz F, Martinez-Lopez J (2009) High resolution melting analysis for JAK2 Exon 14 and Exon 12 mutations: a diagnostic tool for myeloproliferative neoplasms. J Mol Diagn 11:155–161
23. Jones AV, Cross NC, White HE, Green AR, Scott LM (2008) Rapid identification of JAK2 exon 12 mutations using high resolution melting analysis. Haematologica 93:1560–1564
24. Denys B, El Housni H, Nollet F, Verhasselt B, Philippe J (2010) A real-time polymerase chain reaction assay for rapid, sensitive, and specific quantification of the JAK2V617F mutation using a locked nucleic acid-modified oligonucleotide. J Mol Diagn 12:512–519
25. Kroger N, Badbaran A, Holler E, Hahn J, Kobbe G, Bornhauser M, Reiter A, Zabelina T, Zander AR, Fehse B (2007) Monitoring of the JAK2-V617F mutation by highly sensitive quantitative real-time PCR after allogeneic stem cell transplantation in patients with myelofibrosis. Blood 109:1316–1321
26. Larsen TS, Christensen JH, Hasselbalch HC, Pallisgaard N (2007) The JAK2 V617F mutation involves B- and T-lymphocyte lineages in a subgroup of patients with Philadelphia-chromosome negative chronic myeloproliferative disorders. Br J Haematol 136:745–751
27. Lippert E, Boissinot M, Kralovics R, Girodon F, Dobo I, Praloran V, Boiret-Dupre N, Skoda RC, Hermouet S (2006) The JAK2-V617F mutation is frequently present at diagnosis in

patients with essential thrombocythemia and polycythemia vera. Blood 108:1865–1867

28. Zhang SJ, Qiu HX, Li JY, Shi JY, Xu W (2010) The analysis of JAK2 and MPL mutations and JAK2 single nucleotide polymorphisms in MPN patients by MassARRAY assay. Int J Lab Hematol 32:381–386
29. Hermouet S, Dobo I, Lippert E, Boursier MC, Ergand L, Perrault-Hu F, Pineau D (2007) Comparison of whole blood vs purified blood granulocytes for the detection and quantitation of JAK2(V617F). Leukemia 21: 1128–1130
30. Vainchenker W, Delhommeau F, Constantinescu SN, Bernard OA (2011) New mutations and pathogenesis of myeloproliferative neoplasms. Blood 118:1723–1735
31. Heller PG, Lev PR, Salim JP, Kornblihtt LI, Goette NP, Chazarreta CD, Glembotsky AC, Vassallu PS, Marta RF, Molinas FC (2006) JAK2V617F mutation in platelets from essential thrombocythemia patients: correlation with clinical features and analysis of STAT5 phosphorylation status. Eur J Haematol 77: 210–216
32. Sidon P, El Housni H, Dessars B, Heimann P (2006) The JAK2V617F mutation is detectable at very low level in peripheral blood of healthy donors. Leukemia 20:1622
33. Quentmeier H, MacLeod RA, Zaborski M, Drexler HG (2006) JAK2 V617F tyrosine kinase mutation in cell lines derived from myeloproliferative disorders. Leukemia 20: 471–476
34. Pardanani A, Lasho TL, Finke C, Hanson CA, Tefferi A (2007) Prevalence and clinicopathologic correlates of JAK2 exon 12 mutations in JAK2V617F-negative polycythemia vera. Leukemia 21:1960–1963
35. Passamonti F, Elena C, Schnittger S, Skoda RC, Green AR, Girodon F, Kiladjian JJ, McMullin MF, Ruggeri M, Besses C, Vannucchi AM, Lippert E, Gisslinger H, Rumi E, Lehmann T, Ortmann CA, Pietra D, Pascutto C, Haferlach T, Cazzola M (2011) Molecular and clinical features of the myeloproliferative neoplasm associated with JAK2 exon 12 mutations. Blood 117:2813–2816
36. Ugo V, Tondeur S, Menot ML, Bonnin N, Le Gac G, Tonetti C, Mansat-De Mas V, Lecucq L, Kiladjian JJ, Chomienne C, Dosquet C, Parquet N, Darnige L, Porneuf M, Escoffre-Barbe M, Giraudier S, Delabesse E, Cassinat B (2010) Interlaboratory development and validation of a HRM method applied to the detection of JAK2 exon 12 mutations in polycythemia vera patients. PLoS One 5:e8893
37. Schnittger S, Bacher U, Haferlach C, Geer T, Muller P, Mittermuller J, Petrides P, Schlag R, Sandner R, Selbach J, Slawik HR, Tessen HW, Wehmeyer J, Kern W, Haferlach T (2009) Detection of JAK2 exon 12 mutations in 15 patients with JAK2V617F negative polycythemia vera. Haematologica 94:414–418
38. Erali M, Wittwer CT (2010) High resolution melting analysis for gene scanning. Methods 50:250–261
39. Glenn TC (2011) Field guide to next-generation DNA sequencers. Mol Ecol Resour 11: 759–769
40. Meldrum C, Doyle MA, Tothill RW (2011) Next-generation sequencing for cancer diagnostics: a practical perspective. Clin Biochem Rev 32:177–195

Part II

Signal Transducers and Activators of Transcription

Chapter 9

Detection of Activated STAT Proteins

Jean-Patrick Parisien and Curt M. Horvath

Abstract

STAT proteins are activated by diverse cellular stimuli including cytokine and growth factor receptor signaling, proto-oncogene and oncogene expression, and cellular stress mediators. In most cases, canonical STAT activation by a particular treatment or cellular condition results in STAT protein phosphorylation on an activating tyrosine residue near the C terminus. This phosphotyrosine is recognized by SH2 domains in partner STATs, resulting in homo- or hetero-dimerization. The STAT dimers attain the ability to bind specific DNA response element sequences present in the promoters of target genes. Two methods are described for the detection of activated STAT proteins based on (1) acquisition of tyrosine phosphorylation and (2) acquisition of DNA binding ability.

Key words: STAT, Tyrosine phosphorylation, DNA binding, STAT activation

1. Detection of STAT Activation by Immunoblotting

1.1. Introduction

In the canonical models of ligand-activated STAT signal transduction, the STAT proteins are in a quiescent, latent state prior to their activation by tyrosine phosphorylation (1). While the currently available evidence indicates that the latent STATs are likely to be present in alternate oligomeric states and are in some cases able to function as signaling factors and transcription regulators (2), the acquisition of specific ligand-induced tyrosine phosphorylation remains a primary feature of STAT protein activation. Direct detection of phosphorylation on the activating tyrosine residue has been greatly facilitated by the availability of commercially produced phosphorylation-specific peptide antisera that recognize only the modified form of individual STATs. These reagents enable the analysis of STAT proteins in whole cell lysates or nuclear extracts by direct immunoblotting. Prior to the widespread availability of these STAT-specific phospho-tyrosine antisera, it was necessary

Sandra E. Nicholson and Nicos A. Nicola (eds.), *JAK-STAT Signalling: Methods and Protocols*, Methods in Molecular Biology, vol. 967, DOI 10.1007/978-1-62703-242-1_9, © Springer Science+Business Media New York 2013

to first immunoprecipitate the STAT protein and carry out the immunoblotting with secondary antibodies that recognize phosphorylated tyrosine. The ability to detect individual STAT proteins by these assays requires careful optimization of many factors including the primary and secondary antibodies used, the STAT activation and inactivation kinetics in the particular cells or tissues of interest, and the response of endogenous STATs to the ligand under investigation. As such, a general procedure for immunoblotting and immunoprecipitation is provided rather than a detailed protocol for any particular STAT protein.

1.2. Materials

Prepare all solutions using ultrapure water (dH_2O) with a resistivity of 18 MΩ cm at 25°C and analytical grade reagents. Prepare and store all reagents at room temperature (unless indicated otherwise). Follow all safety and waste disposal regulations when disposing of waste materials.

1. 6× DNA Gel loading Buffer: 0.25% bromophenol blue, 0.25% xylene cyanol, 30% glycerol.
2. 10× TBST: 0.5 M Tris–HCl pH 7.4, 2 M NaCl, 0.2% Tween 20.
3. Transfer Apparatus: Trans-Blot® SD (Biorad, Cat# 170-3940).
4. 10× Transfer Buffer: 25 mM Tris, 192 mM glycine.
5. 1× Transfer Buffer (500 mL): 50 ml 10× transfer buffer, 100 ml methanol, 350 ml H_2O.
6. Phospho STAT1 antibody: (Cell Signaling, Cat# 9171L).
7. Phospho STAT2 antibody: (Millipore, Cat# 07-224).
8. Phospho STAT3 antibody: (Cell Signaling, Cat# 9131L).
9. Phospho Tyrosine antibody: (Various Suppliers).
10. Total STAT1 antibody: (Santa Cruz Biotechnology, Cat# sc-345).
11. Total STAT2 antibody: (Santa Cruz Biotechnology, Cat# sc-476).
12. Total STAT3 antibody: (Santa Cruz Biotechnology, Cat# sc-482).
13. ECL Reagent: (Perkin Elmer, Cat# NEL102001EA).
14. Whole Cell Extract Buffer (WCEB): 50 mM Tris–HCl pH 8.0, 280 mM NaCl, 0.5% IGEPAL detergent (octylphenoxypolyethoxyethanol, Sigma-Aldrich), 0.2 mM EDTA, 2 mM EGTA, 10% glycerol, 1 mM DTT (dithiothreitol).
15. Proteins A/G: (Santa Cruz Biotechnology, Cat# sc-2003).

16. 5× Protein Loading Buffer (PLB): 0.312 M Tris–HCl pH 6.8, 10% SDS, 25% 2-mercaptoethanol, 20% glycerol, 0.01% bromophenol blue.

1.3. Methods

1.3.1. Immunoprecipitation

- For IP-Western continue to immunoprecipitation section.
- For direct analysis with anti-phospho-STAT antisera, run appropriate amount of lysate in SDS-PAGE and proceed to immunoblotting section.

1. Lyse cell pellet with 0.5–1 ml complete WCEB (see Note 1).
2. Spin at 20,000 × *g* for 10 min and remove supernatant to new tube.
3. Reserve 10% of supernatants for direct application to gel (Inputs).
4. Add to remaining supernatants the recommended amount of your specific antibody (see Note 2).
5. Rock extracts at 4°C for 1–4 h.
6. Add 30 μl Protein A/G agarose bead suspension or follow the recommended protocol for bead volume based on amount of antibody used.
7. Rock an additional 1 h to overnight at 4°C.
8. Centrifuge beads at 1700 × *g* for 5 min and aspirate supernatant.
9. Wash beads 3× with 1 ml cold WCEB (no inhibitors needed).
10. Add PLB to elute, boil 5 min at 100°C, and separate by SDS-PAGE.

1.3.2. Gel Running and Transfer to Nitrocellulose Membrane (See Note 3)

1. Run SDS-PAGE for appropriate time (generally 1–2 h at 125 V).
2. Disassemble apparatus, cut eight pieces of Whatman paper and briefly soak four pieces in transfer buffer.
3. Place on transfer apparatus (anode is on the bottom).
4. Cut nitrocellulose membrane to size and soak in dH_2O first, then in transfer buffer.
5. Remove gel from plate and place it on top of the nitrocellulose membrane in the transfer buffer.
6. Place nitrocellulose membrane (with gel on top) on the blotting paper. Roll out to make sure there are no bubbles (see Note 4).
7. Soak the remaining four pieces of Whatman paper in transfer buffer and place on top of the gel. Roll out again to remove air bubbles.

8. Transfer at 20 V for 30 min.
9. Make sure markers have transferred to the nitrocellulose.

1.3.3. Immunoblotting

1. After completing SDS-PAGE and electrophoretic transfer, block membranes in TBST containing dry milk powder for from 1 h at room temperature to overnight at 4°C.
2. Wash for 3 × 10 min with TBST.
3. Incubate the membrane with primary antibody for from 1 h at room temperature to overnight at 4°C (see Note 5).
   - *For IP-Western use α-phosphotyrosine antibody.*
   - *For straight Western, use α-phospho STAT antibody.*
   - *It is also essential to reprobe or run a parallel gel for total STAT expression.*
4. Remove the primary antibody and wash for 3 × 10 min with TBST.
5. Incubate the membrane with HRP-secondary antibody for 1 h at room temperature.
6. Remove the secondary antibody and wash for 3 × 10 min with TBST.
7. Prepare the ECL reagents and expose membrane following the manufacturer's recommendations.
8. Develop membrane with X-ray film or digital imaging equipment. Fig. 1

2. Detection of Activated STAT DNA-Binding Complexes by Electrophoretic Mobility Shift Assay

2.1. Introduction

The STAT proteins have the dual functions of signal transduction from ligand-activated receptor kinase complexes followed by nuclear translocation and DNA binding to activate gene transcription. To function as specific transcriptional activators, STAT proteins by themselves or in combination with other proteins must have the ability to recognize specific DNA sequence elements in the promoters of their target genes. In the canonical model of STAT activation, the binding of the STATs to DNA occurs after tyrosine phosphorylation when the proteins form either homodimers (as is the case with STAT1, STAT3, STAT4, STAT5A, STAT5B, and STAT6) or heterodimers (e.g., STAT1-STAT3, STAT1-STAT2) linked by SH2 domain–phosphotyrosine interactions, that bind DNA either alone or in combination with other proteins. The interferon-activated STATs, STAT1 and STAT2, provide excellent examples of both types of binding complexes. Treatment of cells with IFNγ results in activation of a homodimeric IFNγ-activated

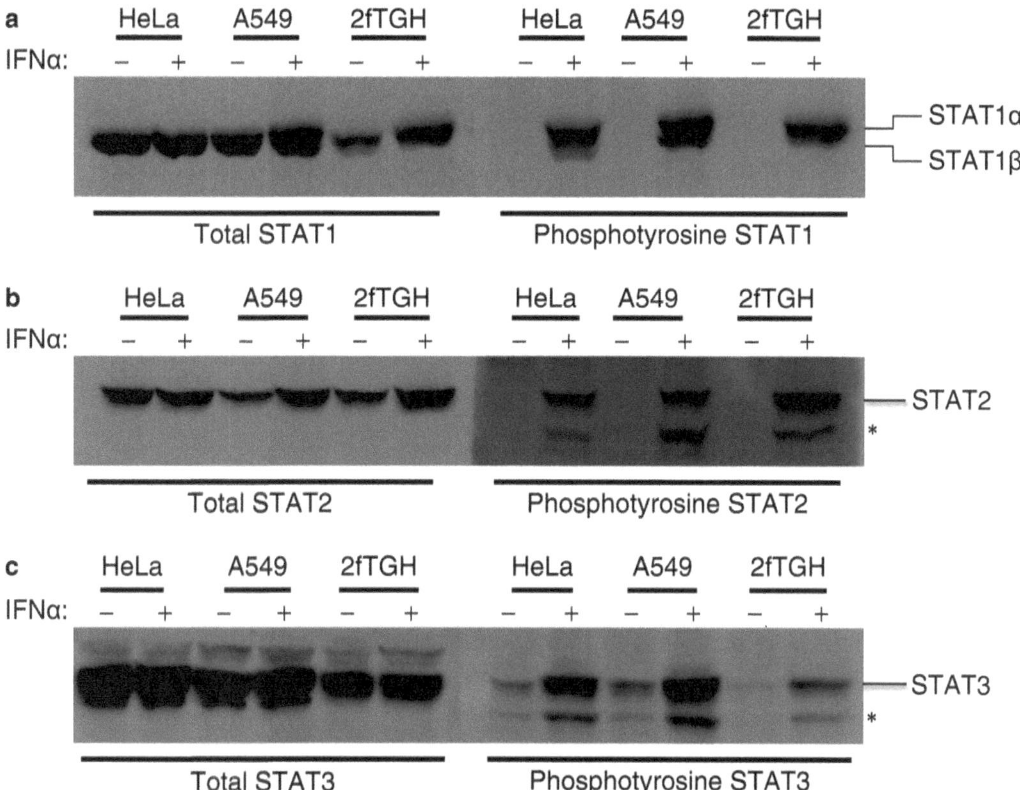

Fig. 1. Examples of detection of STAT tyrosine phosphorylation. Human cell lines HeLa, A549, and 2fTGH were used to demonstrate STAT tyrosine phosphorylation. Cells were treated with 1,000 U/ml of recombinant IFNα (+) for 30 min or left untreated (−) prior to whole cell extract preparation. Gels were loaded with 50 μg total protein per lane and separated before transfer to nitrocellulose membranes and subsequent immunoblotting with STAT-specific antisera. (**a**) Blot was probed for total STAT1 (*left*) and tyrosine phosphorylated STAT1 (*right*). Note the appearance of both STAT1α and STAT1β as a doublet. (**b**) Blot was probed for total STAT2 (*left*) and tyrosine phosphorylated STAT2 (*right*). *Asterisk* indicates a nonspecific or truncated protein recognized in addition to the phosphorylated STAT2. (**c**) Blot was probed for total STAT3 (*left*) and tyrosine phosphorylated STAT3 (*right*). *Asterisk* indicates a nonspecific or truncated protein recognized in addition to the phosphorylated STAT3.

factor (GAF), that binds to a 9 bp inverted repeat DNA sequence elements (referred to as GAS elements for IFNγ-activated sequence) that resemble the consensus, 5′-TTCCGGGAA-3′, or 5′-TTC(N)$_3$GAA-3′. STAT1, STAT3, STAT4, and STAT5 homodimers bind well to variants of the fundamental GAS element consensus, but STAT6 homodimers recognize slightly longer elements matching the consensus 5′-TTC(N)$_4$GAA-3′. A high affinity binding site for STAT1, STAT3, and other STAT dimers is the M67 variant of the c-fos SIE (M67 SIE), 5′-CATTTCCCG TAAATCAT-3′. For interleukin 6 (IL6) treatment, M67 SIE-binding by STAT1 homodimers, STAT3 homodimers, and STAT1:STAT3 heterodimers form distinctly migrating bands in an electrophoretic mobility shift assay (3, 4). In the case of

IFNα/β-activated STAT1 and STAT2, the STAT1–STAT2 heterodimer is joined by the DNA-binding adaptor, IRF9, to form the IFN-stimulated gene transcription factor, ISGF3 (5, 6). The association of the STAT heterodimer with IRF9 results in recognition of a distinct DNA response element, the IFN-stimulated response element (ISRE), which matches the consensus sequence 5′-AGTTTN3TTTCC-3′.

2.2. Materials

2.2.1. Reagents

1. 40% Acrylamide/Bis-acrylamide solution (29:1 acrylamide:Bis): (OmniPur, EMD).
2. 10× TBE Buffer (1 L): Dissolve 108 g Tris and 55 g Boric acid in 900 ml H_2O. Add 40 ml 0.5 M Na_2EDTA (pH 8.0) and adjust volume to 1 L with H_2O.
3. TE buffer: 10 mM Tris–HCl pH 8.0, 1 mM EDTA pH 8.0. To make 1 L: add 10 ml 1 M Tris (pH 8.0, 7.6 or 7.4) and 2 ml 0.5 M Na_2EDTA (pH 8.0) to 800 ml H_2O. Mix and adjust to 1 L with H_2O, sterilize by autoclaving and store at room temperature.
4. 10% APS (ammonium persulfate): Dissolve 1 g APS in 8 ml H_2O. Adjust volume to 10 ml with H_2O. Solution is stable at 4°C for 2 weeks.
5. TEMED (*N,N,N′,N′*-tetramethylethylenediamine): (OmniPur, EMD Cat# 8920).
6. poly (dIdC): Sigma, prepare stock in TE buffer and store at −20°C.
7. 5× Gel Shift Buffer: 100 mM HEPES–HCl, pH 7.9, 20% Ficoll, 5 mM $MgCl_2$, 200 mM KCl, 0.5 mM EGTA, 2.5 mM DTT.
8. 6× DNA Gel loading Buffer: 0.25% bromophenol blue, 0.25% xylene cyanol, 30% glycerol.
9. Whatman™ 3MM chromatography paper: (Cat# 3030-866).
10. 10× TM Buffer: 100 mM Tris–HCL pH 8.0, 60 mM $MgCl_2$, 10 mM DTT.
11. Rainex®, or similar commercial glass polish, a cost effective substitute for dimethyl silane.
12. Sephadex® G50, fine grade: (Sigma-Aldrich, Cat# G5080)

2.2.2. Equipment

1. Savant Slab gel dryer: (Model # SGD 2000).
2. Saran wrap: (SC Johnson Company).
3. Vertical gel electrophoresis apparatus (e.g., PROTEAN® II xi Cell, BioRad Cat #165-1812) with 1.5 mm spacers and combs or any commercially available system that includes glass plates, combs, spacers, and clamps.
4. Syringe with heavy gauge needles.

5. Power supply: (suitable for electrophoresis).
6. X-ray film and cassette or storage phosphor screen.
7. Gel loading pipette tips: (Denville Cat# P3107).
8. Scintillation Counter.

2.3. Methods

Carry out all procedures at room temperature unless otherwise specified.

2.3.1. Gel Preparation

1. Use clean gel plates, spacers, and combs. Remove any fingerprints or other residues from glass plates by coating short plate *only* with Rainex® (see Note 6).
2. Make 5% PAGE in 0.25× TBE in the order shown: (see Note 7).

5% Gel	*70 ml*	*100 ml*
40% Acrylamide:Bis 29:1	8.75 ml	12.5 ml
10× TBE	1.75 ml	2.5 ml
dH_2O	58.25 ml	83.3 ml
10% APS	1.4 ml	2 ml
TEMED (add last)	70 μl	100 μl

3. Mix well and pour slowly to avoid bubble formation. Insert comb immediately and allow gel to polymerize 20–30 min (see Note 8).
4. Mount gel in apparatus and fill top and bottom buffer reservoirs with 0.25× TBE (see Note 9).
5. Pre-run gel in cold room for ≈2 h at a constant 400 V until current reaches <15 mA (see Note 10).
6. While gel is pre running, prepare DNA binding reactions.

2.3.2. Gel Shift Probe

- Double-stranded oligodeoxynucleotides corresponding to the STAT response element are synthesized by annealing individual purified single strands that bear a 5′-dGATC extension. The ends labeled to high specific activity by extension of the overhang with Klenow fragment and ^{32}P-nucleoside triphosphate. Alternatively, the probe can be end-labeled with γ-^{32}P-ATP in a kinase reaction.

  Annealing of Oligos:

  1. Mix top and bottom strands in equimolar ratios with a final concentration of 100 ng/ml (10 mg each single strand, then bring up to 100 ml volume with TE).
  2. Boil 5 min in beaker on hot plate, allow to cool slowly to RT.
  3. Store double-stranded oligo at −20°C.

2.3.3. Labeling to Make Probe

1. Shift probes are designed with 5′-overhanging ends to be filled in with Klenow and α^{32}P-dNTPs. If overhang is designed to contain all bases (GATC), any or all radiolabeled nucleotides can be used to fill in and label the probe; these directions are for α^{32}P-dCTP:
   - 1 μl (100 ng) double-stranded shift probe.
   - 3 μl 10× TM buffer.
   - 5 μl 2 mM dGTP, dATP, and dTTP solution.
   - 5 μl α^{32}P-dCTP.
   - 15 μl H_2O.
   - 1 μl Klenow fragment.

   30 μl total.
2. React at room temperature for 20 min.
3. Add 1 μl 2 mM dGATC to chase for 2 min and add 70 μl TE buffer (100 μl of 1 ng/μl probe solution).

2.3.4. Probe Purification

Several commercial G50 spin columns are available in ready-to-use format. These instructions are for making a home-made spin column from a plugged syringe.

G50 Spin Column Procedure:

1. Plug a 1 ml syringe with glass wool or filter fiber for each probe.
2. Place plugged column into 15 ml conical tube.
3. Fill with G50 Slurry (1:1 G50 in water, autoclaved).
4. Spin at 240 ×*g* for 3 min and top off syringe with G50 slurry.
5. Spin at 240×*g* for 3 min (column bed volume should be 0.8–1 ml).
6. Add 100 μl TE buffer to column and spin at 240×*g* for 3 min.
7. Discard flow through from 15 ml tube.
8. *Put a topless microfuge tube into 15 ml tube to collect the labeled probe, then replace the column.*
9. Aliquot 100 μl probe reaction to top of column spin at 240×*g* for 3 min.
10. Measure radioactivity (cpm) on column (unincorporated) and in tube (labeled probe).
11. Transfer tube contents to fresh microfuge tubes, store at -20°C.
12. Check the radioactive incorporation by scintillation counting. Probes should be $\geq 10^8$ cpm/μg.

2.3.5. DNA Binding Reactions

1. Each 15 μl Reaction contains:
   - 5–10 μg protein/extract—(2 μl nuclear extract).
   - 2 μg poly dIdC—(1 μl at 2 μg/μl).
   - $\geq 10^5$ cpm probe (0.5–1 ng)—(0.5–1 μl).
   - 5× Gel Shift buffer—(3 μl).
   - dH_2O—to 15 μl.
2. Set up master mix without protein and aliquot to each tube.
3. Add protein sample and mix by pipetting.
4. Allow to react at room temperature for 20 min.
5. Stop gel, flush lanes with syringe, and carefully load samples by layering the reaction gently onto the gel (see Note 11).

2.3.6. Gel Electrophoresis

1. Add 1× DNA gel loading buffer to an empty well to monitor electrophoresis. *Do not add to samples* (see Note 12).
2. After loading the samples, run the gel in the cold room for 2–4 h at 300–400 V until the bromophenol blue dye (bottom dye) is near the bottom or off the gel (see Note 13).
3. Pry open the gel plates (see Note 14).
4. Place two sheets of dry Whatman 3MM chromatography paper on top of the gel and carefully lift (see Note 15).
5. Cover with saran wrap and dry on a gel dryer (80°C for 1–2 h as needed).
6. Expose the gel to X-ray film or phosphorimaging screen for 1 h to overnight at −80°C (see Note 16) (Fig. 2).

3. Notes

1. *Optional.* Pre-clear lysates with 20 μl of Protein G/A Agarose + 1 μg normal IgG from the same species as the IP antibody for 30 min at 4°C. This step helps remove proteins that bind nonspecifically to agarose beads thus reducing background.
2. Each antibody must be optimized for the specific condition.
3. The following protocol is for semi-dry transfer.
4. Rolling out bubbles ensures even transfer of proteins to membranes.
5. We typically use at 1:1,000 or follow the manufacturer's recommendations.
6. With Rainex®, gel will not adhere to the short plate but only to the tall plate making gel removal easier after the run.

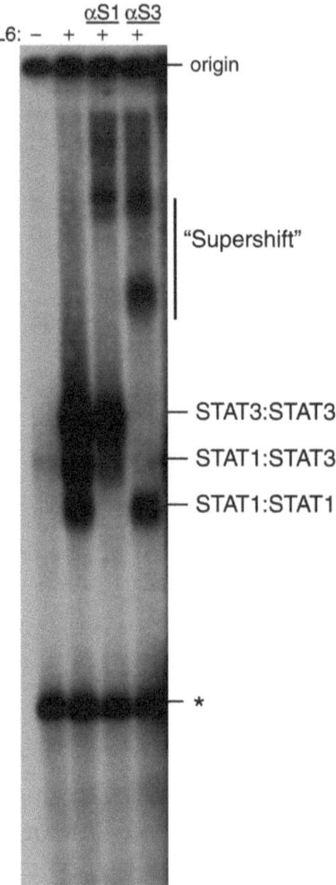

Fig. 2. Example of STAT electrophoretic mobility shift assay. HepG2 cells were treated with interleukin 6 (IL6) for 30 min prior to hypotonic lysis and nuclear extraction. Nuclear extracts from IL6 treated (+) and untreated (–) cells were evaluated in parallel for binding to the M67 SIE probe. Three complexes are formed, consisting of STAT1 and STAT3 homodimers and heterodimers. Addition of STAT1- or STAT3-specific antiserum (αS1, αS3) reveals the protein composition of each species by preventing the normal formation and migration of STAT–DNA complexes to produce the so-called "supershift." *Asterisk* indicates a nonspecific protein in complex with the probe.

7. Gel can also be made the day before, wrapped with wet paper towels and Saran wrap, and stored at 4°C.
8. Bubbles will disrupt electrophoretic migration, therefore it is important to minimize the number and size of bubbles trapped during polymerization. Tilting the plate at 45 ° while pouring helps minimize bubble formation. If bubbles are trapped, gently tap the gel assembly after pouring in order to make bubbles float to the top and remove them before inserting the comb.
9. Use a syringe with a straight needle to rinse debris from the wells. Use another syringe with a 90 ° bent needle to remove bubbles trapped at the bottom of the gel if necessary.

10. The primary reasons for pre-running the gel is to remove all traces of APS and other electrolytes, to lower the ionic strength of the gel and buffer, and to ensure a constant gel running temperature.
11. Do not add dye solutions or loading buffers to the reactions, as they might disrupt DNA–protein interactions. Instead, run a dye sample in a separate lane. The position of the dye can be used to orient the gel after drying.
12. This is to prevent the dyes in the loading buffer from interfering with the binding of proteins to DNA.
13. At this point any unbound probe will have already migrated into the bottom buffer tank, but the much slower migrating STAT–DNA complexes will be well resolved. Note that the tank buffer will contain the ^{32}P-labeled probe DNA.
14. Gel should stick to the untreated taller plate at this point.
15. Sample loading order should be from left to right.
16. In all cases, simple controls should be included such as running the probe alone and always comparing unstimulated nuclear extracts to stimulated. Variations of this basic protocol can be used to qualitatively evaluate DNA-binding species and quantitatively measure binding affinities and on or off rates. These procedures could include competition with unlabeled wild-type and mutant versions of the probe, and antibody-mediated supershifts to characterize the components of the DNA–protein complexes.

References

1. Aaronson DS, Horvath CM (2002) A road map for those who don't know JAK-STAT. Science 296:1653–1655
2. Cheon H, Yang J, Stark GR (2011) The functions of signal transducers and activators of transcriptions 1 and 3 as cytokine-inducible proteins. J Interferon Cytokine Res 31:33–40
3. Zhong Z, Wen Z, Darnell JE Jr (1994) Stat3: A STAT family member activated by tyrosine phosphorylation in response to epidermal growth factor and interleukin-6. Science 264:95–98
4. Sadowski HB, Shuai K, Darnell JE Jr, Gilman MZ (1993) A common nuclear signal transduction pathway activated by growth factor and cytokine receptors. Science 261: 1739–1744
5. Fu X-Y, Schindler C, Improta T, Aebersold R, Darnell JE Jr (1992) The proteins of ISGF3, the interferon α-induced transcriptional activator, define a gene family involved in signal transduction. Proc Natl Acad Sci U S A 89:7840–7843
6. Shuai K, Schindler C, Prezioso VR, Darnell JE Jr (1992) Activation of transcription by IFN-γ: tyrosine phosphorylation of a 91 kD DNA binding protein. Science 259:1808–1812

Chapter 10

Detection of Activated STAT Species Using Electrophoretic Mobility Shift Assay (EMSA) and Potential Pitfalls Arising from the Use of Detergents

Serge Haan and Claude Haan

Abstract

Here we describe the preparation of nuclear extracts and the electrophoretic mobility shift assay (EMSA) for the detection of STAT species. We use the method for the investigation of STAT1 and STAT3 homo- and heterodimers and show how the preparation of the extracts can influence the distribution of the STAT species observed in the EMSA. We show that detergents can massively influence the STAT dimer distribution. Although it is unclear whether they primarily interfere with STAT DNA binding and/or whether they break up or further oligomerize STATs, the observation may also have an impact on the results of other techniques performed with detergent-containing cell lysates (e.g., coimmunoprecipitations of STATs with other proteins).

Key words: EMSA, Nuclear extracts, Cytosolic extracts, STAT dimers

1. Introduction

The electrophoretic mobility shift assay (EMSA) is a technique used to investigate protein–DNA or protein–RNA interactions (1, 2). Generally, EMSA assays are used to investigate whether a given protein or protein complex can bind to a specific DNA/RNA sequence. The recognition of a labeled DNA or RNA sequence by a protein will lead to a reduced mobility of the DNA/RNA fragment during gel-electrophoresis. The EMSA technique has been widely used to study the interaction of STAT transcription factors with various DNA motifs. As well as demonstrating whether a given DNA sequence is bound by STATs, the different mobilities of STAT dimer species such as STAT3/STAT3, STAT1/STAT1, or STAT1/STAT3 will also allow EMSA to give information on the presence and distribution of these DNA-binding-competent

Sandra E. Nicholson and Nicos A. Nicola (eds.), *JAK-STAT Signalling: Methods and Protocols*, Methods in Molecular Biology, vol. 967, DOI 10.1007/978-1-62703-242-1_10, © Springer Science+Business Media New York 2013

STAT species (3). This information cannot be obtained by performing a Western blot analysis where STAT phosphorylation is monitored. In fact, the detection of phosphorylated STAT1 or STAT3 in a Western blot analysis does not always reflect their activity. Notably, it was shown that STAT1 phosphorylation by itself is not a good indicator of the formation of transcriptionally active STAT1 homodimers, especially if only early time points are considered (4, 5). For example, IL-6-type cytokines have been described to activate STAT1 and STAT3. In electrophoretic mobility shift assays of cell lysates STAT3 and STAT1 homodimers as well as STAT3/STAT1 heterodimers have been observed (3, 6). However, IL-6 type cytokines such as IL-6 and oncostatin M do not generally induce STAT1-dependent transcriptional responses (4, 7). Using EMSA analysis, we have previously reported that STAT1, which is phosphorylated after stimulation of cells with IL-6 or oncostatin M, is mostly present in the form of STAT1/STAT3 heterodimers (rather than STAT1/STAT1 homodimers) and that this contributes to the lack of STAT1-dependent responses (5). Importantly, in cells lacking STAT3 and where STAT1/STAT3 heterodimers cannot be formed, interleukin-6 stimulation leads to the formation of STAT1/STAT1 homodimers and induces STAT1-dependent responses (8). Similarly, we described that the IL-6-type cytokine interleukin-27 does not initiate strong STAT3 responses in liver cells, although STAT3 phosphorylation can be detected (9, 10). We also observed for IL-27 that much of the STAT3 is present in STAT1/STAT3 heterodimers, which may explain the lack of STAT3/STAT3-mediated responses in these cells. The EMSA procedure described here allows the detection of DNA-binding competent STAT1 and STAT3 species. We also show how the use of detergents during cell lysis or fractionation can affect the distribution of these species. Importantly, the changes observed in the distribution of STAT species in the presence of different detergents could also affect results of other analyses such as coimmunoprecipitation studies of STATs with other proteins.

2. Materials

2.1. Components for the Preparation of Nuclear Extracts (See Note 1)

1. PBS: 200 mM NaCl, 2.5 mM KCl, 8 mM Na_2HPO_4, 1.5 mM KH_2PO_4.
2. Hypotonic lysis buffer A: 10 mM HEPES–KOH pH 7.8, 1.5 mM $MgCl_2$, 10 mM KCl, 0.5 mM DTT, 0.2 mM phenylmethanesulfonyl fluoride (PMSF), 1 mM Na_3VO_4.
3. Hypertonic buffer C: 20 mM HEPES–KOH pH 7.8, 1.5 mM $MgCl_2$, 420 mM NaCl, 0.2 mM EDTA, 25% (v/v) glycerol, 0.5 mM DTT, 0.2 mM PMSF, 1 mM Na_3VO_4 (see Note 2).

2.2. Components for the Preparation of Oligonucleotides

1. Oligonucleotides, SIE-probe (see Note 3): oligonucleotides are ordered from Eurogentec, Liege, Belgium, or other companies.

   Oligo sense: 5′GAT CCG GGA GGG ATT TAC GGG 3′.

   Oligo antisense: 3′GGC CCT CCC TAA ATG CCC TTT ACG AC 5′.

   The oligonucleotides must not be phosphorylated at the 5′ end!

   Oligonucleotides are dissolved in water to a concentration of 100 pmol/µl and stored at −20°C.
2. 5× annealing buffer: 100 µl of 2 M Tris–HCl, pH 7.5, 100 µl of 1 M $MgCl_2$ and 25 µl of 5 M NaCl are mixed and water is added to a volume of 1 ml. Store at −20°C.

2.3. Components for the Preparation of Labeled Double-Stranded Oligonucleotides

1. Nucleotide-premix: For the preparation of 950 µl mix: Mix 120 µl M buffer (Restriction Enzyme buffer from Roche Applied Science, Penzberg, Germany), 3 µl dGTP (10 mM), 3 µl dCTP (10 mM), 3 µl dTTP (10 mM), 40 µl 1% (w/v) BSA, and 781 µl H_2O. Store in aliquots of 47.5 µl at −20°C.
2. α^{32}P dATP (see Note 4):(alpha-P32)dATP (SCP-203) from Hartmann Analytic GmbH, Braunschweig, Germany.
3. Klenow-Enzyme (1 U/µl).
4. PN buffer (Qiagen, Hilden, Germany).
5. PE buffer (Qiagen).

2.4. Components for EMSA

1. Glycerol (99%).
2. 40% Acrylamide/bisacrylamide solution (19:1) (A1640 from AppliChem, Darmstadt, Germany).
3. Ammonium persulfate (20% w/v solution in water), store at −20°C.
4. *N,N,N,N′*-tetramethyl-ethylenediamine (TEMED). Store at RT and change twice a year (see Note 5).
5. 5× TBE buffer (1 M Trisaminomethane, 0.83 M boric acid, 10 mM EDTA) pH is adjusted to pH 8.3 using HCl; store in the dark at 4°C.
6. BSA solution (10 mg/ml), stored at −20°C.
7. 5× Gel Shift buffer: 50 mM HEPES pH 7.8, 5 mM EDTA, 25 mM $MgCl_2$, 50% glycerol (prepare as follows: use 1 ml of 0.5 M HEPES pH 7.8, 100 µl 0.5 M EDTA, 250 µl 1 M $MgCl_2$ and add H_2O to a total of 4 ml. Adjust pH to 7.8 with 3 M KOH. Then add 5 ml 99% glycerol and adjust total vol ume to 10 ml with H_2O). Store at −20°C in 0.5 ml aliquots.
8. 36 mM PMSF: dissolve 0.063 g PMSF in 10 ml of an ethanol–H_2O mixture (2:1). Store as 1 ml aliquots at −20°C.

9. Dithiothreitol (DTT) solution: 0.5 M in H_2O, stored at −20°C.
10. TE buffer: 10 mM Tris–HCl pH 7.5, 1 mM EDTA.
11. 0.25% xylene cyanol solution (alternatively a xylene cyanol-containing DNA loading buffer can be used: 15% glycerol, 2.5% bromophenol blue, 2.5% xylene cyanol blue).
12. Poly(dI-dC)·Poly(dI-dC) (GE Healthcare, Little Chalfont, UK, Product No 27-7880). Dissolve 1 mg/ml in TE buffer, anneal 5 min at 45°C, and store in 0.2 ml aliquots at −20°C.
13. Methanol.
14. Acetic acid.
15. Gel dryer.
16. Equipment for visualization by autoradiography (we use phosphor screen and a Phosphorimager).

3. Methods

3.1. Preparation of Nuclear Extracts

All steps are performed at 4°C.

1. About $5–10 \times 10^6$ cells are harvested. Adherent cells can usually be scraped off the plate with 1 ml PBS and are transferred to a 1.5 ml tube. Suspension cells are collected by centrifugation in a 15 ml tube ($200 \times g$ for 5 min), resuspended in 1 ml PBS, and transferred to a 1.5 ml tube. The cells are pelleted by centrifugation at $200 \times g$ for 5 min.
2. For the generation of cytoplasmic extracts, the pelleted cells are resuspended in 300 μl hypotonic lysis buffer A and incubated for 20–30 min (see Note 6). After centrifugation at $600 \times g$ for 5 min, the supernatants are collected and centrifuged for another 10 min at $16{,}000 \times g$. The supernatant from this centrifugation constitutes the cytoplasmic extract (see Note 7).
3. The pellet is resuspended in 1–1.5 ml buffer A, incubated for 15–20 min and the cells are pelleted by centrifugation ($16{,}000 \times g$ for 10 min) (see Note 8).
4. To prepare the nuclear extract, the sediment is incubated with 50 μl buffer C for 20 min (resuspend to homogeneity). After centrifugation at $16{,}000 \times g$ for 10 min, the supernatant is collected as nuclear extract (see Note 9).
5. The protein concentration of the extracts is determined by using available techniques (e.g., Nanodrop-spectrophotometer, protein staining by Bradford, or other assay).

6. We usually directly prepare 2–3 aliquots of each sample containing 5 μg total protein in a volume of 9.5 μl buffer A (see Note 10) in order to avoid repeated thawing of the samples if the EMSA needs to be repeated.

3.2. Annealing of the Olignucleotide

For the detection of STAT1 and STAT3 species we usually use the SIE-probe described in Subheading 2. The single strand oligonucleotides can be annealed as follows.

1. 50 μl of each oligonucleotide (100 pmol/μl) are mixed in a 1.5 ml tube with 60 μl H_2O and 40 μl 5× annealing buffer.
2. The vial is placed into a heating block at 95°C for 2 min. The heating block is turned off with the vial remaining in the block. The oligonucleotides will anneal during the slow cooling phase of the block (3–5 h) (see Note 11).
3. The oligonucleotide mix can be stored at −20°C.

3.3. Labeling of Double-Stranded Oligonucleotides

The annealed oligonucleotides undergo a "fill-in" reaction in which the recessed 3′ends are filled. Four ^{32}P dATP molecules are inserted per double-stranded oligonucleotide (see Note 12).

1. "Fill-in" reaction: mix 47.5 μl Nucleotide premix, 2.5 μl of 1 pmol/μl of the annealed oligonucleotide (e.g., SIE-probe), 2.5 μl α ^{32}P dATP, 2.0 μl Klenow-enzyme 1 U/μl, and 5.5 μl H_2O (total volume: 60 μl).
2. Incubate 30 min at 37°C in a heating-block or oven.
3. Purification of the radioactively labeled probe (SIE or other) using the QIAquick Nucleotide Removal Kit (see Note 13): 600 μl of PN buffer is added to 60 μl probe. The mix is applied to the column and centrifuged at 3,300 × g for 1 min. Wash twice with 500 μl PE buffer and centrifuge at 3,300 × g for 1 min. Use a new collection vial (2 ml tube) at each centrifugation step. Finally, centrifuge at 16,000 × g for 1 min and place the column in a 1.5 ml tube. Elute the labeled probe with 50 μl H_2O by incubating for 3 min and subsequent centrifugation at 16,000 × g for 1 min.
4. Measure the radioactivity of the probe (60,000 cpm/μl or higher is desirable).
5. Store the probe at −20°C.

3.4. Electrophoretic Mobility Shift Assay

1. General comment: Different Gel-Chambers and fitting glass plates are available for EMSA. Here, we will describe the composition of the gels that we use and describe the conditions that work well for our self-built set-up. We use xylene cyanol as a color marker and we stop the electrophoresis after the marker has traveled 12 cm (see Note 14). As running buffer for the electrophoresis we use a 0.25 × TBE buffer (1:20 dilution of the 5 × TBE stock).

2. Preparing the gel: For 60 ml of a 5.5% acrylamide gel (see Note 15): Mix 44 ml H_2O with 4.5 ml glycerol, 3 ml 5 × TBE buffer, 8.25 ml acrylamide/bisacrylamide solution and 40 μl TEMED. Add 200 μl of ammonium persulfate solution, cast the gel, and insert the gel comb.
3. Let polymerization occur for a minimum of 60 min.
4. Thaw all the components (stored at −20°C) on ice. Initiate this step early as this takes a while.
5. Preparation of sample buffer (9.5 μl per sample).

   Table 1 shows a pipetting scheme which gives the required volumes for different sample numbers (see Note 16).
6. Dilute the radioactively labeled oligonucleotide to obtain 10,000 cpm/μl. Add 1 μl of labeled oligonucleotide per sample to the sample buffer mix (e.g., 45 μl for 45× of sample buffer) (see Note 17).
7. Add 10.5 μl of the sample buffer/oligonucleotide-mix to the 1.5 ml tubes containing the 9.5 μl sample (prepared in Subheading 3.1) (see Note 18).
8. Incubate for 15 min and load the samples onto the gel. Load 1–2 μl xylene cyanol solution (0.25%) in an empty lane as marker and run the electrophoresis until the marker has migrated 12 cm.
9. The gel is fixed by incubation for 15 min in a 10% methanol/10% acetic acid solution.
10. The gel is placed on a chromatography paper (e.g., Whatman® paper) and dried on a gel dryer. After exposure on a phosphorscreen over night at −80°C, the radioactive bands are visualized using phosphor screens and a phosphorimager (see Note 19).

Table 1
Preparation of the sample buffer

Number of samples (see Note 16)	Required volume in μl							
	1×	5×	10×	15×	20×	25×	35×	45×
H_2O	2.1	10.5	21	31.5	42	52.5	73.5	94.5
BSA	2	10	20	30	40	50	70	90
5× Gel-shift buffer	4	20	40	60	80	100	140	180
DTT	0.2	1	2	3	4	5	7	9
PMSF	0.2	1	2	3	4	5	7	9
Poly(dI-dC)	1	5	10	15	20	25	35	45

4. Notes

1. We strongly discourage the use of total cell lysates for EMSA analysis of STAT1 and STAT3 species. As we reported earlier (5) and as shown in Fig. 1, the STAT1 and STAT3 species occurring after stimulation of cells with cytokines such as Interleukin-6 can significantly differ in cytosolic and nuclear fractions. A total cell lysis or the use of a lysis buffer which does not rupture the nuclei will thus lead to a false species distribution in an EMSA analysis.
2. Na_3VO_4 can be omitted. We did not find differences in the STAT species pattern. However, special care should be taken to perform all steps at 4°C to minimize dephosphorylation by phosphatases.
3. We use a double-stranded, mutated SIE-oligonucleotide from the c-fos promotor (m67SIE). This oligonucleotide binds very well to active STAT1/STAT1 and STAT3/STAT3 homodimers as well as to STAT1/STAT3 heterodimers.
4. We use radioactive ^{32}P because of the high sensitivity of the detection. Alternatively, fluorescently labeled oligonucleotides could be used.

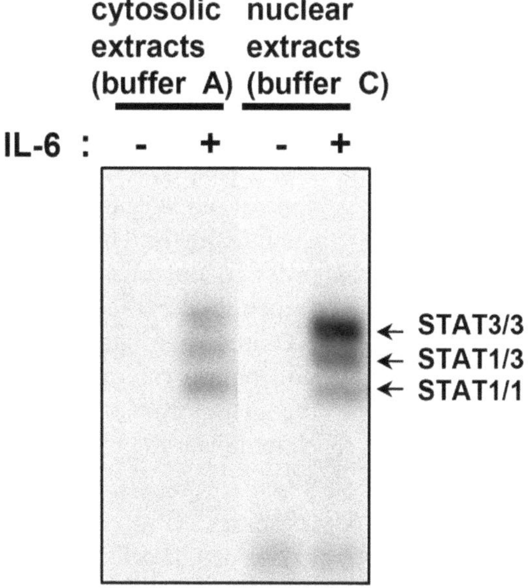

Fig. 1. EMSA analysis of cytosolic and nuclear extracts of HepG2 cells stimulated with Interleukin-6 (20 ng/ml) for 30 min. STAT3/STAT3, STAT3/STAT1, and STAT1/STAT1 species are indicated. As previously reported, the relative distribution of the species is different in the cytosol and the nucleus.

5. TEMED is hygroscopic and water will accumulate with time. This favors its oxidative decomposition and will prevent optimal polymerization of the gels. We replace our TEMED twice a year.
6. Be sure to totally remove the supernatant from the previous step. If the cytoplasmic extracts are not going to be investigated 1–1.5 ml of Lysis buffer A can be used. This will give a more efficient lysis. However if the cytosolic extracts are investigated by EMSA these larger volumes of hypotonic lysis buffer may lead to very weak signals. The optimal volume of hypotonic lysis buffer may have to be adjusted for a given cell line and cell number. Optimal incubation times can also vary with the cell line. For example, we found it useful to incubate primary macrophages for 30 min, whereas cell lines such as HepG2 cells usually only need to be incubated for about 20 min. We strongly recommend controlling the different steps by looking at the cells under a microscope. After the incubation with lysis buffer A you will usually see swollen cells but few ruptured cells and little debris. The cells will rupture during centrifugation. Of course, the rupturing can be optimized by using a loose-fitting Dounce homogenizer. However, care must be taken not to rupture nuclei as the STAT distribution in the cytosolic extracts will be altered if contaminated with nuclear extracts.
7. Ultracentrifugation of this fraction may be considered to increase the purity of the cytosolic extracts. However, the hypotonic lysis will usually not yield large amounts of small membrane fragments.
8. Visual inspection under the microscope after resuspension will usually show a significant pool of non-ruptured cells (except if a Dounce homogenizer was used in the previous step). This visual control allows one to adjust the volume of buffer A required for the initial lysis. This additional incubation step is performed in order to obtain a complete rupture of virtually all the cells. Depending on the cell type a repetition of this step or the use of a Dounce homogenizer may be required. As the STAT species distribution can vary significantly between the cytoplasm and the nuclear fractions (see Fig. 1, (5, 11)) an insufficient lysis will contaminate the nuclear extracts with cytosol.
9. Some protocols suggest adding the detergent IGEPAL (Nonidet NP-40) after the swelling of the cells in buffer A (12). Also, if cells are fractionated, residual ER fragments are usually removed before the lysis of the nuclei in order to further reduce contamination. However, we find that detergents differentially affect the STAT1/STAT1, STAT1/STAT3, and STAT3/STAT3 species and can virtually abrogate the STAT3/STAT3 signal in the EMSA. Figure 2 shows that with the

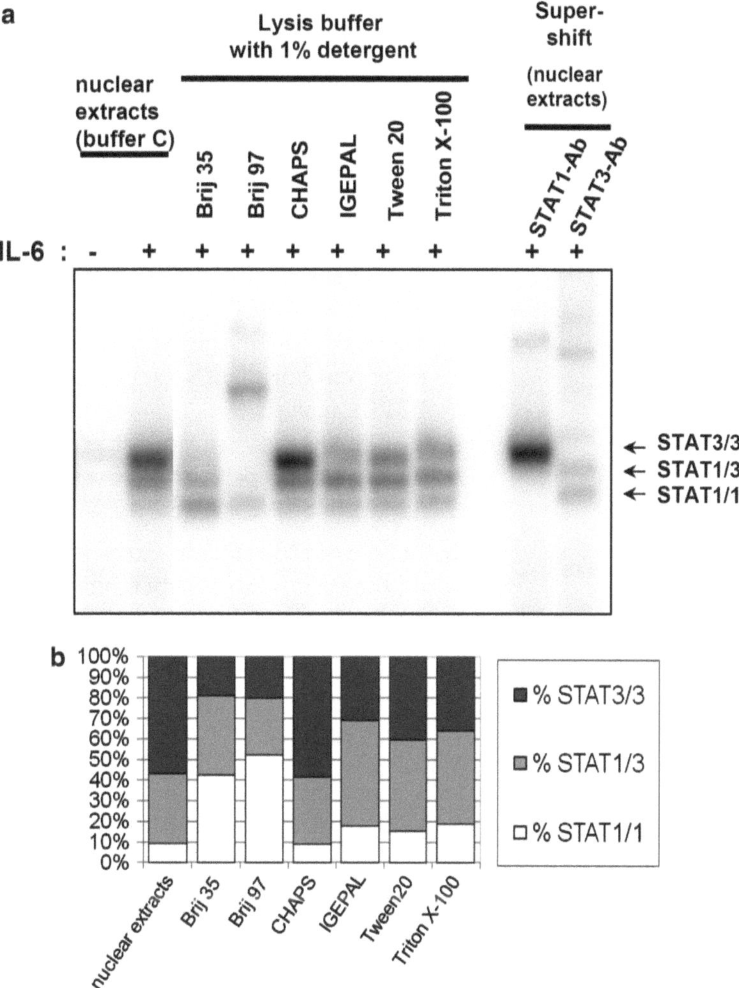

Fig. 2. Effect of detergents on STAT1 and STAT3 species. HepG2 cells were stimulated with interleukin-6 (20 ng/ml) for 20 min. Nuclear extracts (buffer C) were prepared as described above or the cells were lysed using PBS containing 1% of the indicated detergents. (**a**) An EMSA assay was performed as described above. A super-shift with a STAT1 or a STAT3 antibody was performed in order to verify the STAT contents of the EMSA bands. All lanes were run on the same gel and were exposed identically. (**b**) After exposure of a phosphor screen, the band intensities were quantified using a Typhoon Phosphorimager and corresponding software. Only the STAT species running at the right height (see *arrows* above) were considered for the quantification. The total signal was set to 100% in order to compare the species distribution.

exception of CHAPS all other detergents tested here reduce the intensity of the STAT3/STAT3 signal. It must be noted that it is still possible that CHAPS also affects the STAT dimer distributions if the conditions change (e.g., increased incubation time, temperature changes, etc.). This detergent effect explains the sometimes strange STAT species patterns seen in various publications. For example, the use of IGEPAL CA 630 increases the intensity of the STAT1/STAT3 heterodimer.

This could for example be at the origin of the sole STAT1/STAT3 band that was observed by Hückel et al. in a publication where IGEPAL CA-630 was used during the generation of the nuclear extracts (13). The observed effect could be due to interference with STAT3 binding to DNA or by affecting STAT dimerization or oligomerization. An indication of this could be that Brij 97 apparently specifically shifts the STAT3-containing dimers. The results shown in Fig. 3 indicate that Brij 97 shifts STAT3/STAT3 dimers as well as STAT3/STAT1 dimers as STAT1 and STAT3 antibodies lead to a further shifting of the low mobility bands. Finally, we show that the use of detergent during cell fractionations can also affect a subsequent EMSA analysis.

This is why we do not recommend the use of detergents during the preparation of extracts to be used for the determination of STAT species in EMSA.

10. In case of the cytosolic extracts we do not recommend topping up the volume with hypertonic buffer C. We find that buffer C generally reduces the signal intensity obtained in the EMSA

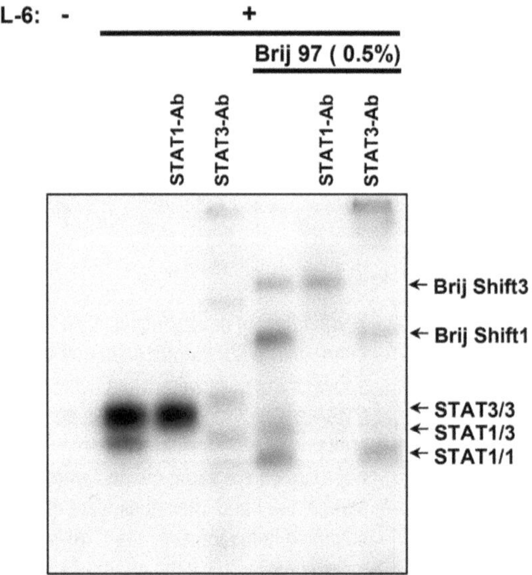

Fig. 3. We performed a single experiment in order to evaluate whether the shifted bands that appear in Brij 97-Lysates contain specific STAT species. HepG2 cells were stimulated with IL-6 (20 ng/ml) for 1 h and nuclear extracts were prepared as described here. The same stimulated sample was used for all the shift experiments. To induce the Brij-Shift, the nuclear extracts were filled up with buffer A containing Brij97 to a final Brij concentration of 0.5% and were incubated for 1 h prior to the addition of the labeled probe. For STAT super-shifts, a STAT1 or STAT3 antibody (1 μg) was added to the samples prior to incubation with the probe. The EMSA shows that the low mobility bands (Brij-Shift 1 and 3) that appear in the Brij samples can be super-shifted with STAT antibodies. We already observed this Brij-Shift at the lowest tested Brij percentage (0.1%, data not shown).

(without changing the species distribution (5)). We thus recommend using the smallest possible volume of buffer C. In order to not affect signal intensities we first add buffer C to the least possible volume (determined by the volume necessary for 5 μg total protein for the least concentrated sample). Then we add buffer A (or H_2O) to reach the final volume of 9.5 μl.

If special care is taken to have equal amounts of cells in every sample the protein concentrations should be very similar and most of the samples only need to be topped up with buffer A (or H_2O).

11. Alternatively, annealing can be performed by heating water to the boiling point in a large beaker (1 L), denaturing the oligonucleotides in the tube for 2 min and then allowing the water to cool down to room temperature.
12. Of course all necessary care must be taken when working with radioactive α ^{32}P dATP. Alternatively, chemiluminescent EMSA kits or others may be used according to the manufacturer's recommendations. We do not have much experience using the nonradioactive probes but in general, they seem to be less sensitive.
13. The double-stranded, labeled oligonucleotide will bind to the column, whereas the single nucleotides will pass through.
14. If the STAT1/STAT1, STAT1/STAT3, and STAT3/STAT3 complexes are to be separately quantified a minimal running distance (xylene cyanol marker) of 9 cm will probably be required in order to achieve a suitable separation. Xylene cyanol (0.25%) (together with 0.25% Bromophenol blue) is part of our 6× DNA loading buffer. We usually load 1–2 μl of 6× DNA loading buffer in the marker lane. Xylene cyanol is the slower moving band.
15. We have long been using 4.5% acrylamide gels to separate the STAT1 and STAT3 species. Although this works as well we generally get better band resolution when using a 5.5% acrylamide gel. We usually prepare the gel the day before and leave it in the cold room at 4°C over night. It this case the gel should be placed into a plastic bag to prevent it from drying.
16. Consider that there will be some loss during pipetting and be sure to prepare enough solution. For example, prepare solution for 45 samples if you have between 35 and 40 samples.
17. This will vary when using nonradioactive oligonucleotide probes. Please follow the providers' indications.
18. In order to ascertain that a band is due to the presence of a specific transcription factor, a so-called super-shift assay can be performed. For this, one additional aliquot (per investigated transcription factor) of a positive control sample must be included. Before incubating the sample with the probe for

15 min, add 0.2–1 μg of an antibody recognizing the respective transcription factor. For super-shifting STAT1 or STAT3 we use the following antibodies: STAT1-E23 (Santa Cruz Biotechnologies) and STAT3-H190 (Santa Cruz Biotechnologies).

19. The visualization procedure will have to be adapted depending on the system used (radioactive or nonradioactive).

Acknowledgments

This work was supported by the University of Luxembourg grants R1F105L01, R1F107L01.

References

1. Garner, M. M., and Revzin, A. (1981) A gel electrophoresis method for quantifying the binding of proteins to specific DNA regions: application to components of the Escherichia coli lactose operon regulatory system. Nucleic Acids Res 9, 3047–3060
2. Fried, M., and Crothers, D. M. (1981) Equilibria and kinetics of lac repressor-operator interactions by polyacrylamide gel electrophoresis. Nucleic Acids Res 9, 6505–6525
3. Wegenka, U. M., Buschmann, J., Lutticken, C., Heinrich, P. C., and Horn, F. (1993) Acute-phase response factor, a nuclear factor binding to acute-phase response elements, is rapidly activated by interleukin-6 at the post-translational level. Molecular and cellular biology 13, 276–288
4. Mahboubi, K., and Pober, J. S. (2002) Activation of signal transducer and activator of transcription 1 (STAT1) is not sufficient for the induction of STAT1-dependent genes in endothelial cells. Comparison of interferon-gamma and oncostatin M. The Journal of biological chemistry 277, 8012–8021
5. Haan, S., Keller, J. F., Behrmann, I., Heinrich, P. C., and Haan, C. (2005) Multiple reasons for an inefficient STAT1 response upon IL-6-type cytokine stimulation. Cellular signalling 17, 1542–1550
6. Guschin, D., Rogers, N., Briscoe, J., Witthuhn, B., Watling, D., Horn, F., Pellegrini, S., Yasukawa, K., Heinrich, P., Stark, G. R., and et al. (1995) A major role for the protein tyrosine kinase JAK1 in the JAK/STAT signal transduction pathway in response to interleukin-6. EMBO J 14, 1421–1429
7. Kortylewski, M., Heinrich, P. C., Mackiewicz, A., Klingmüller, U., Schniertzhauer, U., Nakajima, K., Hirano, T., Horn, F., and Behrmann, I. (1999) Interleukin-6 and oncostatin M-induced growth inhibition of human A375 melanoma cells is STAT-dependent and involves upregulation of the cyclin-dependent kinase inhibitor p27/Kip1. Oncogene 18, 3742–3753
8. Costa-Pereira, A. P., Tininini, S., Strobl, B., Alonzi, T., Schlaak, J. F., Is'harc, H., Gesualdo, I., Newman, S. J., Kerr, I. M., and Poli, V. (2002) Mutational switch of an IL-6 response to an interferon-gamma-like response. Proceedings of the National Academy of Sciences of the United States of America 99, 8043–8047
9. Bender, H., Wiesinger, M. Y., Nordhoff, C., Schoenherr, C., Haan, C., Ludwig, S., Weiskirchen, R., Kato, N., Heinrich, P. C., and Haan, S. (2009) Interleukin-27 displays interferon-gamma-like functions in human hepatoma cells and hepatocytes. Hepatology 50, 585–591
10. Schoenherr, C., Weiskirchen, R., and Haan, S. (2010) Interleukin-27 acts on hepatic stellate cells and induces signal transducer and activator of transcription 1-dependent responses. Cell Commun Signal 8, 19
11. Ndubuisi, M. I., Guo, G. G., Fried, V. A., Etlinger, J. D., and Sehgal, P. B. (1999) Cellular physiology of STAT3: Where's the cytoplasmic monomer?. The Journal of biological chemistry 274, 25499–25509
12. Schreiber, E., Matthias, P., Muller, M. M., and Schaffner, W. (1989) Rapid detection of

octamer binding proteins with 'mini-extracts', prepared from a small number of cells. Nucleic Acids Res 17, 6419

13. Huckel, M., Schurigt, U., Wagner, A. H., Stockigt, R., Petrow, P. K., Thoss, K., Gajda, M., Henzgen, S., Hecker, M., and Brauer, R. (2006) Attenuation of murine antigen-induced arthritis by treatment with a decoy oligodeoxynucleotide inhibiting signal transducer and activator of transcription-1 (STAT-1). Arthritis research & therapy 8, R17

Chapter 11

Flow Cytometric Analysis of STAT Phosphorylation

Jane Murphy and Gabrielle L. Goldberg

Abstract

Flow cytometry can be used to study STAT phosphorylation on a per cell basis. Cells are fixed, permeablized, and stained with antibodies that specifically bind to the phosphorylated form of the STAT protein. This allows the tyrosine phosphorylation of a single STAT to be studied within a heterogeneous cell population and/or the phosphorylation of several STATs within the one cell type.

Key words: STAT phosphorylation, Flow cytometry, Phospho-flow

1. Introduction

The high-affinity binding of a wide variety of growth factors/cytokines to their cognate receptors results in the autophosphorylation and activation of Janus protein tyrosine kinases (JAKs) which in turn phosphorylate and activate signal transducers and activators of transcription (STATs). Phosphorylated STATs subsequently form hetero- and homodimers that translocate to the nucleus and bind specific elements within the promotor regions of STAT responsive genes. While there are several methods for studying protein phosphorylation, few allow for analysis on a per cell basis. Prior to the development of techniques that enabled intracellular staining of phosphorylated proteins, a large number of cells would often have to be cultured in vitro or sorted prior to analysis in order to study a homogeneous population. Over the past decade or so, flow cytometric techniques, antibody production and fluorescent labeling technologies have been developed and refined to allow the study of intracellular protein phosphorylation using flow cytometry of a relatively small number of cells (1). It is now

Sandra E. Nicholson and Nicos A. Nicola (eds.), *JAK-STAT Signalling: Methods and Protocols*, Methods in Molecular Biology, vol. 967,
DOI 10.1007/978-1-62703-242-1_11,

not only possible to study how different cell types in a given sample respond to stimuli but also to study the phosphorylation of many proteins in the one cell, using antibodies conjugated to different fluorophores (2). It is also possible to use this technique to monitor STAT phosphorylation over time. This technique utilizes fluorescent-labeled antibodies that specifically recognize the tyrosine phosphorylated form of a protein and has been rigorously tested against long-used techniques such as western blotting (2).

In this chapter we describe a general protocol that predominantly uses reagents from the Becton Dickinson BD™ Phosflow range, specifically tested for the anti-phosphoSTAT reagents also available from BD Biosciences. It is possible to make in-house fixation and permeablization buffers; however, these will need to vary in composition, depending on the antibodies and fluorochromes used as well as the phosphorylated protein being studied. Here we describe G-CSF- and GM-CSF-induced phosphorylation of STAT-3 and STAT-5, respectively, in human neutrophils, but as mentioned above a wide range of growth factors stimulate phosphorylation of STAT family members.

2. Materials

1. Assay buffer: Phosphate-buffered saline (PBS), 0.1 % bovine serum albumin (BSA), 2 mM EDTA.
2. Lyse/Fix Buffer: BD™ Phosflow Lyse/Fix Buffer 5× (catalogue number: 558049).
3. Fix Buffer: BD™ Phosflow Fix Buffer 1.
4. Permeabilization buffer: BD™ Phosflow Perm Buffer III (catalogue number: 558050).

Antibodies:

5. Anti-human and mouse phospho-STAT3 (pY705) mouse antibody conjugated to phycoerythrin (PE) (BD/Phosflow, catalogue number: 612569).
6. Anti-human and mouse phospho-STAT5a,b (pY694) mouse antibody conjugated to Alexa Fluor® 488 (BD/Phosflow, catalogue number: 612598).
7. IgG2a isotype control mouse antibody conjugated to phycoerythrin (PE) (BD/Phosflow, catalogue number: 340459).
8. Human G-CSF: Also known as Neupogen/Filgrastim (Amgen).
9. Human GM-CSF (ProSpec, catalogue number: CYT-221).

10. Hemocytometer and Trypan blue for counting cells.
11. 96-Well U bottom plate.
12. Vortex mixer.
13. Centrifuge (Heraeus Megafuge 1.0R).
14. Flow cytometer (BD Biosciences, LSRFortessa cell analyser).

3. Methods

1. Count cells of interest (here total leukocytes) in a hemocytometer using Trypan blue to exclude dead cells. Adjust to a concentration of 10^7 cells/mL in assay buffer and plate 10^6 cells in 100 μL of assay buffer in a 96-well U bottom plate.
2. Incubate cells at 37 °C in 10 % CO_2 for 60 min in the absence of any stimuli.
3. If appropriate, add inhibitors (such as blocking antibodies) diluted in 50 μL of assay buffer and pre-incubate for 15 min at 37 °C in 10 % CO_2 before adding the stimuli (here G-CSF and GM-CSF) in 50 μL of assay buffer and incubate (here, for 15 min) at 37 °C in 10 % CO_2 (see Note 1). Spin down cells at 700 × *g* for 3 min and flick out supernatant.

   If cell surface staining is not required proceed to step 5.
4. If further characterization of cells is required, resuspend cells in a cocktail of antibodies to cell surface markers (diluted in assay buffer) and incubate on ice for 30 min (see Notes 2 and 3). Add approximately 200 μL of assay buffer and centrifuge for 3 min at 300 × *g*. Following centrifugation flick out the supernatant and repeat washing step.
5. Vortex well and add 200 μL of 1× Lyse/Fix Buffer (5× stock diluted in deionized or distilled H_2O) if red blood cell lysis is necessary (e.g., when analyzing blood or spleen samples) or Fix Buffer (pre-warmed to 37 °C) and incubate at 37 °C in 10 % CO_2 for 10 min (see Notes 4 and 5).
6. Centrifuge cells for 3 min at 700 × *g*, flick out supernatant. Mix by vortex while slowly adding 200 μL of ice-cold (chilled to -20 °C) permeabilization buffer. Incubate on ice for 30 min. Centrifuge cells at 700 × *g* for 3 min and flick off supernatant. Wash cells 2–3 times in cold assay buffer.
7. Resuspend cells in assay buffer containing the phosphoSTAT antibodies (titrated previously) (see Notes 6 and 7) and incubate on ice for 30 min. Wash cells twice in assay buffer and resuspend in 100–200 μL assay buffer for analysis on a flow cytometer (see Note 8) (Fig. 1).

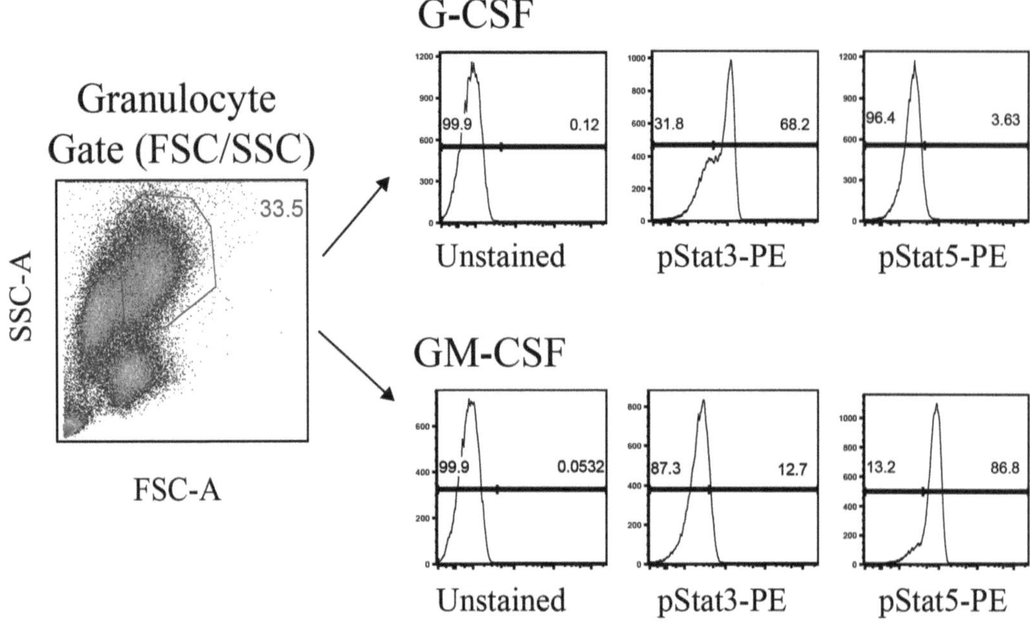

Fig. 1. PhosphoSTAT staining of human neutrophils. Following red blood cell lysis, cells were stimulated with either G-CSF (50 ng/mL) or GM-CSF (50 ng/mL) for 15 min. Cells were treated and stained as described in the protocol and run on the BD Fortessa. Samples were analyzed using Flowjo (TreeStar) and gated on the granulocyte population of the forward scatter/side scatter (FSC/SSC) plot. The *histograms* represent the percentage of granulocytes that are phospho-STAT 3 or 5 positive.

4. Notes

1. The kinetics of STAT phosphorylation will differ depending on the growth factors/receptors used to stimulate cells. It may be necessary to do a time course to establish the optimal stimulation time(s) for your given experiment.
2. Traditional viability stains cannot be used with fixed and permeabilized cells. There are several fixable viability stains on the market. One example is LIVE/DEAD Fixable Dead Cell Stain available from Invitrogen/Molecular probes. It is important to note that cells are incubated with this viability stain prior to cell surface staining.
3. If staining of cell surface antigens is required, it is important to first test all antibodies used for surface staining comparing them on fixed and unfixed samples, as some antigen epitopes and fluorophores may be disrupted by the fixation and permeabilization treatments.
4. It is important to pre-warm the Lyse/Fix buffer and Fix buffers and to pre-chill the permeabilization buffer. Not doing so can lead to substandard staining.

5. Do not store unused diluted BD phosflow buffers. Make up the BD phosflow reagents fresh on the day of the assay. Instability can occur even after a short period of time.
6. Titrate all cell surface and intracellular antibodies. Most antibodies will work at the concentration suggested by the manufacturers (if supplied). Often these antibodies can be diluted further decreasing the cost of the assay, but sometimes need to be used at higher concentrations to resolve the negative and positive populations.
7. Antibodies against human and mouse phospho-STATs 1–6 are commercially available from a number of different sources. Many have now been tested and are known to work for flow cytometry. It is still advisable to test each new antibody and/or conjugate.
8. According to BD specifications, samples should be run immediately, if possible, but may be stored up to 4 h (at 2–8 °C, protected from light)

Acknowledgments

Gabrielle Goldberg is an NHMRC Career Development Fellow. This work was made possible through Victorian State Government Operational Infrastructure Support and Australian Government NHMRC IRIISS.

References

1. Perez OD, Nolan GP (2006) Phosphoproteomic immune analysis by flow cytometry: from mechanism to translational medicine at the single-cell level. Immunol Rev 210:208–228
2. Krutzik PO, Nolan GP (2003) Intracellular phospho-protein staining techniques for flow cytometry: monitoring single cell signaling events. Cytometry A 55(2):61–70

Chapter 12

Acetylation of Endogenous STAT Proteins

Torsten Ginter, Thorsten Heinzel, and Oliver H. Krämer

Abstract

Acetylation of signal transducer and activator of transcription (STAT) proteins has been recognized as a significant mechanism for the regulation of their cellular functions. Site-specific antibodies are available only for a minority of STATs. The detection of acetylated STATs by immunoprecipitation (IP) followed by western blot (WB) will be described in the following chapter. Defined conditions for cell lysis and IP will be elucidated on the basis of STAT1 acetylation.

Key words: Acetylation, HDACi, HAT, IFN, Immunoprecipitation, STAT, STAT1

1. Introduction

Acetylation as a posttranslational modification (PTM) was initially discovered for histones (1). In addition to histones, a rising number of nonhistone proteins (e.g., STATs, p53, and NF-kB) are found to be regulated by acetylation (2, 3). The equilibrium of acetylation and deacetylation is very dynamic and mainly regulated by two enzyme families. Histone acetyltransferases (HATs) transfer acetyl groups from acetyl-CoA molecules to the ε-NH_2 groups of lysine side chains. The histone deacetylases (HDACs) catalyze the removal of the acetyl groups. HDACs need Zn^{2+} or nicotinamide (NAD^+) as cofactors (4). The Zn^{2+}-dependent HDACs are termed as the classical family and are divided in class I (HDAC 1,2,3,8), class II (HDAC 4,5,6,7,9,10), and class IV (HDAC 11). NAD^+-dependent HDACs are designated sirtuins (SIRTs) and comprise class III (SIRT 1–7). In contrast to HDACs, the classification of HATs is less clear (the interested reader is referred to the literature, e.g., (3)). HATs and HDACs are often deregulated in cancer (5), with aberrant protein levels detectable in many tumor types (6, 7). Researchers have developed HDAC inhibitors (HDACi), which could be promising drugs for chemotherapy (8).

Sandra E. Nicholson and Nicos A. Nicola (eds.), *JAK-STAT Signalling: Methods and Protocols*, Methods in Molecular Biology, vol. 967, DOI 10.1007/978-1-62703-242-1_12, © Springer Science+Business Media New York 2013

Since HDACs of the classical family share related active sites, several pan-HDACi were found. Trichostatin A (TSA) a fungal antibiotic, suberoylanilide hydroxamic acid (SAHA), LAQ-824, and LBH589 are representatives of this group. Class I selective inhibitors like MS-275, valproic acid (VPA), and depsipeptide do exist (3), but strictly isoform-specific inhibitors still remain to be generated (9). Due to the need of NAD^+ as a cosubstrate for SIRTs, nicotinamide serves as a noncompetitive inhibitor for this class of enzymes (10). HDACi are essential tools for the detection of acetylated proteins and hence have to be a component of the cell lysis buffer (11).

Transcription factors of the STAT family play key roles for cell survival, differentiation, proliferation, and homeostasis. Cytokines bind to cognate receptors and activate Janus kinases (JAKs), which catalyze tyrosine phosphorylation of STATs at defined sites (12). Tyrosine phosphorylation of STATs allows the nuclear import of preformed cytosolic dimers (13, 14). These act as specific inducers of gene sets relevant for physiological processes (15).

Besides phosphorylation, acetylation has turned out as a PTM that crucially regulates STAT activity. In the last years all STATs, except for STAT4, have been positively tested for acetylation (Table 1). The consequences of this modification for signaling are diverse. For example, acetylation of STAT3 ties in with dimerization and transcriptional activation (16, 17). However, we and others have shown that acetylation of STAT1 is inactivating, as it causes diminished phosphorylation (11, 18–24). It is accordingly accepted that acetylation blocks STAT1-dependent signaling and gene expression (22, 25–31).

Figure 1 shows an example for the detection of STAT1 acetylation in SK-Mel-37 cells after exposure to different treatments. Cytokine-induced acetylation of STAT1 occurs time-dependently (Fig. 2) and ensues the very rapidly induced and transient phosphorylation of STAT1 (18, 19, 22).

Table 1
Acetylation of STAT proteins; HAT, histone acetyltransferase

STAT protein	Lysine moieties required for acetylation	HAT	Tested stimulus	References
STAT1	K410, K413	CBP	IFNα, IFNγ HDACi Cisplatin	(11, 18–24)
STAT2	K390	CBP	IFNα	(23)
STAT3	K685 K49, K87 K679, K707, K709	p300>, CBP p300 >p300	OSM, IFNα IL6 IL6>, OSM, diet	(16, 37–39) (17, 38) (38)
STAT5b	K359, K694, K701	CBP	Prolactin	(40)
STAT6	>unidentified	CBP/p300	IL4	(41)

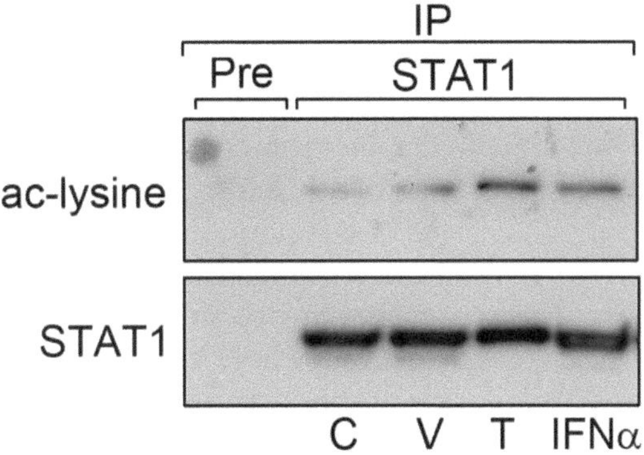

Fig. 1. SK-Mel-37 cells were treated with the HDACi VPA (V; 3 mM) and TSA (T; 100 nM) or the cytokine interferon alpha (IFNα; 1,000 U/ml) for 24 h. Subsequently, cells were immunoprecipitated under acetylation preserving conditions with anti-STAT1 antibody (Santa Cruz, sc-417). Acetylated STAT1 was detected using a pan-specific anti-acetylated lysine antibody (Cell Signaling, # 9441) for WB; C, untreated cells; IP, immunoprecipitation; Pre, pre-immune serum.

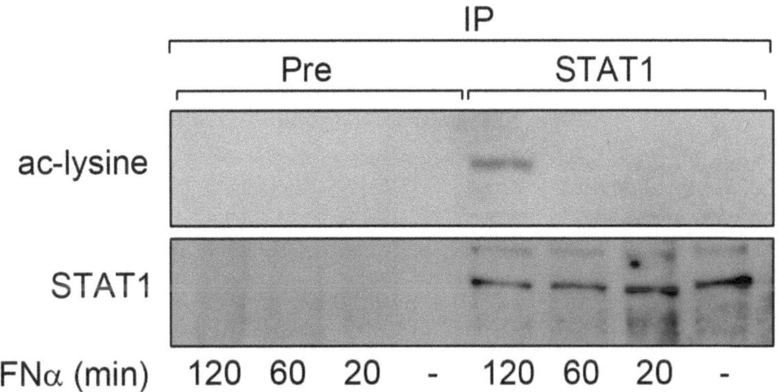

Fig. 2. SK-Mel-37 cells were stimulated with 1,000 U/ml IFNα for the indicated times. Whole cell lysates were immunoprecipitated with STAT1 antibody (Santa Cruz, sc-417) and probed for acetyl-lysine (Cell Signaling, # 9441) by WB analysis; IP, immunoprecipitation; Pre, pre-immune serum.

Several methods may be employed for the detection of acetylated proteins. The most convenient method is a WB analysis using a site-specific antibody. Unfortunately, such agents are still a rarity and their generation is time consuming and expensive. Alternatively, mass spectrometry with enriched acetylated peptides from trypsin-digested whole cell extracts could be performed. However, biologically important acetylation sites of low abundance will remain undetected, because of a high background of non-acetylated peptides (32). This situation is reminiscent of the PTM with SUMO, where only a small portion of protein needs to be modified to evoke important physiological processes (33). Furthermore mass spectrometry requires much technical equipment and expert knowledge. Radioactive labeling of lysines with ^{14}C-acetyl-CoA is an alternative to immunodetection and applicable after in vivo labeling or enzymatic in vitro acetylation of recombinant proteins. The disadvantage of in vivo labeling can be high background signals of other acetylated proteins even upon performance of 1D/2D gel electrophoresis. In vitro labeling of recombinant proteins could be determined easily by scintillation counting (32, 34).

For in vivo analysis of acetylation, an IP of the protein of interest from whole cell lysate, followed by SDS polyacrylamide gel electrophoresis (SDS-PAGE) and WB analysis with pan-specific acetyl-lysine antibodies is often needed (Figs. 1 and 2). This method is most frequently used to analyze protein acetylation and will be described in this chapter using the example of endogenous STAT1 acetylation.

2. Materials

Materials and antibodies listed here are routinely used by our lab and many other groups. However, equipment from other providers should be equally useful.

2.1. Preparation of Whole Cell Extracts

1. RIPA lysis buffer: 0.1–1% sodium dodecyl sulfate (SDS) (w/v) (see Note 1), 1% sodium desoxycholate (w/v), 1% NP-40 (v/v), 50 mM Tris–HCl pH 7.4, 150 mM NaCl, 1 mM EDTA. Freshly add HDACi (TSA and nicotinamide) as indicated (see Notes 2 and 3). For codetection of phosphorylation 1 mM sodium vanadate and 0.5 mM NaF has to be added.
2. Sonification: Branson Sonifier W250D amplitude 40% for 3 s ten times.
3. Dulbecco's Phosphate-Buffered Saline (PBS), e.g., from PAA.

2.2. Immunoprecipitation

1. Antibodies for IP and WB.
   - (a) STAT1 (Santa Cruz; sc-346 [rabbit]).
   - (b) STAT1 (Santa Cruz; sc-417 [mouse]).
   - (c) Pan-specific acetyl-lysine (Cell Signaling; # 9441 [rabbit]).
   - (d) Pan-specific acetyl-lysine (Cell Signaling; # 9681 [mouse]).
   - (e) Acetyl-histone H3 (upstate; 06-599 [rabbit]).
   - (f) Pre-immune serum (Santa Cruz; sc-2025 [mouse]).
   - (g) Pre-immune serum (Santa Cruz; sc-2027 [rabbit]).
   - (h) See Note 4.

2. Protein A Sepharose CL 4B and Protein G Sepharose 4 Fast Flow (GE-Healthcare): Follow the customer instructions to equilibrate and wash the slurry. Resuspend the Sepharose in RIPA buffer and mix protein A Sepharose/protein G Sepharose in a ratio 1:1 (see Note 5).
3. SDS-Laemmli (2×) loading buffer: 116 mM Tris–HCl pH 6.8, 1.4 M β-mercaptoethanol, 10% glycerol, 3.3% SDS (w/v), spatula tip Bromophenol blue.

2.3. SDS Polyacrylamide Gel Electrophoresis

1. Separating gel buffer: 1 M Tris–HCl pH 8.8.
2. Stacking gel buffer: 1 M Tris–HCl pH 6.8.
3. 20% SDS: 20% (w/v) aqueous solution.
4. 10% Ammonium persulfate (APS): 10% (w/v) aqueous solution.
5. Tetramethylethylenediamine (TEMED) (Sigma).
6. 30% Acrylamide/bisacrylamide 37.5/1 (Roth).
7. Butanol (Roth).
8. Mini Protean 3 system—casting stand with corresponding casting frames, combs, and glass plates (spacers included) (Bio-Rad).
9. SDS-Running buffer: 250 mM glycine, 25 mM Tris, 0.1% (w/v) SDS (see Note 6).
10. Protein ladder (Bio-Rad/ Jena Bioscience).

2.4. Protein Transfer—Western Blot

1. Polyvinylidene difluoride (PVDF) membrane (Millipore).
2. Whatman paper (3 M) (VWR Scientific).
3. Ethanol (Roth).
4. Transfer buffer: 250 mM glycine, 25 mM Tris, 0.1% (w/v) SDS, 20% ethanol (see Note 7).
5. Western transfer apparatus (Bio-Rad).

2.5. Antiacetyl-Lysine Detection

1. Antibodies see Subheading 2.2, item 1.
2. Nonfat dry milk (NFDM) (see Note 8).
3. PBS-T: 137 mM NaCl, 8 mM Na_2HPO_4, 2.7 mM KCl, 1.4 mM KH_2PO_4, adjust with HCl to pH 7.25, 0.05% Tween 20.
4. Enhanced Chemoluminecence (ECL) kit (Thermo scientific).
5. Stripping buffer: 62.5 mM Tris–HCl pH 6.8, 2% (w/v) SDS, add freshly 100 mM β-mercaptoethanol (β-ME).

3. Methods

3.1. Preparation of Cell Lysate

Acetylation as a PTM has a very transient nature. The fine-tuned equilibrium of this modification is regulated by the opposing enzymatic activities of HATs and HDACs. Moreover, several PTM can compete for the same lysine residue (3). The challenge is to preserve the acetylation during cell lysis, IP, and WB. Therefore, the experimentator has to take care that the lysis buffer is very stringent to disturb HDAC–substrate complexes required for deacetylation reactions. Additionally, HDACi are added to the buffers and all work steps are done on ice to decelerate remaining HDAC activities.

1. Human or murine cells are cultured in 90 mm dishes to approximately 90% confluence and are stimulated with an appropriate ligand (e.g., 1,000 U/ml IFNα for 3 h) or an HDACi. For each IP at least $3–4 \times 10^6$ cells are required.
2. Every following step is done on ice to slow down HDACs and prevent deacetylation. Before harvesting, cells are washed with ice-cold PBS, containing 200 nM TSA and 5 mM nicotinamide.
3. Cells are lysed immediately in the dish using 1 ml RIPA buffer (containing 1 μM TSA and 5 mM nicotinamide). Take a rubber policeman to scratch the cells from the bottom of the plate. The lysate should appear viscous.
4. Subsequently, the RIPA lysate is sonified directly to reduce viscosity.
5. A centrifugation step at $20,000 \times g$ for 5 min at 4°C is performed to remove cellular debris. It is recommended to directly use the fresh lysate for the IP. Alternatively it can be stored at −80°C or below.

3.2. Immunoprecipitation

The success of an IP depends on several factors. The most important is beyond all questions the reliability of the antibodies. In Subheading 2.2, item 1, antibodies for detection of acetyl-STAT1 are listed (see Note 9). It is necessary to exclude unspecific binding by the usage of a pre-immune serum, i.e., a mock-IP with the same lysate.

1. 500 μl of RIPA lysate are incubated with 1 μg of antibody (e.g., sc-417) and 40 μl of protein A/G sepharose (GE-healthcare) in a ratio 1:1 overnight at 4°C on a rotator. Sepharose slurry was equilibrated in RIPA buffer before use. 500 μl of the same lysate or a pool of several samples is incubated with pre-immune serum in parallel. In addition 5–10% lysate can be taken off for input control (see Note 10).

2. On the next day, IP samples are centrifuged at 5,000×*g* for 1 min at 4°C. Supernatant is aspirated or could be alternatively transferred to a new Eppendorf tube for further analyses (e.g., depletion efficacy of the antibodies).
3. Afterwards the Sepharose beads are washed in 400 μl RIPA (containing freshly added 200 nM TSA and 5 mM nicotinamide) and are centrifuged as before. This step is repeated for three times.
4. Residual RIPA buffer is removed with a precision syringe and beads are boiled in 30 μl SDS-Laemmli (2×) for 5 min at 95°C. Spin the samples briefly down and do SDS-PAGE immediately.

3.3. SDS Polyacrylamide Gel Electrophoresis

1. Cast the 8% acrylamide (v/v) separating gel as stated in Table 2 (see Note 11). Polyacrylamide is built in a radical polymerization. Since APS is the radical former and TEMED the catalyst, both should be added lastly to the mixture. Promptly pour the composite into the assembled gel plate to 3/4 of the volume (see Note 12). Cover the gel surface with 200 μl butanol to ensure proper polymerization.
2. After approximately 30 min the separating gel is polymerized. Pour away the butanol and remove remaining butanol by washing twice with water. Carefully draw off residual water using filter paper. Cast the 5% stacking gel as indicated in Table 2 and quickly insert the comb.
3. Approximately 20 min later stacking gel should be polymerized and the comb could be removed carefully. To clean the slots rinsing with ddH_2O is done twice.

Table 2
Composition of separating and stacking gel

		Separating gel			
Stacking gel		**8%**	**10%**	**12%**	**15%**
Rotiphorese Gel30	415 μl	2 ml	2.5 ml	3 ml	3.8 ml
Stacking gel buffer	312 μl	–	–	–	–
Separating gel buffer	–	2.8 ml	2.8 ml	2.8 ml	2.8 ml
SDS 20% (w/v)	12.5 μl	37.5 μl	37.5 μl	37.5 μl	37.5 μl
H_2O	1.75 ml	2.6 ml	2.1 ml	1.6 ml	0.9 ml
APS 20% (w/v)	12.5 μl	37.5 μl	37.5 μl	37.5 μl	37.5 μl
TEMED	2.5 μl	5 μl	5 μl	5 μl	5 μl

4. The gel could be used directly for electrophoresis, but it is recommended to wrap the gel in wet paper and store it overnight at 4°C to ensure complete polymerization as this results in better separating capabilities (see Note 13).
5. Assemble the gel in an electrophoresis chamber (Bio-Rad) and fill up with SDS-Running buffer.
6. Use a Hamilton syringe to load the samples from Subheading 3.2, step 4 into the slots of the stacking gel and do not forget a protein ladder.
7. Subsequently, run the gel for approximately 1.5 h at 20–30 mA. For orientation the Bromophenol blue from the SDS-Laemmli will mark the running front of the gel.

3.4. Protein Transfer—Western Blot

For the analysis of negatively charged proteins (through SDS attachment) by WB, proteins are transferred by electrical field force to a PVDF membrane. Two major methods exist for protein transfer. Semidry blotting is sufficient for proteins smaller than 70 kDa, but for bigger proteins the transfer efficiency is often poor. An alternative is wet blotting, which provides even convincing transfer results for proteins bigger than 130 kDa. Figure 3 shows a schematic view of a wet blotting configuration.

1. A PVDF membrane (usually 6×9 cm) is equilibrated in ethanol for 1 min. Afterwards the membrane is incubated in transfer buffer for 10 min (see Note 14).
2. In the meantime soak two sponges and two Whatman papers (usually 6×9 cm) with transfer buffer and begin to assemble the sandwich (Fig. 3) in a tray filled with transfer buffer. Use a role (e.g., 15 ml tube) to remove air bubbles between the layers.
3. Take the polyacrylamide gel and cut off the stacking gel. The separating gel is briefly incubated in transfer buffer and then carefully placed onto the PVDF membrane. Avoid air bubbles

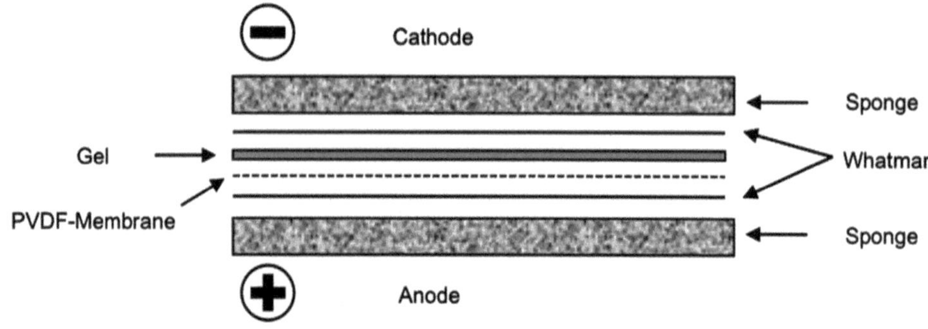

Fig. 3. Schematic configuration of a wet blotting protein transfer assembly.

between the layers, since this results in insufficient protein transfer.

4. Cover the gel with the second piece of Whatman and an additional sponge (Fig. 3).
5. Prepare a wet blot apparatus filled with transfer buffer and insert the assembled sandwich.
6. The transfer should run for 2 h at 150 mA per gel and 4°C (usually in a cold room).
7. Disassemble the apparatus and transfer the membrane to a tray filled with PBS-T. Incubate the membrane in PBS-T on a rocking platform for 5 min.

3.5. Antiacetyl-Lysine Detection

1. Incubate the PVDF membrane in PBS-T with 5% (w/v) nonfat dry milk (NFDM) for 1 h on a shaker to block unspecific binding sites on the membrane.
2. Add the membrane to a tray with PBS-T, 2% (w/v) NFDM, pan-acetyl-lysine antibody (e.g., 9441) diluted 1:1,000 and sodium azide 0.01% (w/v). Incubate over night at 4°C on a shaker (see Note 15).
3. Wash the membrane three times for 10 min in PBS-T on a shaker to reduce unbound antibodies and sodium azide.
4. Subsequently, incubation with the secondary antibody diluted 1:5,000 in PBS-T with 2% NFDM occurs for 1–2 h at room temperature.
5. Afterwards the membrane is washed three times for 10 min with PBS-T to remove unbound secondary antibodies.
6. The signal is detected by autoradiography using a Thermo scientific ECL kit as recommended by the manufacturer. In brief the kit contains two solutions, which are mixed in a ratio of 1:1. 1 ml of this mixture is dispensed homogeneously on the membrane. Put the membrane in a transparent plastic bag and wipe off excessive ECL solution. Maintaining some ECL solution may prevents fast substrate depletion and false-negative signals.
7. After detection of acetylation the identification of total STAT protein level is relevant. For that purpose wash the membrane briefly in PBS-T and afterwards incubate the membrane in stripping buffer for 1 h at room temperature on a shaker (see Note 16).
8. Wash the membrane several times with distilled water until the smell of β-ME has disappeared entirely. Incubate the membrane briefly in PBS-T and continue with steps 1–6.

4. Notes

1. Some antibodies are ineffective at a concentration of 1% SDS. If this is the case reduce the SDS concentration.
2. Please note that, certain protease inhibitors inactivate hydroxamic acid-derived HDACi and therefore can block their effects preserving acetylation (35).
3. It is preferable to prepare a TSA stock of 100 μM dissolved in DMSO and a 1 M aqueous solution of nicotinamide. Both stocks should be stored at −150°C. Before starting the experiment, it is recommended to test the biological activity of the HDACi stocks. This could be done by analyzing global histone acetylation (antibody 06-599). However, this is never a control for the preservation of protein acetylation during IP conditions.
4. Please note that certain antibodies, specifically select for non-acetylated proteins. They do not recognize the acetylated form because the epitopes are masked by acetylation. An example is PAb421, which only identifies non-acetylated forms of p53 (36).
5. Add sodium azide (very toxic) 0.01% (w/v) to prevent microbial contamination and BSA 0.05% (w/v) to diminish unspecific protein binding to the slurry (store at 4°C).
6. A stock of tenfold concentrated SDS Running buffer is preferable.
7. Many protocols recommend methanol, but ethanol serves the same purpose and is less harmful.
8. Some authors reported good results using Roti©-Block (Roth) instead of NFDM.
9. It is strongly recommended to use antibodies from different species for IP and WB, because this reduces background signal.
10. An input control is done to assure equal protein levels before the IP. Furthermore, to detect STAT1 acetylation an efficient IP is absolutely required. Thus, make sure that there is enrichment of protein compared to the input (five- to tenfold).
11. Wear gloves when casting the gel, because acrylamide can be absorbed through skin and is very carcinogenic.
12. Sometimes, despite of much devotion to assemble the casting stand, gels are leaking before polymerization. To prevent this, pipette 300 μl of a sealing gel (1 ml separating gel + 5 μl TEMED and 12.5 μl APS), wait 1–2 min for polymerization and cast the separating gel afterwards.
13. Gels should not be stored longer than 2 weeks, because this results in structurally instable gels.

14. Handle the PVDF membrane always with gloves, since proteins from your skin can attach to the surface and cause high background.
15. Sodium azide prevents microbial contamination, thus the primary antibody solution could be repeatedly used up to months when stored at −20°C. Do not add sodium azide to the secondary antibody, since this will inhibit the antibody-coupled horseradish peroxidase (HRP).
16. Stripping of membranes degrades antibodies as well as proteins of interest on the membrane. For that reason membranes should be stripped as less as possible. Try to alternate the species of secondary antibodies you use to avoid artifacts and background clouding of the band for the acetylated protein.

References

1. Yang XJ, Seto E (2007) HATs and HDACs: from structure, function and regulation to novel strategies for therapy and prevention. Oncogene 26(37):5310–5318
2. Buchwald M, Krämer OH, Heinzel T (2009) HDACi–targets beyond chromatin. Cancer Lett 280(2):160–167
3. Spange S et al (2009) Acetylation of non-histone proteins modulates cellular signalling at multiple levels. Int J Biochem Cell Biol 41(1):185–198
4. Krämer OH (2009) HDAC2: a critical factor in health and disease. Trends Pharmacol Sci 30(12):647–655
5. Bolden JE, Peart MJ, Johnstone RW (2006) Anticancer activities of histone deacetylase inhibitors. Nat Rev Drug Discov 5(9):769–784
6. Sakuma T et al (2006) Aberrant expression of histone deacetylase 6 in oral squamous cell carcinoma. Int J Oncol 29(1):117–124
7. Wilson AJ et al (2006) Histone deacetylase 3 (HDAC3) and other class I HDACs regulate colon cell maturation and p21 expression and are deregulated in human colon cancer. J Biol Chem 281(19):13548–13558
8. Müller S, Krämer OH (2010) Inhibitors of HDACs–effective drugs against cancer? Curr Cancer Drug Targets 10(2):210–228
9. Bradner JE et al (2010) Chemical phylogenetics of histone deacetylases. Nat Chem Biol 6(3):238–243
10. Bitterman KJ et al (2002) Inhibition of silencing and accelerated aging by nicotinamide, a putative negative regulator of yeast sir2 and human SIRT1. J Biol Chem 277(47): 45099–45107
11. Krämer OH et al (2006) Acetylation of Stat1 modulates NF-kappaB activity. Genes Dev 20(4):473–485
12. Reich NC (2007) STAT dynamics. Cytokine Growth Factor Rev 18(5–6):511–518
13. Mertens C, Darnell JE Jr (2007) SnapShot: JAK-STAT signaling. Cell 131(3):612
14. Hu X, Ivashkiv LB (2009) Cross-regulation of signaling pathways by interferon-gamma: implications for immune responses and autoimmune diseases. Immunity 31(4):539–550
15. Kim HS, Lee MS (2007) STAT1 as a key modulator of cell death. Cell Signal 19(3): 454–465
16. Yuan ZL et al (2005) Stat3 dimerization regulated by reversible acetylation of a single lysine residue. Science 307(5707):269–273
17. Ray S, Boldogh I, Brasier AR (2005) STAT3 NH2-terminal acetylation is activated by the hepatic acute-phase response and required for IL-6 induction of angiotensinogen. Gastroenterology 129(5):1616–1632
18. Cudejko C et al (2011) p16INK4a-deficiency promotes IL-4-induced polarization and inhibits its pro-inflammatory signaling in macrophages. Blood 118(9):2556–2566
19. Stronach EA et al (2011) HDAC4-regulated STAT1 activation mediates platinum resistance in ovarian cancer. Cancer Res 71(13): 4412–4422
20. Guo L et al (2007) Stat1 acetylation inhibits inducible nitric oxide synthase expression in interferon-gamma-treated RAW264.7 murine macrophages. Surgery 142(2):156–162
21. Hayashi T et al (2007) IFN-gamma protects cerulein-induced acute pancreatitis by

repressing NF-kappa B activation. J Immunol 178(11):7385–7394

22. Krämer OH et al (2009) A phosphorylation-acetylation switch regulates STAT1 signaling. Genes Dev 23(2):223–235
23. Tang X et al (2007) Acetylation-dependent signal transduction for type I interferon receptor. Cell 131(1):93–105
24. Ginter T et al (2012) Histone deacetylase inhibitors block IFNgamma-induced STAT1 phosphorylation. Cell Signal 24(7):1453–1460
25. Genin P, Morin P, Civas A (2003) Impairment of interferon-induced IRF-7 gene expression due to inhibition of ISGF3 formation by trichostatin A. J Virol 77(12):7113–7119
26. Sakamoto S, Potla R, Larner AC (2004) Histone deacetylase activity is required to recruit RNA polymerase II to the promoters of selected interferon-stimulated early response genes. J Biol Chem 279(39):40362–40367
27. Nusinzon I, Horvath CM (2003) Interferon-stimulated transcription and innate antiviral immunity require deacetylase activity and histone deacetylase 1. Proc Natl Acad Sci USA 100(25):14742–14747
28. Chang HM et al (2004) Induction of interferon-stimulated gene expression and antiviral responses require protein deacetylase activity. Proc Natl Acad Sci U S A 101(26):9578–9583
29. Klampfer L et al (2004) Requirement of histone deacetylase activity for signaling by STAT1. J Biol Chem 279(29):30358–30368
30. Klampfer L et al (2003) Inhibition of interferon gamma signaling by the short chain fatty acid butyrate. Mol Cancer Res 1(11):855–862
31. Krämer OH, Heinzel T (2010) Phosphorylation-acetylation switch in the regulation of STAT1 signaling. Mol Cell Endocrinol 315(1–2):40–48
32. Dormeyer W, Ott M, Schnolzer M (2005) Probing lysine acetylation in proteins: strategies, limitations, and pitfalls of in vitro acetyltransferase assays. Mol Cell Proteomics 4(9):1226–1239
33. Hay RT (2005) SUMO: a history of modification. Mol Cell 18(1):1–12
34. Choudhary C et al (2009) Lysine acetylation targets protein complexes and co-regulates major cellular functions. Science 325(5942): 834–840
35. Sonnemann J et al (2010) Serine proteases in histone deacetylase inhibitor-induced apoptosis. Mol Cancer Ther 9(8):2440–2441, author reply 2441–2
36. Li M et al (2002) Acetylation of p53 inhibits its ubiquitination by Mdm2. J Biol Chem 277(52):50607–50611
37. Wang R, Cherukuri P, Luo J (2005) Activation of Stat3 sequence-specific DNA binding and transcription by p300/CREB-binding protein-mediated acetylation. J Biol Chem 280(12): 11528–11534
38. Nie Y et al (2009) STAT3 inhibition of gluconeogenesis is downregulated by SirT1. Nat Cell Biol 11(4):492–500
39. Nadiminty N et al (2006) Stat3 activation of NF-{kappa}B p100 processing involves CBP/p300-mediated acetylation. Proc Natl Acad Sci U S A 103(19):7264–7269
40. Ma L et al (2010) Acetylation modulates prolactin receptor dimerization. Proc Natl Acad Sci U S A 107(45):19314–19319
41. Shankaranarayanan P et al (2001) Acetylation by histone acetyltransferase CREB-binding protein/p300 of STAT6 is required for transcriptional activation of the 15-lipoxygenase-1 gene. J Biol Chem 276(46): 42753–42760

Chapter 13

Detection and Cellular Localization of Phospho-STAT2 in the Central Nervous System by Immunohistochemical Staining

Reza Khorooshi and Trevor Owens

Abstract

Phosphorylation of signal transducers and activators of transcription (STATs) indicates their involvement in active signaling. Here we describe immunohistochemical staining procedures for detection and identification of the cellular localization of phospho-STAT2 in the central nervous system (CNS) of mice. The procedures include single and dual/triple labeling immunostaining of phospho-STAT2 in the hippocampus of mice following entorhinal cortex lesion.

Key words: Phospho-STAT2, Immunohistochemistry, Central nervous system, Entorhinal cortex lesion, Astrocytes

1. Introduction

Phosphorylation of STAT2 indicates its involvement in active signaling. We present here an immunohistochemical staining method for detection of phospho-STAT2 and identification of its cellular localization in the central nervous system. We developed this protocol in order to analyze glial responses to interferon signaling.

Interferon (IFN)-α and IFN-β belong to the family of type I IFNs that play a crucial role in the innate antiviral response. In addition to their antiviral activities, type I IFNs also have effects on immunomodulation and cell growth control (1). Binding of type I IFNs to their membrane receptor (IFNAR) leads to receptor dimerization, and activation of the receptor-associated janus kinase 1 (JAK1) and tyrosine kinase 2 (Tyk2) followed by phosphorylation of a specific tyrosine residue on IFNAR. This phosphotyrosine provides binding sites for the Src homology-2 (SH2) domain of

Sandra E. Nicholson and Nicos A. Nicola (eds.), *JAK-STAT Signalling: Methods and Protocols*, Methods in Molecular Biology, vol. 967, DOI 10.1007/978-1-62703-242-1_13, © Springer Science+Business Media New York 2013

STAT2. STAT2 is then phosphorylated on tyrosine 689, leading to recruitment and phosphorylation of STAT1 on tyrosine 701. The activated STAT1 and STAT2 proteins then heterodimerize and associate with interferon regulatory factor 9, to form interferon-stimulated gene factor 3 (ISGF3). This ISGF3 complex translocates into the nucleus and activates the expression of IFN-stimulated genes involved in type I IFN function (2–5).

Type I IFNs have therapeutic application in the treatment of Multiple sclerosis (MS) (6), an inflammatory, demyelinating disease of the CNS. Type I IFN signaling also plays a role in the development of Experimental Autoimmune Encephalomyelitis (EAE), an animal model of MS (7–9). It has been shown that lack of type I IFN signaling increases leukocyte infiltration to the CNS and leads to more severe EAE. Our own studies suggest the involvement of type I IFN in control of leukocyte infiltration induced by synaptic degeneration (10). In this model, transection of axonal afferents to the hippocampus leads to synaptic degeneration in the outer molecular layer of the dentate gyrus followed by activation of glial cells, upregulation of inflammatory mediators and leukocyte infiltration (10–14). We showed that axonal transection of the entorhinal cortex neuron projection to the outer molecular layer of the dentate gyrus induces upregulation of STAT1 and STAT2. We showed that STAT2 activation was specific to astrocytes in response to injury-associated signal(s) in the hippocampus (15) using a modified immunostaining protocol that was originally used by Munzberg et al. for detection of phospho-STAT3 (16).

2. Materials

2.1. Animals

Adult female (18–20 g) mice deficient in STAT2 (STAT2-deficient, 129 s6/SvEv background) (17) were kindly provided by Dr. Chris Schindler, Columbia University, New York, NY, and bred in our facility. Wild-type mice (WT) of the 129 s6/SvEv background were purchased from Taconic (Taconic Europe A/S, Ry, Denmark). Mice were provided with free access to normal mouse chow and H_2O and housed under constant light/dark periods at the Laboratory of Biomedicine, University of Southern Denmark, Odense, Denmark. Animal breeding and experiments were conducted according to the guidelines of the National Danish Animal Research Committee.

2.2. Equipment

Instruments for Entorhinal Cortex Lesioning:

- Microdrill (Kjaerulff Fodplejeartikler A/S, Denmark).
- Wire knife and a stereotaxic frame (Kopf Instruments, CA).

- Overhead operation microscope (Zeiss, Germany).

Instruments for sectioning, storing, staining, and analyzing:

- Microtome Cryostat HM 550, series (Brock & Michelsen A/S, Denmark).
- 12-Well cell culture plate, Flat bottom with lid, Polystyrene (Corning Incorporated Costar).
- Cryomold intermediate (Sakura, CA).
- Soft paint brush, size 2 (AV form, Denmark).
- Super frost plus glass slides (Thermo Scientific, Germany).
- Skyline Shaker (CM lab, Denmark).
- Digital camera DP71 mounted on an Olympus BX51 microscope (Olympus, Denmark).

2.3. Solutions, Reagents, and Antibodies Used for Immunohistochemistry

Solutions in the following protocols are made using ultrapure water that was prepared by purifying deionized water to attain a resistivity of 18 MΩ cm. Prepare and store all reagents at room temperature (RT) unless otherwise specified.

2.3.1. Solutions and Reagents

- Methanol.
- Ethanol, 70, 96, and 99 %.
- Triton X-100.
- Hydrogen peroxide.
- 2-Methylbutane (Sigma-Aldrich).
- 3,3′-Diaminobenzidine (DAB; Sigma-Aldrich).
- Xylene.
- DePeX (VWR).
- 10× PBS: Dissolve 80 g of NaCl, 2 g of KCl, 11.5 g of Na_2HPO_4, and 2.4 g of KH_2PO_4, in 800 ml H_2O and adjust pH to 7.4 with HCl. Adjust volume to 1,000 ml with additional deionized H_2O.
- Paraformaldehyde (PFA), 4 %: Dissolve 40 g PFA (Sigma-Aldrich) in approximately 700 ml deionized H_2O and warm up to 60 °C while stirring. Add 1 M NaOH drop wise. When the solution becomes clear, allow it to cool. Add 100 ml 10× PBS. Adjust pH to 7.4. Adjust volume to 1,000 ml with additional deionized H_2O.
- Gelvatol mounting media: Add 1 g polyvinyl alcohol (Sigma-Aldrich) to 2 g glycerol. Mix well with rod and leave at RT for 15–20 min. Add 7 ml 1× PBS, mix well for 10 min with rod, incubate overnight at 37 °C then centrifuge for 20 min at 956×*g*. Take supernatant and store at 4°C.
- 0.03 % Sodium dodecyl sulfate (SDS): Add 300 μl 1 % SDS to 9.7 ml 1× PBS.

- 0.3 % Glycine: Dissolve 30 mg glycine in10 ml 1× PBS.
- 30 % Sucrose: Mix 30 g sucrose in 100 ml 1× PBS.
- de Olmos cryoprotectant solution: Dissolve 9.5 g $Na_2HPO_4 \cdot 2H_2O$ and 2.12 g $NaHPO_4 \cdot H_2O$ in 700 ml H_2O and add 300 ml ethylene glycol, 300 g sucrose, and 10 g polyvinyl pyrrolidone (Sigma-Aldrich) (see Note 1).
- PBST (PBS containing 0.5 % Triton X-100): Mix 500 μl Triton X-100 with 100 ml PBS and mix gently to avoid frothing.
- Blocking solution: 3 % bovine serum albumin (BSA, Sigma-aldrich) in PBST (3 % BSA-PBST). Dissolve 3 g BSA in100 ml PBST.
- 4,6-Diamidino-2-phenylindole (DAPI, Invitrogen A/S, Denmark). To make DAPI stock solution (14.3 mM), dissolve the contents of one vial (10 mg) in 2 mL of deionized water. Aliquot and store at −20 °C. DAPI working solution is 300 nM in PBS. Dissolve 2 μl of DAPI stock solution in 100 ml PBS.

2.3.2. Antibodies

- Primary antibody: Rabbit anti-phospho-STAT2 (Tyr 689) (07-224, Upstate, NY), Goat anti-Glial Fibrillary Acidic Protein (GFAP, Astrocyte Marker, SC-6171, Santa Cruz Biotechnology, Inc. CA)
- Secondary antibody: Biotinylated goat anti-rabbit IgG (Amersham Biosciences, UK), Alexa 568-labeled donkey anti-goat IgG (Invitrogen A/S), Alexa 488-labeled donkey anti-rabbit IgG (Invitrogen A/S).
- Control antibodies: Rabbit IgG (Serotec, UK) and Goat serum.

3. Methods

3.1. Entorhinal Cortex Axonal Lesioning

Axonal lesioning is performed as previously described (15).

1. Anesthetize the mouse by subcutaneous (s.c.) injection of 0.1 ml per 10 g body weight, of a 1:1:2 mixture of Hypnorm™ (fentanyl citrate 0.315 mg/ml and fluanisone 10 mg/ml, Jansen-Cilag, Denmark), Dormicum (5 mg/ml midazolam, Dumex-Alpharma, Denmark), and deionized H_2O, and coat the eyes with Viscotears liquid eye gel (Novartis, Denmark) to protect them from drying.
2. Fix the mouse in a Kopf stereotactic frame, with the nose-bar placed 3.0 mm below zero.
3. Cut the skin above the skull, and drill a burr hole 1.9 mm lateral to lambda and 0.3 mm caudal to the lambdoid suture.

4. Insert a closed wire knife at an angle of 10 ° lateral and rotated 15 ° rostrally and move the knife 3.4 mm ventrally from the dura. Note that the wire knife must be cleaned with ethanol before and after insertion.
5. Unfold wire knife, then retract the knife 3.2 mm to transect the entorhino-dentate perforant path projection, refold, and withdraw from the brain.
6. Close the incision.
7. Supply mice subcutaneously with 0.1 ml 10 % Temgesic™ (Rekitt & Coleman, UK) diluted in isotonic NaCl for pain relief and with 0.8 ml of isotonic NaCl to prevent dehydration.

3.2. Brain Tissue Processing

1. Anesthetize the mice with an i.p. injection of 100 µl/10 g mouse of pentobarbital (200 mg/ml, Glostrup apoteket, Denmark) diluted 10× in isotonic NaCl.
2. Perfuse mice transcardially with 5 mL ice-cold PBS followed by 20 ml 4% PFA. Confirm that the perfusion has been performed successfully (see Note 2).
3. Dissect brains, fix them in 4 % PFA at 4 °C for 2 h and immerse in 30 % sucrose overnight at 4 °C (see Note 3).
4. Place the brain in a cryomold containing Tissue-Tek embedding medium and lower it into liquid-nitrogen-cooled 2-metylebutane, and store at −80 °C until sectioning.
5. Brains are then cut into horizontal, parallel series of 30 µm free-floating sections and stored in 12-well plates containing de Olmos cryoprotectant solution on ice. Store at −14 °C.

3.3. Single Immunohistostaining of PhosphoSTAT2

1. Transfer free-floating sections to 12-well plates containing 1× PBS (see Note 4).
2. Rinse sections in 1× PBS (4×15 min to remove de Olmos solution) at RT on shaker.

2a. Remove PBS (see Note 5).

3. Incubate in 1 % H_2O_2 and 1 % methanol diluted in PBS for 20 min at RT to block endogenous peroxidase (see Note 6).
4. Rinse sections in PBS , for 3×15 min at RT.
5. Incubate with 1 % NaOH and 1 % H_2O_2 for 30 min at RT.
6. Rinse sections for 2×10 min in 1× PBS at RT.
7. Immerse in 0.3 % glycine at RT for 10 min.
8. Rinse sections for 3×10 min in 1× PBS at RT.
9. Incubate in 0.03 % SDS, at RT for 10 min.
10. Remove SDS and incubate in blocking solution for 60 min at RT.
11. Incubate with anti-phospho-STAT2 antibody (1:200) dissolved in blocking solution at RT, overnight on shaker (see Note 7).

12. Rinse sections for 3 × 10 min in PBST at RT.
13. Incubate with secondary biotinylated goat anti-rabbit antibody (1:200, diluted in blocking solution) at RT for 60 min.
14. Rinse sections for 3 × 10 min in PBST at RT.
15. Incubate with streptavidin-conjugated horseradish peroxidase (Dako, Denmark) (1:200, diluted in blocking solution) at RT for 60 min.
16. Rinse sections for 2 × 10 min in PBS at RT.
17. Develop for 5–10 min in 0.05 % DAB and 0.033 % H_2O_2 in PBS (see Note 8).
18. Rinse in PBS for 2 × 10 min at RT.
19. Mount on super frost plus glass slides and allow the sections to dry overnight at 4 °C (see Note 9).
20. Dehydrate succesively in 70, 96, and 99 % ethanol (for 2 × 5 min each) then clear in xylene (for 2 × 5 min) (see Note 10).
21. Coverslip with DePeX.
22. Leave sections for 24 h at RT in fume hood.
23. Images can then be taken and processed, e.g., using Adobe-Photoshop CS4 software program.

3.3.1. Control Reactions

The specificity of phospho-STAT2 staining was assessed by (a) substitution of the primary antibody with normal rabbit IgG, (b) omission of primary antibody in the protocol, and (c) use of STAT2-deficient brain tissue.

Phospho-STAT2 immunostaining was only observed in the denervated hippocampus (Fig. 1b) in comparison to contralateral hippocampus (Fig. 1a). Staining was not detected in lesion-reactive hippocampus from STAT2-deficient mice (Fig. 1c) nor in brain sections stained with normal rabbit IgG (insert, Fig. 1c).

3.4. Double/Triple Staining Immuno-histochemistry

1. Transfer free-floating sections to 12-well plates containing 1× PBS, as described above.
2. Rinse sections in 1× PBS (for 4 × 15 min to remove de Olmos solution) at RT on shaker.
3. Incubate with 1 % NaOH and 1 % H_2O_2 for 30 min at RT.
4. Rinse sections for 2 × 10 min in 1× PBS at RT.
5. Immerse in 0.3 % glycine at RT for 10 min.
6. Rinse sections for 3 × 10 min in 1× PBS at RT.
7. Incubate in 0.03 % SDS, at RT for 10 min.
8. Remove SDS and incubate in blocking solution for 60 min at RT.
9. Incubate with a cocktail containing anti-phospho-STAT2 antibody (1:200) and goat anti-GFAP (1:500) (both diluted in blocking solution) at RT, overnight on shaker.

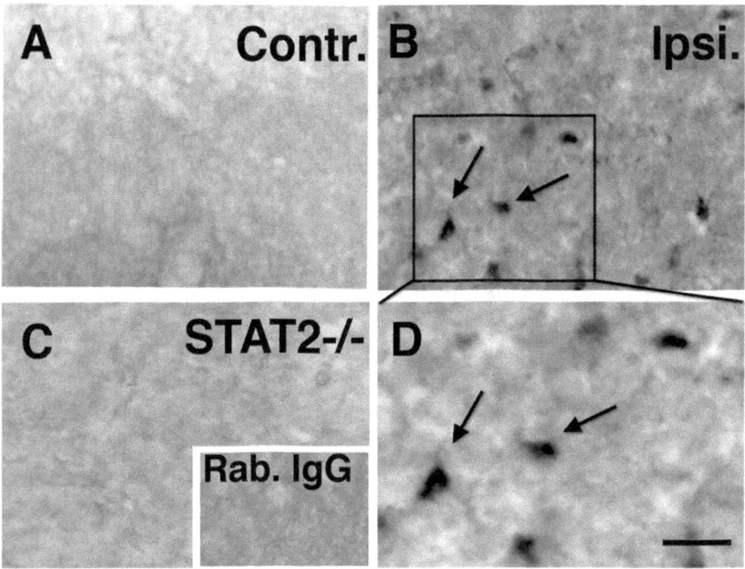

Fig. 1. Phospho-STAT2 staining in the hippocampus of mice in response to synaptic degeneration. Phospho-STAT2 immunostaining was undetectable in the contralateral (unlesioned) dentate gyrus (**a**). The staining for phospho-STAT2 (*arrows*) became detectable in the outer molecular layer, ipsilateral to the lesion, at 1 day post lesion (**b**, **d**). (**d**) is a higher magnification of the area within the *box* in (**b**). Phospho-STAT2 staining was neither detected in denervated hippocampus from STAT2-deficient mice (**c**) nor in lesioned brain sections stained with rabbit IgG [Insert in (**c**)]. Scale bar 50 μm in (**a**–**c**) and 20 μm in (**d**).

10. Rinse sections for 3×15 min in PBST at RT.
11. Incubate sections with a cocktail containing Alexa 568-labeled donkey anti-goat IgG (1:200) and Alexa 488-labeled donkey anti-rabbit IgG (1:200) (both diluted in blocking solution) at RT for 60 min (see Note 11).
12. Rinse sections for 15 min in PBS at RT.
13. Rinse sections at RT for 10 min in PBS-containing DAPI.
14. Rinse sections for 2×15 min in PBS at RT.
15. Mount on super frost plus glass slides and allow to dry overnight at 4 °C (see Note 9).
16. Coverslip with Gelvatol.
17. Images can then be taken and processed, e.g., using Adobe-Photoshop software program.

3.4.1. Control Reactions

The specificity of the phospho-STAT2 and GFAP staining is assessed by (a) substitution of the primary antibodies with normal rabbit IgG or goat serum, (b) omission of primary antibodies in the protocol, and (c) by using STAT2-deficient tissue.

Phospho-STAT2 staining colocalized with astrocyte nuclei (Fig. 2).

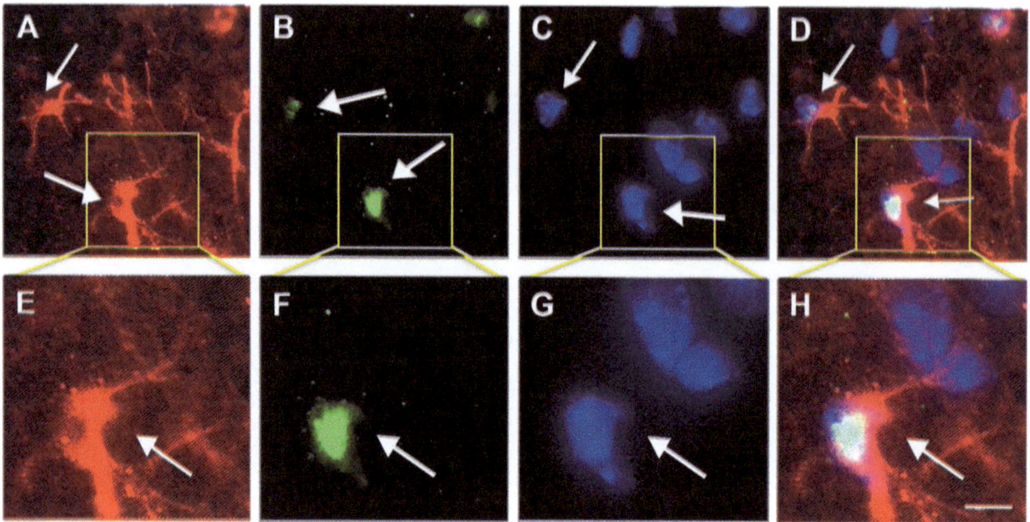

Fig. 2. Triple-labeled, immunofluorescent visualization of cells expressing phospho-STAT2. (**a**) GFAP (*red, arrowed*), (**b**) phospho-STAT2 (*green, arrowed*), (**c**) DAPI (*blue, arrowed*). (**d**) Merged image shows phospho-STAT2 overlap with DAPI in GFAP-positive cells, identifying nuclear localization of phospho-STAT2 in astrocytes. Scale bar 20 μm in (**a–d**) and 10 μm in (**e–h**).

4. Notes

1. Store de Olmos solution in the freezer (−10 to −15 °C) for at least 12 h, before use.
2. A reddish color of the brain indicates a poor perfusion.
3. Note that brains will sink to the bottom of the tube before the freezing process.
4. Sections can be lifted and transferred from one plate to another by a fine, soft paint brush (size 2). Transfer of brain sections should be done very carefully to avoid disruption of sections, which ultimately makes it very difficult to mount them on glass slides.
5. PBS and other solutions should be removed by a pipette very carefully to avoid disruption of sections. Also note that sections should not be allowed to dry.
6. This step should be performed in a fume hood.
7. Incubation with primary antibody can also take place at 4 °C, but for a longer time (e.g., 2× overnight).
8. It is important to monitor staining development with a microscope and to stop the reaction when staining is optimal.
9. Sections can be lifted and placed on slides by using a fine, soft paint brush. Cover sections with a plate to avoid getting dust on the slides.

10. This and the next step should be performed in a fume hood.
11. From this point on all sections should be protected from light exposure.

Acknowledgments

This study was supported by grants from the Danish Agency for Science, Technology and Innovation and from Novo Nordisk Fonden.

We thank Dina Dræby and Pia Nyborg Nielsen for their excellent technical support.

The authors thank Dr. Chris Schindler, Columbia University, New York, NY, for providing the STAT2-deficient mice.

References

1. Theofilopoulos AN, Baccala R, Beutler B, Kono DH (2005) Type I interferons (alpha/beta) in immunity and autoimmunity. Annu Rev Immunol 23:307–336
2. Testoni B, Vollenkle C, Guerrieri F, Gerbal-Chaloin S, Blandino G, Levrero M (2011) Chromatin dynamics of gene activation and repression in response to interferon alpha (IFN(alpha)) reveal new roles for phosphorylated and unphosphorylated forms of the transcription factor STAT2. J Biol Chem 286:20217–20227
3. Stark GR, Kerr IM, Williams BR, Silverman RH, Schreiber RD (1998) How cells respond to interferons. Annu Rev Biochem 67: 227–264
4. Schindler C, Levy DE, Decker T (2007) JAK-STAT signaling: from interferons to cytokines. J Biol Chem 282:20059–20063
5. Levy DE, Darnell JE Jr (2002) Stats: transcriptional control and biological impact. Nat Rev Mol Cell Biol 3:651–662
6. Burks J (2005) Interferon-beta1b for multiple sclerosis. Expert Rev Neurother 5:153–164
7. Teige I, Treschow A, Teige A, Mattsson R, Navikas V, Leanderson T, Holmdahl R, Issazadeh-Navikas S (2003) IFN-beta gene deletion leads to augmented and chronic demyelinating experimental autoimmune encephalomyelitis. J Immunol 170:4776–4784
8. Prinz M, Schmidt H, Mildner A, Knobeloch KP, Hanisch UK, Raasch J, Merkler D, Detje C, Gutcher I, Mages J, Lang R, Martin R, Gold R, Becher B, Bruck W, Kalinke U (2008) Distinct and nonredundant in vivo functions of IFNAR on myeloid cells limit autoimmunity in the central nervous system. Immunity 28:675–686
9. Guo B, Chang EY, Cheng G (2008) The type I IFN induction pathway constrains Th17-mediated autoimmune inflammation in mice. J Clin Invest 118:1680–1690
10. Khorooshi R, Owens T (2010) Injury-Induced Type I IFN Signaling Regulates Inflammatory Responses in the Central Nervous System. J Immunol 185(2):1258–1264
11. Finsen B, Jensen MB, Lomholt ND, Hegelund IV, Poulsen FR, Owens T (1999) Axotomy-induced glial reactions in normal and cytokine transgenic mice. Adv Exp Med Biol 468:157–171
12. Ladeby R, Wirenfeldt M, Dalmau I, Gregersen R, Garcia-Ovejero D, Babcock A, Owens T, Finsen B (2005) Proliferating resident microglia express the stem cell antigen CD34 in response to acute neural injury. Glia 50: 121–131
13. Babcock AA, Wirenfeldt M, Holm T, Nielsen HH, Dissing-Olesen L, Toft-Hansen H, Millward JM, Landmann R, Rivest S, Finsen B, Owens T (2006) Toll-like receptor 2 signaling in response to brain injury: an innate bridge to neuroinflammation. J Neurosci 26: 12826–12837
14. Wirenfeldt M, Dissing-Olesen L, Anne Babcock A, Nielsen M, Meldgaard M, Zimmer J, Azcoitia I, Leslie RG, Dagnaes-Hansen F, Finsen B (2007) Population control of resident and immigrant microglia by mitosis and apoptosis. Am J Pathol 171:617–631

15. Khorooshi R, Babcock AA, Owens T (2008) NF-kappaB-driven STAT2 and CCL2 expression in astrocytes in response to brain injury. J Immunol 181:7284–7291
16. Munzberg H, Huo L, Nillni EA, Hollenberg AN, Bjorbaek C (2003) Role of signal transducer and activator of transcription 3 in regulation of hypothalamic proopiomelanocortin gene expression by leptin. Endocrinology 144:2121–2131
17. Park C, Li S, Cha E, Schindler C (2000) Immune response in Stat2 knockout mice. Immunity 13:795–804

Chapter 14

Nuclear Trafficking of STAT Proteins Visualized by Live Cell Imaging

Velasco Cimica and Nancy C. Reich

Abstract

The ability to observe the dynamic localization of a protein in living cells can provide critical insight to its mode of action and functional molecular interactions. To this purpose, green fluorescent protein (GFP) has served as a powerful tool to tag STAT proteins for microscopic visualization. Live cell imaging with STAT-GFP proteins has contributed to our understanding of signal transduction and the complexities of nuclear transport of STAT proteins. In this report we summarize recent approaches that use GFP-based techniques with live cell imaging to study the mechanisms of STAT nuclear import and export: photoactivation, fluorescence recovery after photobleaching (FRAP), and fluorescence loss in photobleaching (FLIP).

Key words: Signal transducer and activator of transcription, Green fluorescence protein, Live cell imaging, Confocal microscopy, Photoactivation, Fluorescence recovery after photobleaching, Fluorescence loss in photobleaching, Region of interest,

1. Introduction

Eukaryotic cells are highly compartmentalized by lipid bilayer membranes that selectively allow the transport of metabolites and macromolecules in and out of organelles. The most obvious compartment of the cell is the nucleus that separates genomic information from the cytoplasm. Transport of proteins and RNAs in and out of the nucleus takes place through specialized structures called nuclear pore complexes (NPC) that span the nuclear envelope (1–4). Small molecules can diffuse through the NPC; however, the passage of larger proteins is restricted. Nuclear transport of larger molecules such as signal transducers and activators of transcription (STATs) is usually mediated by carrier proteins called karyopherins. The karyopherins and their adapters recognize sequence or conformational motifs on the STAT proteins necessary for their import

Sandra E. Nicholson and Nicos A. Nicola (eds.), *JAK-STAT Signalling: Methods and Protocols*, Methods in Molecular Biology, vol. 967, DOI 10.1007/978-1-62703-242-1_14, © Springer Science+Business Media New York 2013

(nuclear localization signal, NLS) or export (nuclear export signal, NES).

The classical function of STAT proteins is to receive activating signals at the plasma membrane and to transmit those signals to target genes in the nucleus (5, 6). For this reason movement of STATs in and out of the nucleus is central to their function and a key regulatory event. Live cell imaging techniques have greatly impacted our understanding of STAT transcription factors, and have revealed unique properties of the individual STATs (7–14). For instance, although STAT1 nuclear import is conditional and dependent on tyrosine phosphorylation, nuclear import and export of STAT3 is continual and independent of tyrosine phosphorylation (9, 10, 15, 16). In this chapter we will present different techniques used with live cell imaging to analyze the nuclear trafficking of STATs.

Green Fluorescent Protein (GFP). Fluorescence microscopy techniques have been widely used for decades to study the localization and dynamics of proteins within the cell. Proteins can be visualized with immunofluorescence techniques using antibodies linked to fluorescence dyes. However, the immunofluorescence staining procedure usually requires fixation and permeabilization of cells or tissues. Indeed fixation perturbs the structure and alters the architecture of the cell, and allows a very limited spatiotemporal visualization of protein dynamics. The discovery and development of the green fluorescent protein (GFP) from *Aequorea victoria*, has opened new horizons for studying protein dynamics inside living cells (17–20). Virtually any protein can be tagged with the 238 amino acid GFP and expressed inside the cell using simple genetic tools. The resulting tagged protein often maintains its biological properties and so its normal cellular localization, stability, and dynamics can be visualized in living cells. GFP is known to form weak dimers, but this can be prevented with an A206K substitution to eliminate any dimer effects of GFP (21). Molecular engineering has led to improved GFP variants with increased protein stability, brightness, photostability, different spectral properties, and photoactivation properties (Table 1) (20, 22).

Microscopy. In conjunction with the development of fluorescent proteins, confocal microscopy has advanced to offer increased resolution, laser efficiency, digitalized analysis, and time-lapse imaging (23, 24). The main advantage of the confocal microscope is the pinhole aperture in front of the object plane that eliminates out-of-focus images and thereby allows high image resolution. To compensate for reduced light of the final image, lasers are used instead of mercury arc lamps, and sensitive photomultiplier detectors capture the weaker signal. A laser scanning process can be used to move across a plane of the object to produce a visual "slice" of the object, and multiple slices can be combined to produce a three-dimensional image. Two-photon laser scanning microscopes (LSM) use two

Table 1
Properties of fluorescent proteins

Fluorescence protein	Excitation peak (nm)	Emission peak (nm)	Laser for excitation (nm)
DsRed2	556	582	534 Helium/Neon or 568 Argon/Krypton
DsRed Monomer	556	586	534 Helium/Neon or 568 Argon/Krypton
EYFP	513	527	514 Helium/Neon
wt GFP	397	509	405 Diode
EGFP	488	507	488 Argon
ECFP	433	475	405 Diode or 458 Argon
PA-GFP before photoactivation	400	515	405 Diode
PA-GFP after photoactivation	504	517	488 Argon
PA-GFP photoactivation	One-photon laser: 405 Diode, or 413 Krypton. Two-photon laser: 800–820 Ti:Sapphire		

pulsing, lower energy light sources so that the chromophore is only excited by the focus of both photons (25, 26). The confocal pinhole system is not necessary for two-photon microscopy since the excitation provides the three dimensional optical sectioning without absorption above or below the focal plane. This technique allows penetration and precise activation at a plane in the specimen, reduces photobleaching, and is less toxic to living cells.

Photoactivation. A variant of GFP (T203H) was engineered to have low fluorescence when excited by 488 nm light, but 100-fold increased fluorescence emission with 413 nm illumination (27, 28). The photoactivatable-GFP (PA-GFP) is advantageous to study fast moving molecules, and eliminates unwanted bleaching effects that may occur with GFP experiments. Further development of the technique with software using a two-photon laser microscope with 800–820 nm light, has made it possible to define a region of interest (ROI) where activation will be subjected to a limited area of sub-femtoliter volume. Real-time movement of the PA-GFP inside the cell compartment can be captured by time-lapse imaging of emitted green light and has been used to follow dynamic protein and membrane movement (29–32).

Fluorescence Recovery After Photobleaching. Continual light activation of GFP will eventually permanently destroy its fluorescence: a phenomenon called fading or photobleaching. This can be problematic in time-lapse microscopy, but it can also be harnessed to follow the movement of molecules within a cell. The concept of fluorescence recovery after photobleaching (FRAP) is a method to bleach the GFP-tagged molecule in a targeted area of the cell, and subsequently follow the recovery of fluorescence due to movement of unbleached GFP-molecules into the area (31, 33–36). High intensity illumination is used to photobleach GFP in any compartment or ROI of the cell, and the rate of recovery is proportional to the rate of movement of new molecules into that ROI.

Fluorescence Loss in Photobleaching. Another method that uses the photobleaching properties of GFP is fluorescence loss in photobleaching (FLIP). This technique uses repetitive high intensity pulses of light in one ROI to continuously bleach any GFP molecules that enter this targeted area. Fluorescence loss is monitored both in the bleached ROI and in a second ROI selected within the cell. Loss of fluorescence in the second ROI, in addition to the bleached ROI, indicates GFP molecules have moved out of the second ROI into the laser path of the bleached ROI. This technique can be use to demonstrate the movement of GFP tagged molecules from one ROI to another.

2. Materials

2.1. *Microscope.* The experimental procedure is described with a Confocal two-photon Laser Scanning Microscope (CLSM) 510 with Meta non-linear optics (NLO) (Zeiss) equipped with an incubation system for live cell cultures (Zeiss) (see Note 1):

2.1.1. Ring adapter for 35 mm glass bottom dishes with cover consisting of a ring system and CultFoil gas-permeable membrane for avoiding cell culture evaporation.

2.1.2. Incubation chamber S for maintaining temperature, CO_2, and humidity level: Tempcontrol 37-2 digital and Heating Insert P; CTI-Controller 3700 for supplies of heated air with CO_2 mixture connected to a CO_2 gas cylinder; Humidifying System 1.

2.1.3. Lasers: Argon laser (458/477/488/524 nm, 25 mW) (Zeiss), HeNe1 laser (543 nm, 1 mW) (Zeiss), and Chamaleon XR two-photon tunable 705 nm–980 nm laser system (Coherent).

2.1.4. Objectives: 10× Plan-Apochromat (Zeiss) numerical aperture 0.45, and 40× Plan-Neofluar (Zeiss) numerical aperture 1.3 oil immersion.

2.1.5. Anti-vibration table (Vibraplane, Kinetic Systems) within a designated clean room to maintain a dust free, humidity and temperature-controlled environment (Liberty). Adhesive floor mats prevent dust contamination (American Floor Mats).

2.2. *Software.* Microscopy software products are the Zeiss CLSM 520 Meta versions 3.5 to 4.2. Images are analyzed with Image J software (http://rsb.info.nih.gov/ij) to plot curves and are presented using Adobe Photoshop graphic software. Curve fitting and statistics are performed using GraphPad Prism.

2.3. *Glass Bottom Dishes.* For live cell imaging, tissue culture cells are plated on 35 mm glass bottom dishes (MatTek Corporation).

2.4. *Cells.* A monolayer adherent cell is best used that has distinct nuclear morphology by bright field microscopy, and biological features specific to your needs. Cell lines such as HT1080, HeLa, or Hep3B (ATCC repository) are human lines that are responsive to type I interferon or epidermal growth factor and can be transfected with DNA.

2.5. *Cell Culture.* Cells are maintained in Dulbecco's modified Eagle's Media (DMEM) supplemented with 8% fetal bovine serum (see Note 2); 0.5% trypsin-EDTA solution is used for passaging cells; standard sterile cell culture plates and pipettes; laminar flow cabinet to maintain sterility.

2.6. *Plasmid DNA Expressing STAT-GFP.* To generate STAT proteins tagged with enhanced green fluorescent proteins, mammalian expression plasmid vectors from Clontech can be used: pEGFP-C1, pECFP-C1, and pE\YFP-C1 to tag the STAT at the amino terminus, or pEGFP-N1, pECFP-N1, and pE\YFP-N1 to tag the STAT at the carboxyl terminus. (see Note 3). The photoactivatable GFP (PA-GFP) was a gift from Dr. Jennifer Lippincott-Schwartz (National Institutes of Health) (27). Commercially available vectors encoding PA-GFP are available from AddGene.

2.7. *Transfection.* The transfection agent used is Trans-IT-LT1 (Mirus) (see Note 4).

3. Methods

3.1. *Cells.* Standard tissue culture is used to maintain the cells. When cultures near 70–90% coverage of the tissue culture plate, they should be passaged 1:5 or 1:10 into new plates by trypsinization. For live cell imaging, the trypsinized cells are counted with a hemocytometer in order to plate 100,000 cells into one 35 mm glass bottom dish in 2 ml complete media.

3.2. *Transient Transfection.* The day following plating on a 35 mm glass bottom dish, DNA transfection is performed with Trans-IT-LT1 agent according to the manufacturer's instructions. 2.4 μg of plasmid DNA encoding STAT-GFP is mixed with 0.2 ml serum-free DMEM in a sterile plastic tube. 7.2 μl of Trans-IT reagent is then added to the DMEM-DNA mixture and incubated for 20 min before adding to the cells. 24 h post-transfection, the media is removed and fresh DMEM/ serum is added to the plates. 48 h post-transfection, live cell imaging is performed.

3.3. *Microscope and Chamber Set-Up for Live Cell Imaging.* The temperature and CO_2 controllers are warmed up before the experiment for 10–30 min at 37°C and 5% CO_2. A drop of oil is placed over the 40× Neofluar objective lens prior to securing the glass bottom cell culture dish inside the environmental chamber with the ring adapter. A CultFoil gas permeable membrane is placed over the dish to prevent media evaporation and the incubation chamber lid is closed. The gas bypass is opened in order to provide CO_2 to the specimen.

3.4. *Start the System.* Switch on the mercury arc lamp power supply when the confocal system is *off*. The mercury lamp startup produces electrical transients which can damage the confocal and microscope systems if they are running. Turn on the remote control button in order to switch on the computer, microscope system and lasers. For live cell imaging initiate the CLSM 510 program for specimen observation and experimental procedure. Please refer to the manufacturer's (Zeiss) manual for software instructions.

3.5. *Photoactivation.* The most relevant factors for a successful photoactivation experiment with the CLSM are the level of STAT-PA-GFP expression, the laser power and number of pulse iterations, and proper focus on the area subjected to photoactivation. The use of a two-photon laser system offers the advantage of photoactivating the protein in a restricted cellular volume (femtoliter) with light that is less cytotoxic than the one photon laser. The peak wavelength of excitation for enhanced GFP (EGFP) with a one photon laser is 488 nm (Table 1). However, with the coordinated pulsing of the two-photon laser, longer wavelengths about twice that of the one-photon laser are used (800 nm) (37). Fluorescent emission spectra are exactly the same.

3.5.1. *Turn on the Lasers (Acquire-Lasers).* Turn on Argon laser output set at 50%, turn on the HeNe laser (if using DsRed; see Note 5), and turn on the two-photon laser set to 800 nm.

3.5.2. *Focus.* Adjust the 40× Neofluor oil immersion objective to meet the glass bottom culture dish. Transmitted light (mercury arc lamp) is used to focus the cells with normal illumination. Identify a cell of interest by moving the dish using the joystick control of the motorized stage. Use reflected light with the FITC/GFP filter to identify a fluorescent cell expressing STAT-PA-GFP. The fluorescence of STAT-PA-GFP is very low and identification requires careful observation. (see Note 5).

3.5.3. *Acquire Pre-activated Image.* To identify detailed structures of the cell of interest, the CLSM acquires both a normal light image and a fluorescent image (Argon laser 488 nm with 5% transmission) using the approximate scan control settings: frame scanning with 512×512 pixels frame size, scan speed 7–9, 12 bit imaging, 2–3 times zoom, pinhole set with 1 Airy unit. The focus is optimized with the fluorescent image, and subsequently only fluorescence images are obtained using a long-pass 505 filter for green fluorescence. The detector gain, amplifier offset and pinhole settings should be determined empirically in order to obtain a fluorescence image with low background, and maximal fluorescence signal without pixel saturation. Use the "Palette" software option for optimizing the best condition (refer to the manual instructions). Three fluorescence images are captured at time zero.

3.5.4. *Photoactivate the ROI and Capture Images.* After acquiring images of the chosen cell, a Region Of Interest (ROI) for photoactivation within the cell is determined. Drawing tools are used to define the location, size, and shape of the ROI. The typical experiment will capture continual time-lapse fluorescence images before and after photoactivation. The laser settings for photoactivation must be determined empirically, but usually the process consists of 50–100 continual laser iterations in the ROI with 10–20% transmission power using the two-photon laser set at 800 nm. Following photoactivation the fluorescence images should be acquired as quickly as possible because the evident spatial redistribution of the STAT-PA-GFP will occur immediately after photoactivation. The photoactivated ROI is expected to emit 10–100-fold greater fluorescence than before excitation. Real time detection of the movement of the STAT PA-GFP inside the cell compartments is obtained by time-lapse imaging. (see Note 6)

An example of a photoactivation experiment performed with STAT3-PA-EGFP is shown in Fig. 1 (16). The experiment was designed to follow the

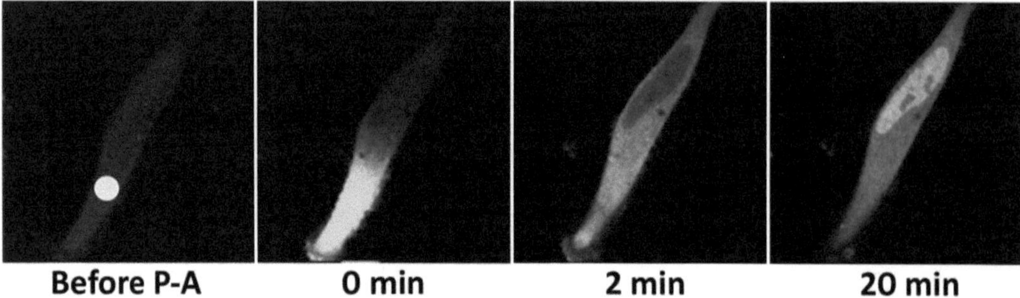

Fig. 1. *Photoactivation of STAT3-PA-GFP.* A ROI (*white dot*) within the cytoplasm of a HeLa cell expressing STAT3-PA-GFP was subjected to high intensity two-photon laser pulses. Time-Lapse imaging of the cell after photoactivation captures STAT3-PA-GFP rapid movement throughout the cytoplasm and accumulation in the nucleus by 20 min.

movement of STAT3 from the cytoplasm to the nucleus. An ROI in the cytoplasm (dot) was selected for activation and subjected to two-photon laser pulses at 800 nm. Time lapse imaging detects the activated STAT3-PA-EGFP rapid movement throughout the cytoplasm within 2 min, and then accumulation in the nucleus by 20 min.

3.6. *Fluorescence Recovery After Photobleaching.* FRAP is another method used to measure protein dynamics in living cells. EGFP is relatively stable using normal imaging with 488 nm light excitation, but with high intensity illumination it can be photobleached in an ROI. After photobleaching there is a dynamic exchange of unbleached molecules with bleached molecules inside the ROI and the kinetics of repopulation into the ROI can be measured (Fig. 2). This technique does not require a two-photon laser; use the Argon laser 488 nm for excitation and filter BP 500-550 IR for emission.

3.6.1. *Turn on One Photon Laser.* Turn on Argon laser output set at 50%.

3.6.2. *Focus.* Similar to Subheading 3.5.2., focus with transmitted light and the 40× Neofluor oil immersion objective. Identify a cell of interest by moving the dish in the *x* and *y* axes. Use reflected light with the FITC/GFP filter to focus fluorescent cells expressing STAT-EGFP.

3.6.3. *Acquire Pre-photobleached Image.* Similar to Subheading 3.5.3, scan the specimen with normal light and Argon laser 488 nm light, and optimize focus with fluorescent image using a long-pass 505 filter for green fluorescence. Select a Region Of Interest (ROI) to photobleach using the drawing tools. For FRAP of an entire HeLa cell nucleus, a circular ROI of approximately 250 μm^2 can be used. Capture three fluorescence images at time zero (488 nm with 5% transmission).

3.6.4. *Photobleach the ROI.* Bleach the spot ROI with Argon laser 488 nm at 50% power output, 100% transmission for 30–60 s and acquire images with time until the recovery is complete. Depending on the area of the ROI, a significant portion of the cellular STAT-EGFP may be photobleached and thus reduce the total amount of fluorescent molecules. This should not significantly affect the interpretation of the kinetics of recovery in the ROI; however, the final fluorescence intensity is expected to be lower than at time zero.

3.6.5. *Image Analysis.* The software and computer-controlled functions of the CLSM provide both flexibility for experimental design and measurement of fluorescence redistribution. Fluorescence intensity of the average pixels in the ROI is quantified before and subsequent to the photobleaching. A second ROI can be selected and fluorescence intensity in this second ROI can be measured to determine the influence of the photobleached ROI.

The raw fluorescence intensity measurements are plotted with time to generate a graphic representation of the ROI photobleaching, normalizing the pre-bleached value to 1. The STATs are mobile proteins and the curve is expected to indicate three phases: the initial fluorescence decreases with photobleaching of the ROI, followed by an increase in recovery with the ingress of fluorescent proteins, and finally a plateau indicating recovery to steady state. Curve fitting analysis allows the measurement of a half-time of recovery parameter that is inversely proportional to the mobility of the protein.

An example of nuclear spot FRAP with STAT3-EGFP is shown in Fig. 2a. An ROI was selected in a cell expressing STAT3-EGFP encompassing the entire nucleus and subjected to photobleaching. Time-lapse imaging displays the recovery of fluorescence in the ROI through 800 s. A graphic representation of the FRAP is shown to the right with time zero normalized to 1 and the half-time of full recovery noted. This approach can be used to measure the kinetics of STAT3-EGFP nuclear import.

Another variation of FRAP called strip-FRAP can be used to study protein mobility and is shown in Fig. 2b. In this example a very small ROI of 31 $\mu\mu^2$ in the shape of a strip was selected within the nucleus of

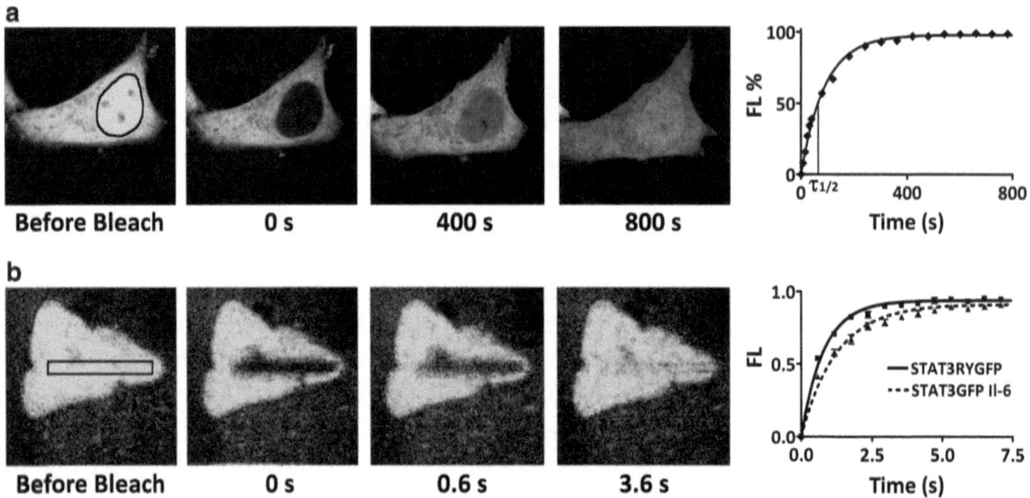

Fig. 2. *Nuclear Fluorescence Recovery After Photobleaching of STAT3-EGFP.* (**a**) To evaluate STAT3 nuclear import, a nuclear FRAP experiment was performed with HeLa cells expressing STAT3-EGFP. A ROI encompassing the nucleus of the cell (*black line*) was subjected to a high intensity laser to photobleach nuclear STAT3-EGFP. Time-lapse imaging of fluorescence recovery in the nucleus was measured, and the increase in nuclear fluorescence intensity (FL) is shown graphically to the *right*. The half-time of fluorescence recovery is indicated (τ1/2). (**b**) To measure the intra-nuclear mobility of STAT3-EGFP, nuclear strip FRAP experiments were performed with Hep3B cells expressing STAT3-EGFP and treated with IL-6. Images of a strip ROI (*black rectangle*) inside an irregularly shaped nucleus that was subjected to a high intensity laser to bleach STAT3-EGFP is shown. Curve fitting analysis of the multiple experiments generated the graph to the *right*. Experiments were also performed with a tyrosine and SH2 domain double mutant of STAT3 (STAT3-RY-EGFP)(image not shown) and curve fitting analysis was performed. A graphical representation comparing the kinetics of nuclear strip recovery for the two proteins is shown on the *right* (τ1/2 STAT3GFP treated with IL-6 = 0.869 s; τ1/2 STAT3RYGFP = 0.557 s).

cells expressing STAT3-EGFP. The cells were stimulated with the interleukin-6 (IL-6) cytokine to activate STAT3 DNA-binding prior to photobleaching. A short photobleaching time of 0.25 s was used with the 488 nm laser at output power 50% and transmission 100%. The time of recovery within the nucleus was less than 1 s indicating high mobility. Multiple experiments were used to generate a graph by using curve fitting analysis. Another strip FRAP analysis was performed with a mutant of STAT3 lacking the phosphorylated tyrosine 705, STAT3-RY-GFP (15). Multiple experiments were used to generate a graph by curve fitting analysis of the STAT3-RY-GFP, and a comparison of the nuclear dynamics of the two proteins is shown. The slower mobility of the activated wt STAT3-GFP may be due to association with chromatin.

3.7. *Fluorescence Loss in Photobleaching.* The FLIP technique differs from FRAP in that an ROI is repetitively photobleached during the course of the experiment. The continual bleaching will deplete the cell of any fluorescent molecules that move into the ROI, and may cause a loss of fluorescence in a separate unbleached ROI. The loss of fluorescence in the second unbleached ROI is measured and indicates movement of molecules from the unbleached ROI to the bleached ROI. This is a valuable method to demonstrate connectivity between compartments.

3.7.1. *Turn on the One Photon Laser.* Turn on Argon laser output set at 50%.

3.7.2. *Focus.* Similar to Subheading 3.5.2., focus with transmitted light and identify a cell of interest by moving the dish in the x and y axes. Use reflected light with the FITC/GFP filter to focus fluorescent cells expressing STAT-EGFP.

3.7.3. *Acquire Pre-photobleached Image.* Similar to Subheading 3.5.3, scan the specimen with normal light and laser 488 nm light, and optimize focus with the fluorescent image. Select two ROIs within a cell, the first ROI will be photobleached and the second ROI will be evaluated for the loss of fluorescence intensity. Capture three fluorescence images at time zero.

3.7.4. *Photobleach the ROI.* Photobleach the first ROI with Argon laser 488 nm at 50% power output, 100% transmission for 15 s and acquire an image that includes the second ROI. Repeat cycles of the photobleaching and image capture for multiple iterations. The cycling of photobleaching and image capture is designed to evaluate fluorescence loss in the second ROI. Generalized effects of laser pulses on the specimen can be determined by fluorescence acquisition with a control cell in the field of view.

Cytoplasmic FLIP can be used to evaluate STAT-EGFP nuclear export. Figure 3 displays a FLIP experiment with cells expressing STAT3-EGFP. An ROI to

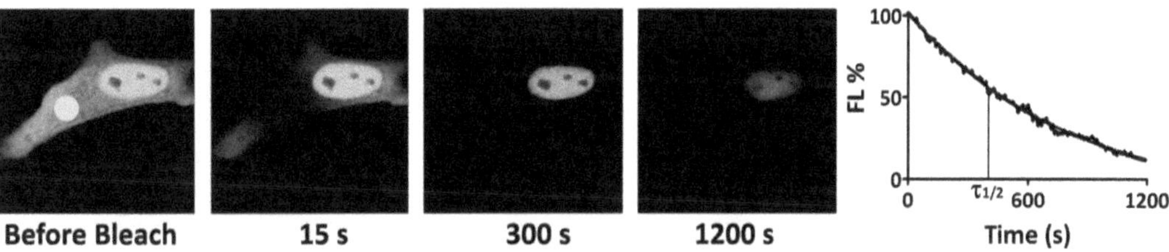

Fig. 3. *Cytosolic FLIP with STAT3-EGFP.* To evaluate STAT3 nuclear export, a ROI inside the cytosol of a HeLa cell expressing STAT3-EGFP was subjected to continuous laser photobleaching. Time-lapse imaging demonstrated a rapid decay of cytoplasmic fluorescence followed by a decay in nuclear fluorescence. The graph to the right displays the fluorescence intensity (FL) decay in the nucleus with time. Nuclear half time of fluorescence decay is indicated ($\tau 1/2$).

be bleached was selected in the cytoplasm of the cell (solid dot) and results are shown with cycles of photobleaching for 15 s and image capture. The entire cytoplasmic compartment of STAT3-EGFP was photobleached within 300 s, followed by a decrease in fluorescence intensity in the nucleus. A graphic depiction is shown of fluorescence intensity in the second ROI within the nuclear compartment. The loss of fluorescence in the nucleus can be used to evaluate STAT3-EGFP nuclear export with time.

4. Notes

The specifics of photoperturbation and image capture paradigms that are used with PA, FRAP, and FLIP are unique to each technique; however, the basic concept of photoperturbation of a GFP-tagged protein in living cells is similar. A simple illustration comparing the sequence of photoperturbation and image capture for PA, FRAP, and FLIP is shown in Fig. 4. Each of these methods can be used to evaluate the nuclear trafficking of STAT proteins.

1. Imaging was performed at the Stony Brook University Central Microscopy Imaging Center.
2. Phenol red-free media is an option; however, we use media containing phenol red since it has not hindered our imaging.

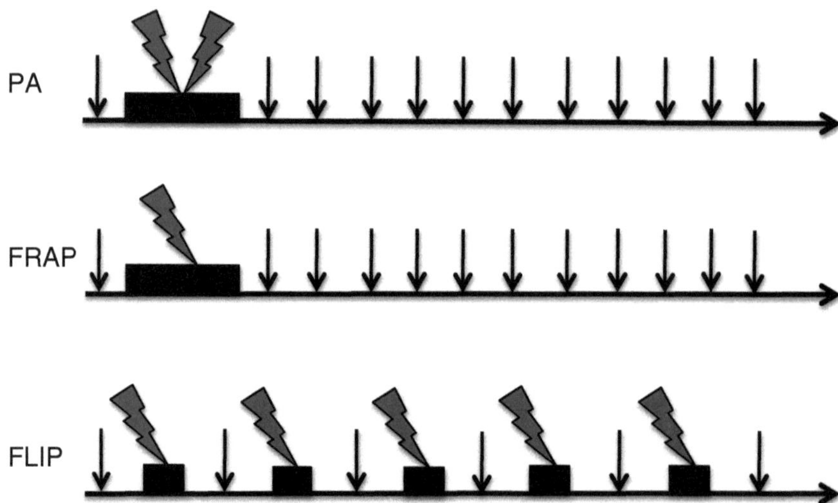

Fig. 4. *Conceptual protocol for photoperturbation and imaging with PA, FRAP, and FLIP.* The general protocols for the different photoperturbation experiments with GFP-tagged STAT proteins are illustrated for comparison. A time continuum is represented by a *horizontal arrow*; *black boxes* indicate laser illumination of the ROI; *small vertical arrows* indicate image capture.

3. Our studies use STAT proteins tagged at the carboxyl terminus since tagging at the amino terminus can inhibit tyrosine phosphorylation in response to cytokines.
4. Any method to introduce the gene expressing STAT-EGFP is acceptable including transfection agents such as Lipofectamine (Invitrogen), or calcium-phosphate-DNA co-precipitates, or infections with recombinant virus.
5. A challenge inherent with PA-GFP is the low green fluorescence signal prior to photoactivation. Co-expression of STAT-PA-GFP with DsRed can help with cell identification. DsRed-expressing cells can be detected with the HeNe1 laser (543 nm). Alternatively photoconvertible fluorescent proteins can be used.
6. A caveat with any live cell imaging technique is that the cell can move out of the plane of focus. This is not a significant parameter with short time imaging, but can be problematic with a longer time course. Automated tracking may be available with the microscope system, otherwise the focus must be checked during the experiment and may require manual adjustment.

Acknowledgments

This work was supported by grants from NIH (R56AI095268 and RO1CA122910) and a Carol M. Baldwin Breast Cancer Research Award to NCR.

References

1. Chook YM, Blobel G (2001) Karyopherins and nuclear import. Curr Opin Struct Biol 11:703–715
2. Pemberton LF, Paschal BM (2005) Mechanisms of receptor-mediated nuclear import and nuclear export. Traffic 6:187–198
3. Macara IG (2001) Transport into and out of the nucleus. Microbiol Mol Biol Rev 65: 570–594
4. Cook A, Bono F, Jinek M, Conti E (2007) Structural biology of nucleocytoplasmic transport. Annu Rev Biochem 76:647–671
5. Levy DE, Darnell JE Jr (2002) Stats: transcriptional control and biological impact. Nat Rev Mol Cell Biol 3:651–662
6. Darnell JE Jr, Kerr IM, Stark GR (1994) Jak-STAT pathways and transcriptional activation in response to IFNs and other extracellular signaling proteins. Science 264:1415–1421
7. Lillemeier BF, Koster M, Kerr IM (2001) STAT1 from the cell membrane to the DNA. EMBO J 20:2508–2517
8. Koster M, Hauser H (1999) Dynamic redistribution of STAT1 protein in IFN signaling visualized by GFP fusion proteins. Eur J Biochem 260:137–144
9. McBride KM, McDonald C, Reich NC (2000) Nuclear export signal located within theDNA-binding domain of the STAT1transcription factor. EMBO J 19:6196–6206
10. McBride KM, Banninger G, McDonald C, Reich NC (2002) Regulated nuclear import of the STAT1 transcription factor by direct binding of importin-alpha. EMBO J 21:1754–1763

11. Meyer T, Vinkemeier U (2004) Nucleocytoplasmic shuttling of STAT transcription factors. Eur J Biochem 271:4606–4612
12. Banninger G, Reich NC (2004) STAT2 nuclear trafficking. J Biol Chem 279:39199–39206
13. Iyer J, Reich NC (2008) Constitutive nuclear import of latent and activated STAT5a by its coiled coil domain. FASEB J 22:391–400
14. Chen HC, Reich NC (2010) Live cell imaging reveals continuous STAT6 nuclear trafficking. J Immunol 185:64–70
15. Liu L, McBride KM, Reich NC (2005) STAT3 nuclear import is independent of tyrosine phosphorylation and mediated by importin-alpha3. Proc Natl Acad Sci U S A 102:8150–8155
16. Cimica V, Chen HC, Iyer JK, Reich NC (2011) Dynamics of the STAT3 transcription factor: nuclear import dependent on Ran and importin-beta1. PLoS One 6:e20188
17. Giepmans BN, Adams SR, Ellisman MH, Tsien RY (2006) The fluorescent toolbox for assessing protein location and function. Science 312:217–224
18. Wang Y, Shyy JY, Chien S (2008) Fluorescence proteins, live-cell imaging, and mechanobiology: seeing is believing. Annu Rev Biomed Eng 10:1–38
19. Tsien RY (1998) The green fluorescent protein. Annu Rev Biochem 67:509–544
20. Chalfie M, Tu Y, Euskirchen G, Ward WW, Prasher DC (1994) Green fluorescent protein as a marker for gene expression. 263: 802–805
21. Zacharias DA, Violin JD, Newton AC, Tsien RY (2002) Partitioning of lipid-modified monomeric GFPs into membrane microdomains of live cells. Science 296:913–916
22. Schmid JA, Birbach A (2007) Fluorescent proteins and fluorescence resonance energy transfer (FRET) as tools in signaling research. Thromb Haemost 97:378–384
23. Smith CL (2008) Basic confocal microscopy. Curr Protoc Mol Biol 81:14.11.1–14.11.18
24. Stephens DJ, Allan VJ (2003) Light microscopy techniques for live cell imaging. Science 300:82–86
25. Oheim M, Michael DJ, Geisbauer M, Madsen D, Chow RH (2006) Principles of two-photon excitation fluorescence microscopy and other nonlinear imaging approaches. Adv Drug Deliv Rev 58:788–808
26. Ustione A, Piston DW (2011) A simple introduction to multiphoton microscopy. J Microsc 243:221–226
27. Patterson GH, Lippincott-Schwartz J (2002) A photoactivatable GFP for selective photolabeling of proteins and cells. Science 297:1873–1877
28. Henderson JN, Gepshtein R, Heenan JR, Kallio K, Huppert D, Remington SJ (2009) Structure and mechanism of the photoactivatable green fluorescent protein. J Am Chem Soc 131:4176–4177
29. Snapp EL, Lajoie P (2011) Activating photoactivatable proteins with laser light to visualize membrane systems and membrane traffic in living cells. Cold Spring Harb Protoc 2011:1368–1369
30. Chen Y, MacDonald PJ, Skinner JP, Patterson GH, Muller JD (2006) Probing nucleocytoplasmic transport by two-photon activation of PA-GFP. Microsc Res Tech 69:220–226
31. Lippincott-Schwartz J, Altan-Bonnet N, Patterson GH (2003) Photobleaching and photoactivation: following protein dynamics in living cells. Nat Cell Biol 5:S7–S14
32. Bancaud A, Huet S, Rabut G, Ellenberg J (2010) Fluorescence perturbation techniques to study mobility and molecular dynamics of proteins in live cells: FRAP, photoactivation, photoconversion, and FLIP. Cold Spring Harb Protoc 2010(12):pdb.top90
33. Goodwin JS, Kenworthy AK (2005) Photobleaching approaches to investigate diffusional mobility and trafficking of Ras in living cells. Methods 37:154–164
34. Houtsmuller AB, Vermeulen W (2001) Macromolecular dynamics in living cell nuclei revealed by fluorescence redistribution after photobleaching. Histochem Cell Biol 115:13–21
35. van Royen ME, Farla P, Mattern KA, Geverts B, Trapman J, Houtsmuller AB (2009) Fluorescence recovery after photobleaching (FRAP) to study nuclear protein dynamics in living cells. Methods Mol Biol 464:363–385
36. Snapp EL, Lajoie P (2011) Photobleaching regions of living cells to monitor membrane traffic. Cold Spring Harb Protoc 2011:1366–1367
37. Drobizhev M, Makarov NS, Tillo SE, Hughes TE, Rebane A (2011) Two-photon absorption properties of fluorescent proteins. Nat Methods 8:393–399

Chapter 15

Characterization of STAT Self-Association by Analytical Ultracentrifugation

Nikola Wenta and Uwe Vinkemeier

Abstract

Multiple experimental tools have demonstrated that cytokine-induced STAT activation entails the transition of dimer conformations rather than de novo dimerization. In this chapter, we describe the utilization of analytical ultracentrifugation (AUC) as a powerful technique for the quantitative analysis of hydro- and thermodynamic properties of STAT proteins in solution. These studies provided a quantitative understanding of dimer stability and conformational transitions associated with the activation of STAT1.

Key words: Signal transducer and activator of transcription, Analytical ultracentrifugation, Sedimentation velocity, Sedimentation equilibrium, UltraScan, 2-Dimensional spectrum analysis, Genetic algorithm, Protein-protein interaction, Self-association, Dimerization

1. Introduction

It was believed for a long time that signal transducers and activators of transcription (STATs) are monomeric in resting cells, and that dimerization is a consequence of STAT activation (1). The activation process involves the phosphorylation of a conserved tyrosine residue at the STAT C-terminus, enabling STAT proteins to homo- or heterodimerize via phospho-tyrosine:SH2 domain interactions. This dimer conformation, which is called "parallel" (2, 3), enables high affinity binding of STATs to cognate DNA sequences and the transcriptional activation of target genes. Accumulating structural and functional evidence, however, indicates that STATs already exist as dimers prior to activation (4–11). Crystals of unphosphorylated STAT dimers revealed a completely different conformation, termed "anti-parallel", where the monomers interact via N-domains and reciprocal interactions between coiled-coil and DNA-binding domains (12–14). However, these

Sandra E. Nicholson and Nicos A. Nicola (eds.), *JAK-STAT Signalling: Methods and Protocols*, Methods in Molecular Biology, vol. 967, DOI 10.1007/978-1-62703-242-1_15, © Springer Science+Business Media New York 2013

experiments do not provide information on the stability and behavior of STAT oligomers in solution. As these data are required to fully grasp the dynamics of cytokine signal processing, we turned to analytical ultracentrifugation (12, 15).

In the following, we provide protocols for the application of this very powerful technique to generate precise and accurate quantitative data on thermo- and hydrodynamic properties of STAT proteins. These protocols allowed us to elucidate the contributions of tyrosine phosphorylation and different interaction interfaces on the stability and shape of STAT dimers (15, 16).

Analytical ultracentrifugation (AUC) is a method for quantitatively describing properties of macromolecules in solution. The first high-speed ultracentrifuge was developed by Theodor Svedberg and co-workers as an oil turbine-driven instrument in the 1920s and used for the analysis of gold nanoparticles (17, 18). In 1926, Svedberg was awarded the Nobel Prize in Chemistry "for his work on disperse systems". In his honor, the ratio of maximum sedimentation velocity of a particle and the applied centrifugational acceleration is expressed in Svedberg-units. The next big invention, the design of an electrically driven ultracentrifuge in the 1940s—the Beckman "Model E" (19)—was more reliable and easier to operate than Svedberg's instrument, and made AUC available to a broader range of scientists all over the world. However, the application of this method declined, until in 1992 Beckman made the Optima XL-A commercially available. This instrument and its successor, the XL-I, together with the availability of cheaper and faster computers for data recording and analysis, led to a revival of AUC and paved the way to not only scientific, but also methodological advancements that are still going on today. Currently, the Open AUC Project (20) attempts to integrate the newest developments of instrumentation, detectors, data acquisition and analysis software, and collaborative tools into a new open platform with a newly designed AUC instrument.

Analytical Ultracentrifugation provides first principle information about thermodynamic and hydrodynamic properties of particles sedimenting in a centrifugal field that can be precisely modeled based on the laws of physics. The following section provides a brief theoretical overview essential for understanding the two major approaches of AUC, namely, Sedimentation Velocity (SV) and Sedimentation Equilibrium (SE).

For an ideal monodisperse particle with concentration c and sedimentation coefficient s, sedimenting at constant speed in a centrifugal field, the rate of mass transport $\mathrm{d}m$ over time $\mathrm{d}t$, $J_s(r)$, through a sector-shaped cell is defined as:

$$J_s(r) = \frac{\mathrm{d}m}{\mathrm{d}t} = c\varphi rls\omega^2 r \quad (1)$$

where ϕrl is the surface of the sector-shaped cell with angle ϕ, length l and radius r, and $\omega^2 r$ is the centrifugal acceleration.

The sedimentation coefficient s with the unit [1 Svedberg (S) = 10^{-13} s] of a particle with mass m and a specific partial volume $\bar{v}$, sedimenting in a solution of density ρ, is defined as:

$$\frac{m_p(1-\bar{v}_p\rho_s)}{f} \equiv s \tag{2}$$

The friction factor f according to Einstein (21) is given as:

$$f = \frac{k_B T}{D} \tag{3}$$

where k_B is the Boltzmann constant, T is the absolute temperature and D is the diffusion coefficient. The combination of Eqs. 2 and 3 yield the Svedberg equation:

$$\frac{s}{D} = \frac{m_p(1-\bar{v}_p\rho_s)}{k_B T} \tag{4}$$

According to Stokes (22), f for a spherical particle with hydrodynamic radius a in a solution of viscosity η is defined as:

$$f = 6\pi a\eta \tag{5}$$

The combination of Eqs. 3 and 5 yields the Stokes–Einstein relation:

$$D = \frac{k_B T}{6\pi a\eta} \tag{6}$$

which allows for the calculation of the buoyancy-corrected molecular weight of a sedimenting spherical particle according to the Svedberg Eq. 4:

$$m_p = \frac{s}{D}\frac{6\pi a\eta}{(1-\bar{v}_p\rho_s)} \tag{7}$$

The diffusion rate $J_d(r)$ is, according to Fick's first law (23), defined as:

$$J_d(r) = -D\varphi rl\frac{\partial c}{\partial r} \tag{8}$$

which yields the total transport rate J_{tot}:

$$J_{tot} = \varphi rl(cs\omega^2 r - D\frac{\partial c}{\partial r}) \tag{9}$$

and eventually yields the Lamm equation (24), which describes the change of concentration of a particle during sedimentation over time:

$$\frac{\partial c}{\partial t} = \frac{\partial}{\partial r}\left[r\left(cs\omega^2 r - D\frac{\partial c}{\partial r} \right)\right] \tag{10}$$

While sedimentation velocity experiments monitor concentration profiles that yield hydrodynamic information, that is sedimentation coefficient and shape in the form of the frictional ratio f/f_0 (the ratio of friction of a macromolecule and an ideal sphere of the same molecular weight), the data from sedimentation equilibrium experiments are obtained at equilibrium of sedimentation and diffusion. Thus, net transport of particles is zero, and only thermodynamical information as for example molecular weight and equilibrium dissociation constants (K_d) of a particle can be obtained. This leads to the description of the concentration distribution of a particle in equilibrium:

$$c(r) = c(r_m)\exp\left[\frac{s}{D}\frac{\omega^2(r^2 - r_0^2)}{2}\right] \tag{11}$$

2. Materials

1. 1–2 mg purified STAT proteins (15, 25) in phosphate buffered saline, pH 7.4 (see Note 1).
2. Optima XL-A or XL-I analytical ultracentrifuge (Beckman-Coulter, Brea, CA, USA).
3. Two-sector Epon-charcoal centerpieces for sedimentation velocity experiments.
4. Six-sector Epon-charcoal centerpieces for sedimentation equilibrium experiments.
5. Quartz windows (3.8 g per window; for absorption optics).
6. Sapphire windows (5.7 g per window; for interference optics).
7. UltraScan Analysis Software ((26, 27), see Note 2).
8. WinMATCH software (28).

3. Methods

3.1. Design of Sedimentation Velocity and Equilibrium AUC Experiments

1. Choice of experiment type: Depending on the problem under investigation, two complementary approaches of AUC can be performed: sedimentation velocity and equilibrium AUC. Sedimentation velocity (SV) can be used as an initial diagnostic

method for assessing sample composition, stability and aggregation within a few hours. The main strength of SV as a hydrodynamic method lies in the determination of hydrodynamic parameters such as sedimentation coefficient and protein shape. SV can also yield indications for concentration-dependent self-interaction that can be applied as prior knowledge to the design of SE experiments. Even reaction kinetics in the range of 10/s to 10^{-4}/s can be obtained (29), depending on the molecular weight of the protein.

2. Sample requirements: The sample should be at least 95% pure and in dialysis equilibrium with the reference buffer. This can be easily achieved by gel filtration or dialysis. Gel filtration also adds an additional purification step and separates the sample from aggregates and proteolysis products. 1–2 mg protein in 1 ml buffer should be sufficient for an initial characterization of the protein by AUC. It is recommended to dilute the protein, allowing it to fill the cells as completely as possible, rather than using high concentrations with subsequent low information content due to a smaller observable column height. Alternatively, 3 mm-centerpieces can be used, which yield only 1/4 signal intensity of a standard 1.2 cm-centerpiece.

3. Buffer considerations: When using absorbance optics, the buffer should not absorb at the wavelength at which the experiment is performed, as it would reduce the exploitable dynamic concentration range (0.1–1.0 OD at 280 nm; 0.1–1.5 OD at 230 nm). If reducing agents are required in the buffer, their contribution to the absorbance profile should be taken into account, e.g., by measuring the buffer extinction against water at the wavelength used in the experiment. Tris(2-carboxyethyl) phosphine (TCEP) is an ideal reducing agent as it exhibits only low absorbance at 280 nm and its extinction coefficient does not alter with oxidation state as is the case for dithiothreitol (DTT), for example. When using interference optics, high salt concentrations in the buffer can contribute strongly to the sedimentation profile at high speeds and quench the protein signal dramatically if reference and sample buffer do not match exactly. In this case it is advisable to record protein sedimentation profiles against water in the reference sectors, as UltraScan can then easily deconvolute the signal contribution of each solute to the sedimentation profile.

4. Choice of the optical detection system (ODS): To date, three major optical systems are commercially available and compatible with Beckman's AUC. The XL-A is equipped with an absorbance (ABS) detector only, while the XL-I is equipped with a combined ABS and interference (IF) detector, thus allowing data collection in IF, ABS and intensity mode. Both Beckman centrifuge models can additionally be modified to be

equipped with a fluorescence detection system (Aviv Biomedical, Inc.). A prototype multiwavelength detector has also recently been developed (30, 31), and blueprints and data acquisition software are freely available as part of the Open AUC Project (20).

Absorbance detectors take several minutes to scan a cell at a good radial resolution (0.001–0.003 cm) and can detect, depending on extinction coefficient and wavelength used (wavelength range from 190 to 700 nm), sample concentrations in the range of μg/ml to mg/ml. Of note, recording the data in intensity mode not only allows for increasing the number of samples in one SV run, but also reduces the stochastic noise level by a factor of square root of two. Interference detectors take about 1 min to scan a cell at a radial resolution of 0.001 cm and detect sample concentrations of about 0.1 mg/ml up to several mg/ml. IF is particularly useful when analyzing non-absorbing samples or using strongly absorbing buffers. Fluorescence detectors offer the highest level of selectivity as the sample has to emit fluorescence signals from 505–565 nm after excitation at 488 nm. This also allows detection of samples at very low concentrations in the range of ng/ml to pg/ml. The detector can scan all cells in any rotor within 1.5 min at a moderate radial resolution of 0.005 cm.

5. Choice of speed: The speed for SV should allow the recording of at least 40–50 scans (see Note 3) per cell before complete sedimentation of the protein to the cell bottom (see Note 4). UltraScan contains an ASTFEM (Adaptive Space-Time Finite Element Method; (32) simulation module that accurately simulates the sedimentation profile of a protein with given shape and molecular weight depending on speed, temperature and buffer conditions, and thus helps the user to find an appropriate maximum speed that allows collection of a sufficient number of scans. Additional limiting factors are scan times per cell that depend on sample volumes, rotor, ODS and radial resolution. Sedimentation at high speeds is not influenced by diffusion and thus allows the precise measurement of sedimentation coefficients. At lower speeds, sedimentation is influenced more strongly by diffusion, which allows the extraction of shape information, while the *s* value is not as well defined.

   SE requires a thermodynamic equilibrium gradient that can be established by stepwise increase of speed from 1 to 3.5 σ ($\sigma = \omega^2 m(1-\bar{v})/2RT$; (33). UltraScan contains an Equilibrium Time simulator that calculates an optimal speed setup and the time increments for a protein depending on shape, molecular weight, buffer conditions and temperature to establish equilibrium conditions (see Note 5). This setup can be entered directly into the method editor of the ultracentrifuge. Notably, the duration of an SE experiment is only restricted by protein stability and the time needed to establish equilibrium conditions. A typical experiment with three speed steps can therefore take about 3–4 days. This allows for data recording with high radial resolution, many scan repetitions, at up to three wavelengths, for multiple samples over a broad concentration range and at multiple speeds. However, the time required to reach equilibrium can be shortened by using smaller sample volumes which reduce precision due to a reduced column height.

3.2. Performing SV-AUC Experiments

1. Pre-cool the AUC with empty rotor and optical detection system mounted to the desired temperature for at least 1 h (see Note 6).
2. Meanwhile, clean and assemble AUC cells (see Note 7), depending on the optical detection system used, with appropriate windows and two-sector centerpieces (see Note 8). Once the AUC has reached vacuum, intensity wavelength scans at 6.5 cm of empty rotor holes should be performed at 3,000 rpm (see Note 9).
3. Prepare protein dilutions in reference buffer and fill AUC cells through their filling holes with the recommended maximum volumes of protein solution and buffer (see Notes 10 and 11), and seal the cells. The top screw ring should be tightened up to about 120 in.-pounds (13.6 Nm).
4. Weigh the completely assembled cells and make sure that opposing cells have a weight difference of no more than 0.5 g. If a counterbalance is used, it should have an equal or up to 0.5 g lower weight than the opposing cell.
5. Stop the AUC, release the vacuum and dismount optical system and rotor. Insert AUC cells into the rotor and align them according to the scribe lines on the rotor and cell bottom. Place the rotor back into the AUC, mount the optical detection system and set the AUC to the desired temperature and 3,000 rpm (see Note 12) until the pressure drops below 100 microns (see Note 13).
6. During temperature equilibration, the following diagnostics and calibrations (see manual) should be performed: (a) checking the radial calibration of the AUC, using the counterbalance; (b) recording a wavelength spectrum (190–300 nm) for each cell at a radial position of 6.5 cm (see Note 14); (c) recording intensity scans at the desired experimental wavelength for each cell from radial positions 5.9 cm to 7.1 cm (see Note 15). If using interference optics, a laser delay calibration is advisable.
7. Stop the AUC, release the vacuum and dismount ODS and rotor. Shake the rotor gently to remix the protein samples and place rotor and ODS back into the AUC. Set rotor speed to 0 rpm for at least 1 h.

8. Meanwhile, set up experimental method parameters in the Beckman data acquisition software, such as (a) rotor type; (b) speed; (c) temperature; experiment type (Velocity); (d) ODS, and for ABS additionally wavelength and operation mode; (e) radial data range; (f) cell designations; (g) radial resolution (0.003 cm for ABS), and continuous step motor mode; (h) time between subsequent scans, and number of recorded scans.
9. Once a pressure of 0 microns and the desired temperature are reached (see Note 16), start the data acquisition (see Note 12).
10. Once the experiment has finished and the AUC stopped, release the vacuum and dismount optical system and rotor. The samples can be recovered after shaking the cells and analyzed by SDS-PAGE for degradation during the AUC run. Copy experimental data into the UltraScan data folder for analysis.

3.3. Analysis of SV Data with UltraScan

1. For UltraScan II, internet connection and registration for UltraScan LIMS database (Laboratory Information Management System) are required, which provides access to supercomputers, where all calculation steps are performed. Ultrascan III can perform initial calculations (up to step 12) locally, but requires registration for later stages of data analysis.
2. Commit experimental data to database: associate raw data (see Notes 17 and 18) with information about used samples (e.g., amino acid or nucleic acid sequence), buffers (see Note 19), rotor, and centerpieces.
3. Upload datasets to UltraScan LIMS database. Retrieve data from database to the computer.
4. UltraScan will guide you through the data editing process: edit meniscus and bottom positions, define useful data range, remove radially invariant noise, exclude scans if necessary (see Note 20), and define absorption plateau and baseline. Upload edited data (stored in the UltraScan results folder) to database.
5. Perform a model-independent analysis method, e.g., van Holde–Weischet (vHW; (34, 35)) or dc/dt (36) analysis, to determine a useful s value range (see Note 21).
6. Login to your UltraScan LIMS portal and start a 2-dimensional spectrum analysis (2DSA; (37)). Enter an email address for receiving results. Then select an experiment and add cells to be included for processing to the analysis queue (see Note 22).
7. Perform 2DSA with default settings and appropriate ranges for s (as determined in step 5) and f/f_0 values, and include fits of meniscus position and time-invariant (TI) noise.

8. After retrieving the results as email attachments, fit and update the meniscus position (contained in email body) for each dataset after applying the UltraScan Meniscus Fitter module. Download the model for finite element (FE) solutions and TI noise file with the lowest RMSD, and save them in the UltraScan results folder. Apply the 2DSA models in the FE Model Viewer module, and then subtract TI noise from each dataset. Re-upload and overwrite the initial datasets in UltraScan LIMS database with the noise- and meniscus-corrected datasets.
9. Refine the 2DSA analyses as described in steps 6 and 7, but this time with TI and radially invariant (RI) noise fit included and fit of meniscus position excluded.
10. After retrieving the results as email attachments, download the model for FE solutions and TI and RI noise files and save them in the UltraScan results folder. Apply the 2DSA model in the FE Model Viewer module, save the 3D plot, and then subtract TI and RI noise from each dataset. Re-upload and overwrite the former datasets in UltraScan LIMS database with the noise-corrected datasets (see Note 23).
11. Perform vHW analysis for all protein concentrations. A concentration-dependent increase of *s* value indicates protein self-interaction.
12. Initialize the genetic algorithm (GA; (38)) analysis using saved pseudo-3D plots.
13. Login to your UltraScan LIMS portal and start a GA analysis request. Enter an email address for receiving results. Then select an experiment and add cells to be included for processing to the analysis queue (see Note 22).
14. Perform a GA analysis with default settings, upload the GA initialization file from the UltraScan results folder, and submit analysis request.
15. After retrieving the results as email attachments, download the model file for GA-FE solutions and save it in the UltraScan results folder. Apply the model in the FE Model Viewer module, and save the 3D plot. Initialize the GA once again using saved pseudo-3D plots.
16. Perform a GA analysis with 50 Monte Carlo iterations (39) and default settings again, upload the GA initialization file from the results folder, and submit analysis request.
17. After retrieving the results as email attachments, download the model file for GA-FE solutions and save it in the UltraScan results folder. Apply the model in the FE Model Viewer module, and save the final 3D plot.

3.4. Performing Sedimentation Equilibrium Experiments

1. Clean and assemble AUC cells completely (see Note 7), depending on the optical detection system used, with the appropriate windows (see Note 24) and six-sector centerpieces (see Note 25). Afterwards, disassemble top screw ring and top window.
2. Prepare protein dilutions in reference buffer and fill AUC cells with the recommended maximum volumes of protein solution and buffer (see Notes 26 and 27), and seal the cells.
3. Weigh the completely assembled cells and make sure that opposing cells have a weight difference of no more than 0.5 g. If a counterbalance is used, it should have an equal or up to 0.5 g lower weight than the opposing cell.
4. Insert AUC cells into the rotor and align them according to the scribe lines on the rotor and cell bottom. Place the rotor back into the AUC, mount the ODS and set the AUC to the desired temperature and 3,000 rpm (see Note 12) until the pressure drops below 100 microns (see Note 13).
5. Record wavelength spectra (190–300 nm) for each cell at the radial positions of 6.0, 6.5 and 7.0 cm at 3,000 rpm (see Note 28).
6. Set up experimental method parameters in the Beckman data acquisition software, such as (a) rotor type; (b) speed profile (see Note 5); (c) temperature; (d) experiment type (Equilibrium); (e) ODS, and for ABS additionally up to three wavelengths; (f) radial data range for the whole cell; (g) cell designations; (h) radial resolution (0.001 cm), 20 averages, and motor in step mode; (i) time between subsequent scans (8 h).
7. Start the data acquisition (see Notes 12 and 29). Once every 8 h, check with WinMatch if sedimentation profiles are in equilibrium (see Note 30). Proceed to next higher speed and check for equilibrium again.
8. Once the experiment has finished and the AUC stopped, release the vacuum and dismount optical system and rotor. The samples can be recovered after shaking the cells and analyzed by SDS-PAGE for degradation during the AUC run. Copy experimental data into the UltraScan data folder for analysis.

3.5. Analysis of SE Data with UltraScan

1. The analysis of SE data with UltraScan does not require access to UltraScan LIMS database or supercomputers, but it might be advantageous to make use of a spare computer with UltraScan installed on it for performing Monte Carlo analyses.
2. Fit all recorded wavelength scans (at least three) globally using the Extinction Coefficient Calculator module in order to associate measured wavelength profiles with the protein's

extinction coefficient at 280 nm (calculated from amino acid sequence) and save the results.

3. UltraScan will guide you through the data editing process (see Note 17): select centerpiece and rotor, and add a unique run ID. Edit meniscus position for each sector, define useful data range, and exclude scans that are not in equilibrium, if necessary.

4. Perform an initial global equilibrium fit: (a) select scans to be included; (b) select model "Fixed Molecular Weight Distribution" (FMWD; see Note 31) over a broad molecular weight range, and use $\overline{\nu}$ from protein sequence; (c) perform "Scan Diagnostics" and read the diagnostic messages carefully. If necessary, exclude data points that do not seem to contribute to an exponential concentration curve; included scans are marked with a green triangle—if including scans associated with a warning message, they will be marked with a red triangle; (d) initialize parameters: $\overline{\nu}$ (from protein sequence), buffer density (from buffer composition), and extinction coefficient profile (from step 2); (e) float parameters; (f) check scans for fit; (g) apply fitting control, with auto-converge turned on, if available.

5. Perform diagnostics: (a) Ln(C) vs. R^2: the plots will be nonlinear if multiple components are present; (b) MW vs. R^2: the plots will overlay if the protein is noninteracting; (c) MW vs. c: shows the relative concentration of detected molecular weights. Save the fit with default suffix.

6. Perform additional fits for at least the following models: (a) 1 ideal component; (b) 2 ideal noninteracting components; (c) 3 ideal noninteracting components. These fits should be applied with floated and constrained molecular weights.

   If these fits with floated molecular weights yield expected molecular weights and RMSD comparable to FMWD, the protein is actually noninteracting. If expected molecular weights were fixed and RMSD deviates significantly from FMWD, interacting models should be tested analogous to step 4 (see Note 32).

7. Decide which model fits the data best: (a) the fit that yields the expected molecular weights with the least number of fitted parameters is considered the "best" fit (see Note 33); (b) the residuals for the "best model" have to be randomly distributed; (c) for self-interacting models, the dissociation constant obtained has to match the investigated concentration range of the sample, with at least 10% of each oligomeric state present; (d) models for more than two interacting species usually are poorly defined and can easily be overfitted. (e) if two models fit the data equally well, a Monte Carlo analysis will reveal how well the parameters are actually defined.

8. Perform a Monte Carlo analysis with at least 10,000 iterations on the best fit model to obtain 95% confidence intervals for molecular weight and K_d.

3.6. Examples of Analyses of Experimental SV and SE Data for STAT Proteins

1. Initial Sedimentation Velocity analysis of phosphorylated STAT1β, a ~84 kDa protein that forms dimers. Figure 1 shows the analysis of a sedimentation profile recorded in absorbance mode at 230 nm for 5 μM phosphorylated STAT1β in PBS at 20°C and 50,000 rpm. Van Holde–Weischet analysis after 2DSA and TI and RI noise removal reveals one major species with a sedimentation of 9.5 S.
2. Characterization of the influence of phosphorylation on the self-association of STAT proteins by sedimentation equilibrium analyses (15). Table 1 shows an overview example of the results for various models applied to SE data for phosphorylated STAT1β. 108 data sets at equilibrium were collected in PBS at

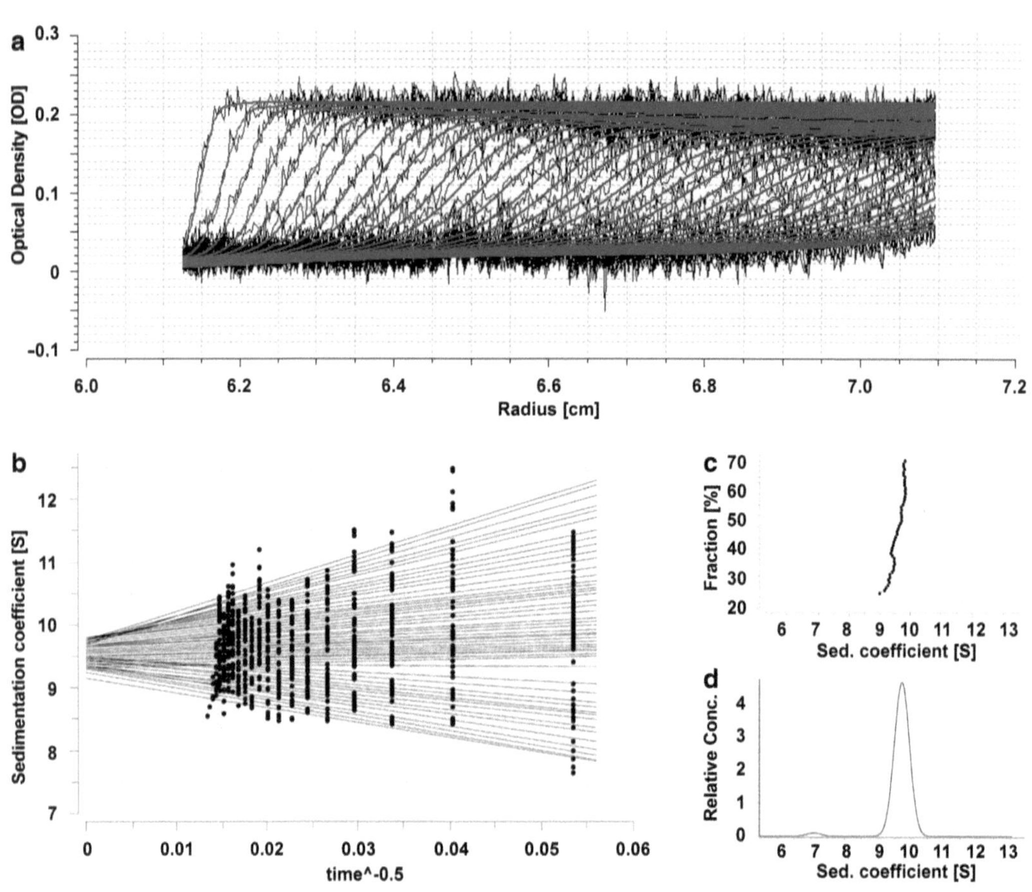

Fig. 1. Analysis of phosphorylated STAT1β by Sedimentation Velocity experiments. (**a**) Raw experimental data (*black*) was analyzed by 2DSA analysis for TI and RI noise removal. The fit model is shown in *gray*. (**b**) The noise-corrected data was then analyzed with the van Holde–Weischet method. The corresponding integral distribution (**c**) and g(s) plots (**d**) show a homogeneous sample with one major sedimenting species at 9.5 S.

Table 1
Results of global fits for SE data of phosphorylated STAT1β

Model	MW (kDa)	Modus	K_d (M)	RMSD (OD)
Fixed molecular weight distribution	10–2,000	500 Slots	NA[a]	1.128E-02
1 Ideal component	153.3	Floated	NA	1.304E-02
	84.1	Fixed	NA	3.472E-02
	168.3	Fixed	NA	1.423E-02
2 Ideal components, noninteracting	122.5/204.6	Floated/floated	NA	1.161E-02
	84.1/183.7	Fixed/floated	NA	1.172E-02
	93.0/168.3	Floated/fixed	NA	1.183E-02
	84.1/168.3	Fixed/fixed	NA	1.186E-02
Monomer–Dimer	84.0	Floated	3.73E-08	1.274E-02
	84.1	Fixed	3.83E-08	1.274E-02
Monomer–Trimer	150.1	Floated	3.18E-05	1.282E-02
	84.1	Fixed	5.08E-07	1.808E-02
Monomer–Tetramer	151.8	Floated	2.93E-05	1.294E-02
	84.1	Fixed	7.89E-07	2.560E-02
Monomer–Dimer–Trimer	83.9	Floated	4.05E-08[b]	1.273E-02
	84.1	Fixed	3.85E-08[b]	1.274E-02
Monomer–Dimer–Tetramer	83.9	Floated	3.90E-08[b]	1.271E-02
	84.1	Fixed	4.09E-08[b]	1.272E-02
Monomer–Dimer–Hexamer	84.0	Floated	3.75E-08[b]	1.274E-02
	84.1	Fixed	3.82E-08[b]	1.274E-02

Shown are values of fixed or floated molecular weights (MW), K_d and root mean square deviation (RMSD) for various noninteracting and interacting models

[a] NA: not applicable

[b] K_d of the monomer–dimer component only

4°C, 4 speeds (8,000, 13,000, 18,000, and 23,000 rpm) with loading concentrations ranging from 0.1 to 1.0 absorption units at the three wavelengths used (220, 225 and 230 nm) and globally analyzed with UltraScan. The best fit for phosphorylated STAT1β was obtained for a monomer–dimer equilibrium model, with a monomer molecular weight of about 84,000 Da and a K_d of 37.3 nM (Fig. 2). The best fit models, including 95% confidence ranges for molecular weight, K_d and RMSD values after Monte Carlo analyses are summarized in Table 2. These results demonstrate that all phosphorylated STAT1 variants form equally stable dimers with K_d values of ~50 nM. Surprisingly, unphosphorylated STAT1α and STAT1β exist as dimers and are thermodynamically as stable as their phosphorylated counterparts. On the other hand, unphosphorylated STAT1 variants that are missing the N-terminal domain (STAT1ΔN and STAT1tc) form dimers with a dramatically

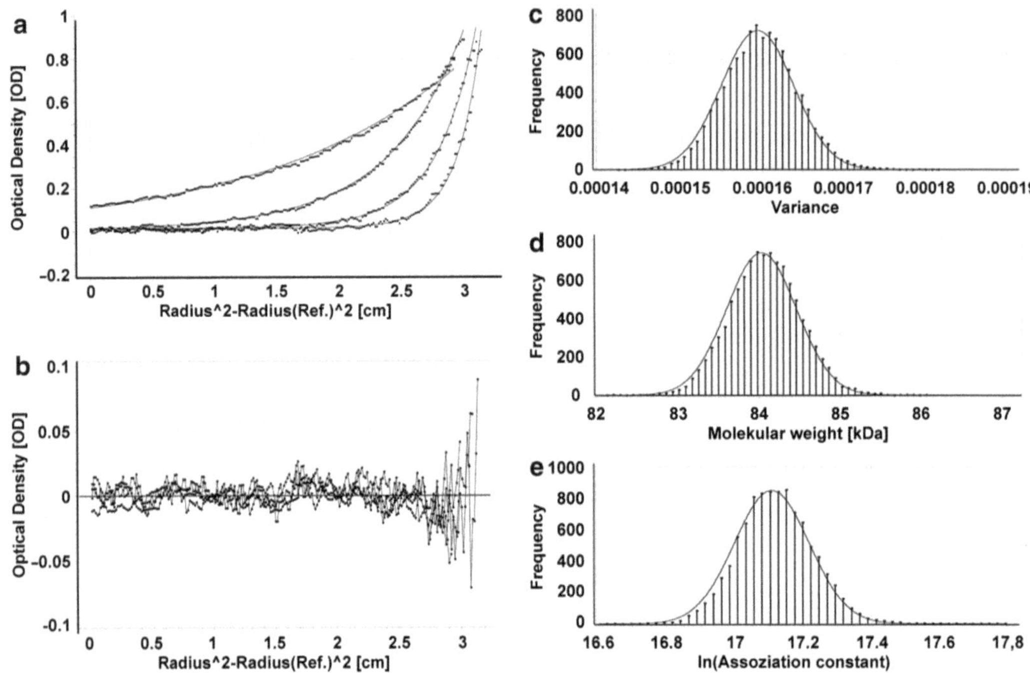

Fig. 2. Results of the global nonlinear fit on SE data of phosphorylated STAT1β to a monomer–dimer model. For clarity, only the equilibrium profiles and fits for one single protein concentration at $\lambda = 230$ nm and four different speeds (**a**) and the corresponding residuals (**b**) are shown. The 95% confidence intervals after 10,000 Monte Carlo iterations are shown for variance (**c**), monomer molecular weight (**d**), and natural logarithm of the association constant (**e**).

impaired capability to dimerize by a factor of 500. This indicates that phosphorylated and unphosphorylated STAT1 dimers are stabilized by mutually exclusive interaction interfaces: SH2-domain interactions for phosphoproteins, and N-terminal domain-mediated interactions for the unphosphorylated STAT1 proteins. Notably, homotypic N-terminal domain interactions are not conserved in the STAT family, because the N-terminal domain dissociation constants of STAT1, STAT3, and STAT4 differed by more than three orders of magnitude. In conclusion, STAT1 constantly oscillated between different dimer conformations, whereby the abundance of the dimer conformations was determined by tyrosine phosphorylation.

3. Characterization of the shape of phosphorylated and unphosphorylated STAT1α by 2-dimensional spectrum analysis/genetic algorithm (16). Individual sedimentation profiles for phosphorylated (0.7 mg/ml) STAT1α and unphosphorylated (0.8 mg/ml) STAT1α were recorded in buffer GF (2 mM Tris (pH 8), 150 mM NaCl, 1 mM EDTA, 1 mM TCEP) at 10°C and 30,000 rpm in absorbance mode. The data were separately analyzed using subsequent 2-dimensional spectrum and genetic algorithm analyses. The combined

Table 2
Summary of best fit results for all investigated STAT proteins analyzed by sedimentation equilibrium

Protein	Model	Monte Carlo 95% confidence range		
		MW (kDa)	K_d (M)	RMSD (10^{-3} OD)
STAT1-N	Monomer–Dimer	15.5 (+1.9/−1.9) 17.8 (fixed)	5.4 (+1.5/−1.1)10^{-6} 23.3 (+0.7/−0.6)10^{-6}	5.033 (+0.203/−0.211) 5.156 (+0.800/−0.227)
STAT3-N	Monomer–Dimer	17.0 (+2.1/−2.1) 17.3 (fixed)	2.9 (+1.6/−0.3)10^{-3} 3.7 (+0.7/−0.6)10^{-3}	6.067 (+0.226/−0.234) 6.015 (+0.223/−0.231)
STAT4-N	Monomer–Dimer	18.9 (+0.2/−0.2) 18.0 (fixed)	4.5 (+0.4/−0.4)10^{-6} 2.7 (+0.1/−0.1)10^{-6}	7.618 (+0.188/−0.193) 7.701 (+0.193/−0.199)
Unphosphorylated STAT1α	Monomer–Dimer	92.2 (+1.1/−1.1) 89.9 (fixed)	68.5 (+12.2/−10.3)10^{-9} 45.2 (+3.4/−3.1)10^{-9}	8.562 (+0.324/−0.336) 8.589 (+0.317/−0.328)
Unphosphorylated STAT1β	Monomer–Dimer	83.4 (+4.5/−3.3) 84.1 (fixed)	46.7 (+26.5/−17.2)10^{-9} 52.1 (+4.4/−4.5)10^{-9}	5.663 (+0.258/−0.271) 5.662 (+0.256/−0.268)
Unphosphorylated STAT1ΔN	Monomer–Dimer	72.5 (+3.5/−4.1) 73.9 (fixed)	2.4 (+1.6/−0.9)10^{-6} 2.9 (+0.3/−0.2)10^{-6}	8.228 (+0.735/−0.808) 8.455 (+0.378/−0.396)
Unphosphorylated STAT1tc	Monomer–Dimer	67.5 (+4.1/−3.1) 69.8 (fixed)	3.4 (+2.0/−1.3)10^{-6} 4.5 (+0.3/−0.3)10^{-6}	7.709 (+0.325/−0.340) 7.716 (+0.322/−0.336)
Phosphorylated STAT1α	Monomer–Dimer–Tetramer	89.7 (+1.2/−1.3) 89.9 (fixed)	29.2 (+9.0/−6.9)10^{-9a} 31.2 (+12.4/−8.9)10^{-9a}	8.444 (+0.313/−0.326) 8.466 (+0.329/−0.341)
Phosphorylated STAT1α	Dimer–Tetramer	190.5 (+3.6/−3.6) 179.8 (fixed)	136.9 (+28.2/−23.5)10^{-9b} 75.7 (+4.6/−4.3)10^{-9b}	8.644 (+0.296/−0.308) 8.721 (+0.287/−0.296)
Phosphorylated STAT1β	Monomer–Dimer	84.0 (+0.1/−0.1) 84.1 (fixed)	37.1 (+9.1/−7.3)10^{-9} 38.2 (+4.5/−4.0)·10^{-9}	12.620 (+0.340/−0.350) 12.620 (+0.340/−0.350)

(continued)

Table 2 (continued)

Protein	Model	Monte Carlo 95% confidence range		
		MW (kDa)	K_d (M)	RMSD (10^{-3} OD)
Phosphorylated STAT1ΔN	Monomer–Dimer	77.6 (+4.4/−3.0)	98.8 (+77.7/−43.5)10^{-9}	9.605 (+0.559/−0.594)
		73.9 (fixed)	48.1 (+6.7/−5.9)10^{-9}	9.636 (+0.547/−0.580)
Phosphorylated STAT1tc	Monomer–Dimer	74.2 (+5.8/−4.4)	47.6 (+59.5/−26.2)10^{-9}	9.181 (+0.361/−0.376)
		69.8 (fixed)	19.0 (+5.1/−4.1)10^{-9}	9.276 (+0.352/−0.366)

Shown are the values and 95% confidence ranges obtained for fixed or floated molecular weights, K_d and RMSD after 10,000 Monte Carlo iterations
[a]Only K_d of monomer–dimer component is shown
[b]Only K_d of dimer–tetramer component is shown
Reproduced from ref. 15, Table S1

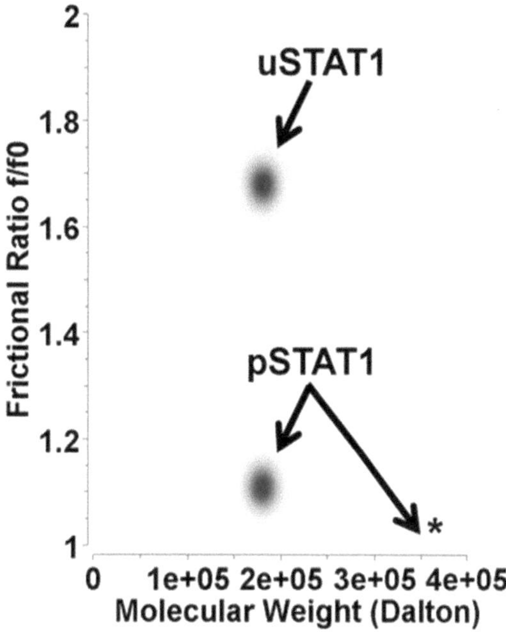

Fig. 3. Analysis of sedimentation velocity data of phosphorylated (pSTAT1) and unphosphorylated (uSTAT1) STAT1α. Shown is a combined molecular weight/frictional ratio distribution from separate analyses of uSTAT1 and pSTAT1 data after 50 Monte Carlo iterations of genetic algorithm analysis. Note the identical molecular masses of uSTAT1 and pSTAT1 dimers (~180 kDa), but their different frictional ratios. Also note an additional high-molecular-weight peak for pSTAT1 consistent with tetramerization (*asterisk*). Reproduced from ref. 16, Fig. 1c.

representation of molecular weights and frictional ratios for phosphorylated and unphosphorylated STAT1α, obtained after 50 Monte Carlo iterations of genetic algorithm analysis is shown in Fig. 3. The outcome confirms the results from SE experiments that STAT1α exists as a dimer before and after activation. However, the dimers differ in shape. Phosphorylated STAT1α dimers have a globular shape ($f/f_0 \sim 1.2$), while unphosphorylated STAT1α dimers have a rather elongated shape ($f/f_0 \sim 1.7$). These results are in agreement with existing crystal structures of phosphorylated and unphosphorylated STAT proteins (10–14).

4. Notes

1. The protein should be of high purity (>95%, as judged by visual inspection of Coomassie-stained SDS gels) and show no signs of precipitation or degradation in any selected buffer for several hours at 20°C or several days at 4°C. If necessary, reducing

agents can be added (Attention: buffer components and reducing agents may absorb at $\lambda = 280$ nm or other wavelengths and reduce the available dynamic detection range!). The protein has to be in dialysis equilibrium with the reference/dilution buffer; this can be achieved by collecting dialysis buffer after dialysis or the flow-through from gel filtration or ultrafiltration.

2. The analyses described in this chapter were performed with UltraScan-II. However, the steps described in the data analysis chapters can be performed in the same order with the same outcome with UltraScan-III. New users are encouraged to use the major rewrite, UltraScan-III, which contains several enhancements and will eventually replace the former version. UltraScan-II is still available for download, but will be maintained only to assure access to legacy data because UltraScan-III is not backward compatible with UltraScan-II.
3. Superfluous scans can later be discarded during analysis.
4. The 8-hole An 50-Ti rotor is rated for a maximum speed of 50,000 rpm; the 4-hole An 60-Ti rotor is rated for a maximum speed of 60,000 rpm.
5. In order to ensure that equilibrium will be reached, you should perform the simulation for the largest expected component present for the inner channel of the centerpiece, and then add 20% on top of the estimated time required to reach equilibrium.
6. The time required for rotor pre-cooling can be reduced by manually setting the speed to 0 rpm.
7. The "Analytical Ultracentrifuge User Guide—Volume 1: Hardware" by Karel L. Planken and Virgil Schirf is freely available at: http://wiki.bcf.uthscsa.edu/aucmanual/.
8. Depending on the optical detection system used, a counterbalance in hole 1 will be necessary for absorption optics in normal and intensity mode, which reduces the number of available holes for AUC cells to three in a 4-hole An-60 Ti rotor; and to seven in an 8-hole An-50 Ti rotor, respectively. A counterbalance is not required when using interference optics only.
9. The intensity values should not vary between individual holes, and the spectra should show a maximum peak at 230 nm (otherwise a calibration by Beckman service is advisable). A reference spectrum can be found in the manual.
10. It is crucial to remember that depending on the optical detection system used, different recommended maximum loading volumes apply: (a) when using absorption optics, for all cells fill the reference sectors with 450 μl of buffer and the sample sectors with up to 440 μl of protein solution; (b) when using

absorbance optics in intensity mode, for all cells fill one reference sector with 450 μl of water and all other sample and reference sectors with up to 440 μl of protein solution; (c) when using interference optics, for all cells fill the sample sectors with up to 440 μl of protein solution and the reference sectors with 440 μl of buffer or water. Alternatively, meniscus-matching centerpieces (Spin Analytical, Inc.) can be used for optimal performance with sample volumes of 340 μl in sample and 350 μl in reference sectors. These centerpieces require a short pre-run at 5,000 rpm for menisci to match. The rotor will then be stopped, gently shaken to mix the samples, and inserted back into the centrifuge.

11. If the top screw ring is facing towards you and filling holes upwards, the reference sector is on the left and the sample sector on the right.

12. The development of vacuum should be carefully monitored; sudden increases in pressure indicate leaking cells.

13. Above a pressure of 100 microns, the temperature of the chamber is measured with a thermistor. Below a pressure of 100 microns, a radiometer kicks in and directly measures the rotor temperature. This will lead to a sudden apparent increase of the displayed temperature.

14. During an SV experiment, only one wavelength and/or interference optics should be used. The absorption values at the wavelength of choice for all protein solutions used should be in the dynamic range between 0.1 and 1.0 absorption units at 280 nm or up to 1.5 at 230 nm. Remember that the usual optical path length for AUC cells is 1.2 cm when preparing protein dilutions. 3 mm centerpieces can be used to reduce the optical path length and thus allow extension of the dynamic range by an additional factor of four.

15. The intensities throughout each cell should not vary by more than 10%. If necessary, cleaning the lamp and slit assembly might help improve intensity readings.

16. Due to adiabatic rotor stretching and cooling, the rotor can already be accelerated to the desired speed and the experiment started when the displayed temperature is 0.3°C above the desired experimental temperature.

17. If using UltraScan III, first convert and save raw data into UltraScan III format.

18. If raw data is in intensity mode, convert into pseudo-absorbance mode.

19. Sample $\overline{v}$ and hydrodynamic corrections for buffer density and viscosity can be automatically calculated by UltraScan. If untabulated buffer components were used, density and viscosity should be experimentally determined and entered manually.

20. If the last scans contain only baseline information, they can be excluded from analysis in order to reduce computational requirements. Additionally, a scan exclusion profile can be applied to obtain about 50 non-baseline scans that are sufficient to perform a high-resolution analysis.
21. dc/dt is particularly useful for data with considerable time-invariant noise contribution, as is often the case for intensity data.
22. If you have selected multiple cells, the data can be fit globally or separately.
23. Optionally, a Monte Carlo analysis of the 2DSA fit with 50 iterations can be performed in parallel to the Genetic Algorithm analysis.
24. SE experiments are usually performed with ABS optics; interference optics can be used optionally, at least for higher protein concentrations. Intensity mode cannot be used.
25. Add the top window after you have filled the samples, seal the cells and tighten the screw ring up to about 120 in.-pounds (13.6 Nm).
26. The maximum recommended loading volumes for all cells are 110 μl of protein solution in the sample sectors and 120 μl of buffer in the reference sectors.
27. If the cell housing top is facing upwards and the part number at the cell housing is facing to the left, the top 3 holes correspond to the sample sectors and the bottom 3 holes to the reference sectors.
28. The absorption values at given wavelengths for all protein dilutions should be in the dynamic range between 0.1 and 1.0 absorption units. Take into account that the usual optical path length for AUC cells is 1.2 cm when preparing protein dilutions.
29. There is no need to wait for temperature equilibration before starting the AUC run. The temperature will have equilibrated long before the equilibrium state will be reached.
30. It is sufficient to check only the inner sectors for equilibrium, because with the centrifugal forces being weaker in the inner than in the outer sector, it requires more time for the inner sector to reach equilibrium. If two subsequent scans do not differ by more than the random noise level, equilibrium is reached.
31. The Fixed Molecular Weight Distribution model is the most unconstrained model available and always results in the lowest achievable RMSD.

32. In the case of a monomer–dimer equilibrium, also higher oligomerization states on the basis of a momoner–dimer equilibrium, e.g., 1-2-3mer, 1-2-4mer, and 1-2-6mer models should be tested; these will correspond to 1-2-Xmer either with no Xmer present (RMSD does not change) or with an actual Xmer present (RMSD becomes lower).
33. Try to constrain as many known parameters as possible. This implies measuring a parameter rather than fitting it wherever possible.

Acknowledgment

We thank Borries Demeler (University of Texas Health Science Center at San Antonio) for critical reading and comments. This work was supported by BBSRC grant BB/GO019290/1.

References

1. Chen X, Vinkemeier U, Zhao Y, Jeruzalmi D, Darnell JE Jr, Kuriyan J (1998) Crystal structure of a tyrosine phosphorylated STAT-1 dimer bound to DNA. Cell 93:827–839
2. Becker S, Groner B, Muller CW (1998) Three-dimensional structure of the Stat3beta homodimer bound to DNA. Nature 394:145–151
3. Shuai K, Horvath CM, Huang LH, Qureshi SA, Cowburn D, Darnell JE Jr (1994) Interferon activation of the transcription factor Stat91 involves dimerization through SH2-phosphotyrosyl peptide interactions. Cell 76:821–828
4. Stancato LF, David M, Carter-Su C, Larner AC, Pratt WB (1996) Preassociation of STAT1 with STAT2 and STAT3 in separate signalling complexes prior to cytokine stimulation. J Biol Chem 271:4134–4137
5. Novak U, Ji H, Kanagasundaram V, Simpson R, Paradiso L (1998) STAT3 forms stable homodimers in the presence of divalent cations prior to activation. Biochem Biophys Res Commun 247:558–563
6. Ndubuisi MI, Guo GG, Fried VA, Etlinger JD, Sehgal PB (1999) Cellular physiology of STAT3: where's the cytoplasmic monomer? J Biol Chem 274:25499–25509
7. Haan S, Kortylewski M, Behrmann I, Muller-Esterl W, Heinrich PC, Schaper F (2000) Cytoplasmic STAT proteins associate prior to activation. Biochem J 345(Pt 3):417–421
8. Braunstein J, Brutsaert S, Olson R, Schindler C (2003) STATs dimerize in the absence of phosphorylation. J Biol Chem 278:34133–34140
9. Ota N, Brett TJ, Murphy TL, Fremont DH, Murphy KM (2004) N-domain-dependent nonphosphorylated STAT4 dimers required for cytokine-driven activation. Nat Immunol 5:208–215
10. Zhong M, Henriksen MA, Takeuchi K, Schaefer O, Liu B, ten Hoeve J, Ren Z, Mao X, Chen X, Shuai K, Darnell JE Jr (2005) Implications of an antiparallel dimeric structure of nonphosphorylated STAT1 for the activation-inactivation cycle. Proc Natl Acad Sci USA 102: 3966–3971
11. Mertens C, Zhong M, Krishnaraj R, Zou W, Chen X, Darnell JE Jr (2006) Dephosphorylation of phosphotyrosine on STAT1 dimers requires extensive spatial reorientation of the monomers facilitated by the N-terminal domain. Genes Dev 20:3372–3381
12. Mao X, Ren Z, Parker GN, Sondermann H, Pastorello MA, Wang W, McMurray JS, Demeler B, Darnell JE Jr, Chen X (2005) Structural bases of unphosphorylated STAT1 association and receptor binding. Mol Cell 17:761–771
13. Neculai D, Neculai AM, Verrier S, Straub K, Klumpp K, Pfitzner E, Becker S (2005) Structure of the unphosphorylated STAT5a dimer. J Biol Chem 280:40782–40787
14. Ren Z, Mao X, Mertens C, Krishnaraj R, Qin J, Mandal PK, Romanowski MJ, McMurray JS, Chen X (2008) Crystal structure of unphosphorylated STAT3 core fragment. Biochem Biophys Res Commun 374:1–5

15. Wenta N, Strauss H, Meyer S, Vinkemeier U (2008) Tyrosine phosphorylation regulates the partitioning of STAT1 between different dimer conformations. Proc Natl Acad Sci USA 105:9238–9243
16. Nardozzi J, Wenta N, Yasuhara N, Vinkemeier U, Cingolani G (2010) Molecular basis for the recognition of phosphorylated STAT1 by importin alpha5. J Mol Biol 402:83–100
17. Svedberg T, Nichols JB (1923) Determination of size and distribution of size of particle by centrifugal methods. J Am Chem Soc 45:2910–2917
18. Svedberg T, Rinde H (1924) The ultra-centrifuge, a new instrument for the determination of size and distribution of size of particles in amicroscopic colloids. J Am Chem Soc 46:2677–2693
19. Pickels EG (1950) Mach Des 22:102–107
20. Colfen H, Laue TM, Wohlleben W, Schilling K, Karabudak E, Langhorst BW, Brookes E, Dubbs B, Zollars D, Rocco M, Demeler B (2010) The open AUC project. Eur Biophys J 39:347–359
21. Einstein A (1905) Über die von der Molekularkinetischen Theorie der Wärme geforderte Bewegung von in ruhenden Flüssigkeiten suspendierten Teilchen. Ann Phys 17:182–193
22. Stokes GG (1850) On the effect of the internal friction of fluids on the motion of pendulums. Trans Cambridge Phil Soc 9:8–106
23. Fick A (1855) Über diffusion. Ann Phys Chem 94:59–86
24. Lamm O (1929) Die Differenzialgleichung der Ultrazentrifugierung. Ark Mat Astron Fysik 21B:1–4
25. Vinkemeier U, Cohen SL, Moarefi I, Chait BT, Kuriyan J, Darnell JE Jr (1996) DNA binding of in vitro activated Stat1 alpha, Stat1 beta and truncated Stat1: interaction between NH2-terminal domains stabilizes binding of two dimers to tandem DNA sites. EMBO J 15:5616–5626
26. Demeler B (2005) UltraScan—a comprehensive data analysis software package for analytical ultracentrifugation experiments. In: Scott DJ, Harding SE, Rowe AJ (eds) Modern analytical ultracentrifugation: techniques and methods. Royal Society of Chemistry, UK, pp 210–229
27. UltraScan is freely available for download as source code (GPL) and binary packages for Windows, Linux and Macintosh platform. UltraScan-II: http://www.ultrascan2.uthscsa.edu/download.php; UltraScan-III: http://www.ultrascan3.uthscsa.edu/download.php. as source code (GPL) and binary packages for Windows, Linux and Macintosh platform
28. WinMATCH binary package for Windows can be freely downloaded at http://www.biotech.uconn.edu/auf/ftp/WINMATCH.ZIP (tested on October 4th, 2012).
29. Demeler B, Brookes E, Wang R, Schirf V, Kim CA (2010) Characterization of reversible associations by sedimentation velocity with UltraScan. Macromol Biosci 10:775–782
30. Bhattacharyya SK, Maciejewska P, Borger L, Stadler M, Gulsun AM, Cicek HB, Colfen H (2006) Development of fast fiber based UV-Vis multiwavelength detector for an ultracentrifuge. Prog Colloid Polym Sci 131:9–22
31. Strauss HM, Karabudak E, Bhattacharyya S, Kretzschmar A, Wohlleben W, Colfen H (2008) Performance of a fast fiber based UV/Vis multiwavelength detector for the analytical ultracentrifuge. Colloid Polym Sci 286:121–128
32. Cao W, Demeler B (2005) Modeling analytical ultracentrifugation experiments with an adaptive space-time finite element solution of the Lamm equation. Biophys J 89:1589–1602
33. Yphantis DA (1964) Equilibrium ultracentrifugation of dilute solutions. Biochemistry 3:297–317
34. van Holde KE, Weischet WO (1978) Boundary analysis of sedimentation velocity experiments with monodisperse and paucidisperse solutes. Biopolymers 17:1387–1403
35. Demeler B, van Holde KE (2004) Sedimentation velocity analysis of highly heterogeneous systems. Anal Biochem 335:279–288
36. Stafford WF 3rd (1992) Boundary analysis in sedimentation transport experiments: a procedure for obtaining sedimentation coefficient distributions using the time derivative of the concentration profile. Anal Biochem 203: 295–301
37. Brookes E, Cao W, Demeler B (2009) A two-dimensional spectrum analysis for sedimentation velocity experiments of mixtures with heterogeneity in molecular weight and shape. Eur Biophys J 39:405–414
38. Brookes E, Demeler B (2006) Genetic algorithm optimization for obtaining accurate molecular weight distributions from sedimentation velocity experiments. Prog Colloid Polym Sci 131:33–40
39. Demeler B, Brookes E (2007) Monte Carlo analysis of sedimentation experiments. Prog Colloid Polym Sci 286:129–137

Chapter 16

Constitutively Active STAT5 Constructs

Lynn M. Heltemes-Harris and Michael A. Farrar

Abstract

The transcription factor *S*ignal *T*randucer and *A*ctivator of *T*ranscription 5 (STAT5) plays an important role in many biological processes. To study STAT5 biology, several different constructs have been designed that render STAT5 constitutively active. These constructs have now been used to generate animal models that allow for targeted expression of constitutively active STAT5 including a model where STAT5 is expressed in developing B and T cells. Herein we briefly describe the design of constitutively active STAT5 constructs and recent advances in their use.

Key words: Constitutively active, STAT5, Phosphoflow, $CD4^+$, Tregs,

1. Introduction

The STAT family of transcription factors consists of seven distinct members that have wide ranging biological activities (1). The transcription factor STAT5 exists in two isoforms called STAT5a and STAT5b. These two proteins exhibit 95% sequence identity and are encoded by two closely linked genes on murine chromosome 11 or human chromosome 17 (2–4). The STAT5a and STAT5b proteins consist of an N-terminal domain that is involved in promoting STAT5 dimerization, a DNA binding domain, an SH2 domain, a C-terminal regulatory tyrosine residue, and a C-terminal transactivation domain (Fig. 1). The N-terminal domain plays an important role in directing the generation of specific STAT dimers (5). The DNA binding domain interacts with a conserved DNA binding sequence consisting of TTCXXXGAA (6). The SH2 domain of STATs is what drives the initial interaction of STAT proteins with phosphorylated tyrosine residues in the cytoplasmic tails of cytokine receptors that activate specific STATs (7). This allows for cytokine receptor-associated kinases (most typically belonging to the Jak family) to phosphorylate either Tyr-694

Sandra E. Nicholson and Nicos A. Nicola (eds.), *JAK-STAT Signalling: Methods and Protocols*, Methods in Molecular Biology, vol. 967, DOI 10.1007/978-1-62703-242-1_16, © Springer Science+Business Media New York 2013

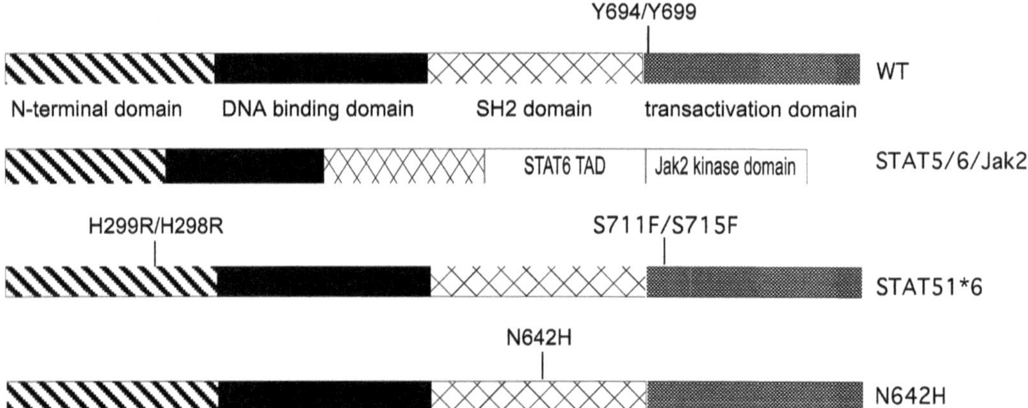

Fig. 1. Schematic outline of the various domains of STAT5. The *top* construct depicts wild-type (WT) STAT5 and includes an N-terminal domain involved in promoting STAT5 homodimerization, a DNA binding domain, an SH2 domain, and a C-terminal transactivation domain. The relative positions of Tyr-694 (STAT5a) or Tyr-699 (STAT5b), the residue which is phosphorylated and promotes the formation of a distinct STAT5 homodimer and hence activation, is shown. Schematic descriptions of the three distinct constitutively active forms of STAT5 that have been developed so far are also shown. The respective names of the constructs are listed on the right. The STAT5/6/Jak2 construct consists of the first 750 amino acids of ovine STAT5 fused to the transactivation domain from STAT6 and the kinase domain from Jak2. The relative positions of point mutations in the STAT51*6 and N642H constructs that promote constitutive STAT5 activation are indicated; the mutation listed first at each position refers to locations in STAT5a, while the second mutation refers to that location in STAT5b.

(STAT5a) or Tyr-699 (STAT5b). These phosphorylated tyrosine residues then interact with neighboring STAT5 proteins via reciprocal phospho-tyrosine SH2 interactions that result in the release of STAT5 from the activated receptor complex and the rearrangement of STAT5 as a head to tail homodimer. This new homodimeric form of STAT5 can then translocate to the nucleus where it binds to STAT5-regulated genes.

STAT5 proteins are relatively ubiquitously expressed and have been shown to play an important role downstream of a number of distinct receptors. STAT5-deficient mice exhibit a perinatal lethal phenotype with defects in multiple cell types including lymphocytes, erythrocytes, and hepatocytes (4). Mice expressing a hypomorphic form of STAT5 survive to birth on a mixed genetic background but exhibit defects in growth hormone responses, T cell proliferation and lactation (8, 9). The availability of mice expressing floxed alleles of the Stat5a/Stat5b locus has allowed studies to examine the effect of STAT5 loss-of-function mutations on these various developmental processes (4). Conversely, the development of constitutively active gain-of-function mutations for STAT5 has also been extremely useful for elucidating STAT5 function. Although this latter approach has been quite useful, there are a number of issues that must be considered when designing studies using constitutively active STAT5 constructs. In this review we describe the most commonly used constitutively active constructs and describe recent advances in their use.

2. Materials

2.1. Mice

1. *Stat5b-CA* (10).
2. *Foxp3-DTR* (11).

2.2. Reagents

1. 16% Paraformaldehyde.
2. 100% Methanol.
3. Complete RPMI medium: 10% fetal bovine serum (FBS), 2 mM L-glutamine, 5 IU penicillin, 50 μg/ml streptomycin, 55 μM beta-mercaptoethanol, 10 mM HEPES, 1 mM Nonessential amino acids.
4. FACS Buffer: PBS with 3% FBS, 0.02% Sodium Azide, 2 mM EDTA.
5. Anti-phospho-STAT5 (BD Biosciences #612567, clone 47).
6. Anti-mouse IgG1 (Isotype).
7. Phosphate-Buffered Saline (PBS).
8. Diptheria Toxin (Sigma).
9. MACS Buffer: PBS with 1% FBS, 2 mM EDTA.
10. $CD4^+$ Mouse T Cell Isolation Kit (Miltenyi).

3. Methods

3.1. Design of Constitutively Active STAT5 Constructs

To better understand the role of STAT5 in biological systems, multiple forms of constitutively active STAT5 have been created. These constructs are illustrated in Fig. 1. The first such construct involves the generation of a hybrid protein consisting of the first 750 amino acids of ovine STAT5 fused to the transactivation domain from STAT6 and the kinase domain from Jak2 (12). The second construct is one in which two point mutations are introduced into Stat5a; these include a histidine to arginine substitution at position 299 and a serine to phenylalanine substitution at position 711. STAT5b can be rendered constitutively active via similar substitutions (His298→Arg and Ser715→Phe). These constructs are sometimes referred to as STAT5a1*6 or STAT5b1*6 (13). Subsequent studies have found that the His→Arg mutation is not needed to render STAT5 constitutively active. In fact, constructs incorporating the His→Arg mutation are only functional when they can heterodimerize with a wild-type version of STAT5 (14). Finally, STAT5 can also be rendered constitutively active by mutation of Asp642 to histidine (15). Of these three constructs, the STAT5a1*6 and STAT5b1*6 versions are the ones that have been used the most frequently and which will be the focus of this review.

Several concerns need to be considered when using STAT5a1*6 and STAT5b1*6 constructs including how to deliver the constructs, the level of expression following delivery and the endogenous expression of STAT5 protein. Constitutively active STAT5a/b1*6 constructs have been delivered using standard retroviral protocols. In certain cell types though, expression of these constructs may pose problems. Hematopoietic cells transduced with constitutively active STAT5 constructs give rise to mixed multi-lineage leukemia following transfer back into a host animal (14, 16). This was overcome in one situation by using an inducible system to turn on STAT5 following transfer of the cells into the host (16). Another concern involves the level of expression and activation of STAT5 constructs. Studies designed to assess STAT5 function in a whole animal model provided important insight on how expression levels affect outcome. For example, we generated a series of transgenic mice expressing STAT5b1*6 throughout B and T cell development. Most of the founder lines expressed super-physiological levels of STAT5 and died 4–6 weeks following birth from massive expansion of progenitor B cells and $CD8^+$ T cells (10). In contrast, two lines expressed STAT5b1*6 at levels that were ~50% of that observed for the wild-type endogenous STAT5 proteins. This lower expression level resulted in a hyperactivatable form of STAT5b that exhibited modest constitutive phosphorylation in lymphocytes. Bissell and colleagues observed similar results when using constitutively active STAT5 constructs to examine STAT5 function in mammary epithelial cells (17). In this study, they found that the STAT5 construct required cytokine signaling for measurable activity, suggesting that the construct was acting as a hyperactive rather than truly constitutively active form of STAT5. The expression level of STAT5 can be examined using conventional western blotting approaches. For western blotting studies we have successfully used antibodies from Santa Cruz (clone C17, catalog number sc-835). The activation level of STAT5a/b1*6 constructs can be determined by examining the status of STAT5 phosphorylation either by western blotting or by flow cytometry.

When using constitutively active STAT5 constructs it is critical to know whether the construct is in fact constitutively active in your system of study. This is best assessed using intracellular flow cytometry to detect the phosphorylated and hence active form of STAT5. We have previously used a protocol developed by Teague and colleagues to track STAT5 phosphorylation (described in detail in ref. 18). More recently we have begun using a modified protocol from the Nolan laboratory to track STAT5 phosphorylation, which is described below. The protocol described involves ex vivo cytokine stimulation; if one wants to examine the phosphorylation status of cells freshly isolated from tissues the cytokine stimulation step can be omitted (i.e., start at step 2).

1. For IL-7 cytokine stimulation, place ~2–4 × 10^6 cells into 100 μl of complete RPMI medium. Make one tube for stimulation and one tube without stimulation. Add recombinant IL-7 at a concentration of 5–10 ng/ml to one tube. Incubate both tubes at 37°C for 30 min.
2. Add 900 μl of complete RPMI medium to bring tube to a final volume of 1 ml.
3. Terminate signaling and fix cells by adding 110 μl of 16% paraformaldehyde (fresh). Incubate at room temperature for 10 min (see Note 1).
4. Add 2 ml of room temperature PBS. Centrifuge at 450 × *g* for 5 min, aspirate, and remove residual volume.
5. Vortex cells (medium to low setting on vortex mixer) and permeablize by adding 1 ml of 4°C 100% methanol (see Note 2). You may also vortex, add methanol, and vortex after addition of methanol.
6. Incubate for 10 min at 4°C. At this point, cells can be used immediately or stored at −80°C.
7. Wash cells at least three times with large volumes of FACS buffer (see Note 3).
8. Resuspend cells in 100 μl FACS buffer containing antibodies for the surface markers you are interested in and either pSTAT5 or the isotype control. Stain for 30 min at room temperature in the dark.
9. Wash cells two times with FACS buffer and run on a flow cytometer.

To further assess the role of STAT5 in vivo, several groups including ours have generated animal models making use of the constitutively active STAT5 constructs. For example, we have generated mice expressing the STAT5b1*6 construct (referred to as *Stat5b-CA* in our papers) throughout B and T lymphocyte development (10). As mentioned previously, several founder lines expressed high levels of STAT5 and were not viable. In contrast, two founder lines expressed levels of STAT5b-CA that were ~50% that of endogenous STAT5a and STAT5b. Though there was no outwardly obvious phenotype, these mice did display perturbed lymphocyte development as they exhibited increased numbers of progenitor B cells, $CD8^+$ memory T cells, $CD4^+$ regulatory T cells, γδ T cells, and NK T cells. By crossing these mice to *il7r−/−* or *il2rβ−/−* mice, we were able to establish a key role for STAT5 in the development of progenitor B cells and Treg cells (19–21). These models provide a method for studying the physiological affects of STAT5 activation in a whole animal model.

3.2. Limiting STAT5 Activation to Appropriate Target Cells

Our STAT5b-CA mice have increased numbers of regulatory T cells (Tregs) in addition to increased numbers of progenitor B cells and CD8$^+$ memory-like T cells. The fact that multiple lymphocyte populations are perturbed in these mice can make it difficult to interpret results obtained in the *Stat5b-CA* mouse. One solution to this problem is to make mice expressing the STAT5b-CA construct in a more tissue restricted fashion. However, generating such mice that express appropriate levels of STAT5b-CA is not trivial and there is no guarantee of success. An alternative approach that makes use of existing Stat5b-CA transgenic mice was recently developed by Rowe and colleagues to specifically study the effect of increased STAT5 activation in Tregs (22). To accomplish this goal they made use of transgenic mice in which the diptheria toxin receptor was selectively expressed on Tregs (Foxp3-DTR mice) (11). Addition of diptheria toxin selectively ablated host Tregs. Purified CD4$^+$ T cells from Stat5b-CA mice were then transferred into these Foxp3-DTR mice. This resulted in expansion of the transferred Stat5b-CA CD4$^+$Foxp3$^+$ Treg cells until they reached a population size that was comparable to that observed in unmanipulated Stat5b-CA mice. In contrast, CD4$^+$Foxp3$^-$ T cells (non-Tregs) did not successfully engraft thereby leaving a host mouse in which the Stat5b-CA construct was selectively expressed in CD4$^+$Foxp3$^+$ regulatory T cells. A detailed protocol is provided below.

1. Remove spleen from STAT5b-CA mouse and create a single cell suspension in FACS Buffer.
2. Use the Miltenyi CD4$^+$ Purification Kit to isolate CD4$^+$ cells (~90% purity).
3. Resuspend cells in PBS for transfer into mice.
4. Inject 5×10^5 purified CD4$^+$ T cells per mouse intravenously.
5. Inject first dose of diphtheria toxin (intraperitoneal) at 50 μg/kg per mouse.
6. Each subsequent day inject 5 μg/kg of diphtheria toxin (intraperitoneal) per mouse.

This procedure results in full reconstitution of the CD4$^+$Foxp3$^+$ Treg compartment within 2 weeks. In this setting, the STAT5b-CA transgene is selectively expressed in Tregs, while the CD4$^+$Foxp3$^-$ effector population (and all other lymphocyte populations) are derived from wild-type cells. Thus, this method provides a novel mechanism to study the role of STAT5 activation in Tregs in a more physiological setting. This strategy could presumably be applied to other strains of STAT5 transgenic mice that have been developed by Ghysdael and colleagues (23). Similar strategies could potentially be applied to study the selective effect of STAT5 activation in other lymphocyte cell subsets.

4. Notes

1. A color change will take place due to pH alteration.
2. We add the methanol while vortexing (on low to medium setting) the cells. This is done by holding the tube with one hand and vortex while adding methanol in a drop-wise manner with a pipetman.
3. It is extremely important to take care to wash out all of the methanol. If concerned, you may do extra washes, as residual methanol will interfere with staining.

Acknowledgments

We wish to thank Sing Sing Way and Jared Rowe for helpful advice and comments. MAF is supported by a Leukemia & Lymphoma Society Scholar Award.

References

1. Leonard WJ, O'Shea JJ (1998) Jaks and STATs: biological implications. Annu Rev Immunol 16:293–322
2. Copeland NG, Gilbert DJ, Schindler C, Zhong Z, Wen Z, Darnell JE Jr, Mui AL, Miyajima A, Quelle FW, Ihle JN et al (1995) Distribution of the mammalian Stat gene family in mouse chromosomes. Genomics 29:225–228
3. Lin JX, Mietz J, Modi WS, John S, Leonard WJ (1996) Cloning of human Stat5B. Reconstitution of interleukin-2-induced Stat5A and Stat5B DNA binding activity in COS-7 cells. J Biol Chem 271:10738–10744
4. Cui Y, Riedlinger G, Miyoshi K, Tang W, Li C, Deng CX, Robinson GW, Hennighausen L (2004) Inactivation of Stat5 in mouse mammary epithelium during pregnancy reveals distinct functions in cell proliferation, survival, and differentiation. Mol Cell Biol 24:8037–8047
5. Ota N, Brett TJ, Murphy TL, Fremont DH, Murphy KM (2004) N-domain-dependent nonphosphorylated STAT4 dimers required for cytokine-driven activation. Nat Immunol 5:208–215
6. Darnell JE Jr (1997) STATs and gene regulation. Science 277:1630–1635
7. Greenlund AC, Farrar MA, Viviano BL, Schreiber RD (1994) Ligand-induced IFN gamma receptor tyrosine phosphorylation couples the receptor to its signal transduction system (p91). EMBO J 13:1591–1600
8. Moriggl R, Topham DJ, Teglund S, Sexl V, McKay C, Wang D, Hoffmeyer A, van Deursen J, Sangster MY, Bunting KD, Grosveld GC, Ihle JN (1999) Stat5 is required for IL-2-induced cell cycle progression of peripheral T cells. Immunity 10:249–259
9. Teglund S, McKay C, Schuetz E, van Deursen JM, Stravopodis D, Wang D, Brown M, Bodner S, Grosveld G, Ihle JN (1998) Stat5a and Stat5b proteins have essential and nonessential, or redundant, roles in cytokine responses. Cell 93:841–850
10. Burchill MA, Goetz CA, Prlic M, O'Neil JJ, Harmon IR, Bensinger SJ, Turka LA, Brennan P, Jameson SC, Farrar MA (2003) Distinct effects of STAT5 activation on CD4+ and CD8+ T cell homeostasis: development of CD4+CD25+ regulatory T cells versus CD8+ memory T cells. J Immunol 171:5853–5864
11. Kim JM, Rasmussen JP, Rudensky AY (2007) Regulatory T cells prevent catastrophic autoimmunity throughout the lifespan of mice. Nat Immunol 8:191–197
12. Berchtold S, Moriggl R, Gouilleux F, Silvennoinen O, Beisenherz C, Pfitzner E, Wissler M, Stocklin E, Groner B (1997) Cytokine receptor-independent, constitutively active variants of STAT5. J Biol Chem 272:30237–30243

13. Onishi M, Nosaka T, Misawa K, Mui AL, Gorman D, McMahon M, Miyajima A, Kitamura T (1998) Identification and characterization of a constitutively active STAT5 mutant that promotes cell proliferation. Mol Cell Biol 18:3871–3879
14. Moriggl R, Sexl V, Kenner L, Duntsch C, Stangl K, Gingras S, Hoffmeyer A, Bauer A, Piekorz R, Wang D, Bunting KD, Wagner EF, Sonneck K, Valent P, Ihle JN, Beug H (2005) Stat5 tetramer formation is associated with leukemogenesis. Cancer Cell 7: 87–99
15. Ariyoshi K, Nosaka T, Yamada K, Onishi M, Oka Y, Miyajima A, Kitamura T (2000) Constitutive activation of STAT5 by a point mutation in the SH2 domain. J Biol Chem 275:24407–24413
16. Antov A, Yang L, Vig M, Baltimore D, Van Parijs L (2003) Essential role for STAT5 signaling in CD25+CD4+ regulatory T cell homeostasis and the maintenance of self-tolerance. J Immunol 171:3435–3441
17. Xu R, Nelson CM, Muschler JL, Veiseh M, Vonderhaar BK, Bissell MJ (2009) Sustained activation of STAT5 is essential for chromatin remodeling and maintenance of mammary-specific function. J Cell Biol 184:57–66
18. Farrar MA (2010) Design and use of constitutively active STAT5 constructs. Methods Enzymol 485:583–596
19. Burchill MA, Yang J, Vogtenhuber C, Blazar BR, Farrar MA (2007) IL-2 receptor beta-dependent STAT5 activation is required for the development of Foxp3+ regulatory T cells. J Immunol 178:280–290
20. Goetz CA, Harmon IR, O'Neil JJ, Burchill MA, Farrar MA (2004) STAT5 activation underlies IL7 receptor-dependent B cell development. J Immunol 172:4770–4778
21. Burchill MA, Yang J, Vang KB, Moon JJ, Chu HH, Lio CW, Vegoe AL, Hsieh CS, Jenkins MK, Farrar MA (2008) Linked T cell receptor and cytokine signaling govern the development of the regulatory T cell repertoire. Immunity 28:112–121
22. Rowe JH, Ertelt JM, Aguilera MN, Farrar MA, Way SS (2011) Foxp3(+) regulatory T cell expansion required for sustaining pregnancy compromises host defense against prenatal bacterial pathogens. Cell Host Microbe 10:54–64
23. Joliot V, Cormier F, Medyouf H, Alcalde H, Ghysdael J (2006) Constitutive STAT5 activation specifically cooperates with the loss of p53 function in B-cell lymphomagenesis. Oncogene 25:4573–4584

Part III

Suppressor of Cytokine Signalling Proteins

Chapter 17

Analysis of Suppressor of Cytokine Signalling (SOCS) Gene Expression by Real-Time Quantitative PCR

Tatiana B. Kolesnik and Sandra E. Nicholson

Abstract

The Suppressor of Cytokine Signalling (SOCS) proteins are a family of negative regulators characterized by a central SH2 domain and C-terminal SOCS box motif. Cytokine Inducible SH2-containing protein (CIS), SOCS1, 2 and 3 are rapidly upregulated in response to cytokine stimulation and act to inhibit JAK/STAT signalling by a variety of mechanisms. The expression of SOCS proteins provides a level of specificity in the control of signalling, with SOCS proteins differentially upregulated in response to individual cytokines and in various cell-types. Real-time reverse transcription (RT) quantitative polymerase chain reaction (RT-qPCR) is an established technique for quantifying mRNA in biological samples, measuring the relative expression of genes of interest and identifying single nucleotide polymorphisms. Here we describe the use of SYBR® Green I RT-qPCR to quantify the relative expression level of *SOCS* mRNA in murine bone marrow-derived macrophages (BMDM). The approach can be universally applied to different cell types and various tissues.

Key words: RT-quantitative PCR, SOCS, Suppressor of cytokine signalling, Macrophages, Reference gene, Standard curve, Ct value, Relative expression

1. Introduction

The CIS/SOCS family of intracellular proteins are represented by eight members (CIS and SOCS1-7) (1, 2) and are involved in the negative regulation of Janus Kinase/Signal Transducer and Activator of Transcription (JAK/STAT) signalling pathways. They are expressed in response to multiple cytokines and growth factors, and have an important role in regulating many biological systems, including hemopoietic development and the innate and adaptive immune response, particularly in the context of bacterial and viral infections. Thus far, CIS, SOCS1, SOCS2, and SOCS3 are the

Sandra E. Nicholson and Nicos A. Nicola (eds.), *JAK-STAT Signalling: Methods and Protocols*, Methods in Molecular Biology, vol. 967, DOI 10.1007/978-1-62703-242-1_17, © Springer Science+Business Media New York 2013

best-characterized family members. STAT binding to specific elements in the promoter regions of *SOCS* genes results in a rapid increase in both mRNA and protein, with the SOCS proteins often acting to inhibit the initiating cytokine as part of a classic negative-feedback loop (1, 3). CIS, SOCS1, SOCS2, and SOCS3 inhibit signalling by binding to phosphorylated tyrosine residues within the receptor cytoplasmic domains to block further STAT recruitment (CIS/SOCS2) (4, 5), or alternatively bind directly to activated JAK kinases, inhibiting catalytic activity (SOCS1/3) (6–8). In addition, the SOCS proteins act as adaptors to bring the SOCS box-associated E3 ubiquitin ligase complex in contact with an SH2-bound substrate, targeting bound proteins for ubiquitination and degradation via the proteasome (9–11).

The expression level of SOCS proteins varies in different cells and tissues, with for instance, SOCS1 protein only detectable in the thymus under normal (non-stimulatory) conditions. If mice are injected with cytokines such as granulocyte colony stimulating factor (G-CSF), interferon (IFN) γ or interleukin (IL)-6, then expression of SOCS mRNA is easily detected in various tissues (1, 3). In BMDM, *SOCS2*, *SOCS4*, and *SOCS5* are expressed at a low to moderate level, whereas *CIS*, *SOCS1*, and *SOCS3* are undetectable in the absence of stimulation. Treatment of BMDM with various stimuli results in the rapid upregulation of SOCS mRNA, often within short timeframes (0.5–4 h). SOCS1 is induced in response to IFNγ, SOCS3 in response to IL-6 family members and Toll-like receptor (TLR) ligands such as lipopolysaccharide, whilst SOCS2 and CIS are upregulated in response to IL-4 (3, 12, 13). Regulation of the *SOCS* genes provides a biological context to understand their function and is of a great interest in different aspects of biomedical research including cancer, inflammation, autoimmunity and response to infection (14–16).

Following the discovery of PCR (1985) (17, 18) and then qPCR (1993) (19), the latest technique has evolved and at present combines PCR with many fluorescent detection systems to quantify the mRNA level (absolute or relative). In a basic PCR the amplified product is detected in the end of the cycling program, whereas in RT-qPCR, the DNA can be detected after each cycle and quantified by measuring the associated fluorescent signal. The intensity of fluorescence increases in proportion with the number of cycles. The results can be presented as an Amplification Plot where the Ct value defines the cycle number at which the amplified product reaches the threshold. The threshold is the point at which sample fluorescence can be detected above background fluorescence. The *Ct* value is related to the initial amount of DNA and inversely proportional to the expression level of the gene. If the *Ct* value is low, it means the fluorescence reaches the threshold early, and the amount of target in the sample is high. The *Ct* is therefore dependent on the threshold level and it is important to maintain a consistent threshold value for the same gene from one run to another (20).

In this chapter we describe RT-qPCR analysis of *SOCS* genes using the SYBR® Green I method; *SOCS1* expression in IL-4-stimulated BMDM is shown as an example. The following main aspects are outlined:

1. The principles of successful primer and amplicon design.
2. RNA purification and cDNA synthesis.
3. Preparation of the standard curve and sample analysis.
4. Standard curve quantification and normalization against a reference gene (RG).

2. Materials

Prepare all solutions using nuclease-free water. RNA purification, cDNA synthesis, and PCR setup can be performed at room temperature (unless indicated). Fine pipettes P2, P10 and a multichannel P10 should be used to reduce variation between samples. Use filter tips only.

Reagents and equipment:

1. 70% Ethanol (EtOH).
2. 14.3 M ß-mercaptoethanol (ß-ME).
3. RNA purification kit, includes RLT buffer and columns (see Note 1).
4. SYBR® Green I master mix (see Note 2).
5. Forward and Reverse primers (5 μM stock solutions).
6. Reverse transcriptase (RT) (50 U/μl) and buffer (see Note 3).
7. RNase-Free DNase (30 U/μl).
8. Ribonuclease inhibitor (RNasin) (40 U/μl).
9. dNTP mix (dATP, dGTP, dCTP, and dTTP) (10 mM each).
10. Oligo $(dT)_{14–20}$ (50 μM).
11. *Reaction Mix I*: 1 μl of Oligo $(dT)_{14\text{-}20}$ (50 μM) and 1 μl of dNTP mix.
12. *Reaction Mix II*: 4 μl 5× RT buffer, 2 μl of 0.1M DTT, 1 μl of RNAsin, and 1 μl of RT.
13. Standard amplicons at 0.01 pmol/4 μl (0.0025 μM in water).
14. Eight dilutions of the Standard amplicons from 10^{-2} pmol/4 μl to 10^{-9} pmol/4 μl.
15. 0.1 M Dithiothreitol (DTT).
16. Sterile 1.5 mL Eppendorf tubes.

17. Microcentrifuge (for 1.5 and 2 ml tubes) and macrocentrifuge (suitable for 384- or 96-well plates).
18. Optical adhesive film (transparent tape) for sealing the plates.
19. Plate stand.
20. Heating blocks set up with the temperature 50 °C, 65 °C, and 80 °C.
21. 384- and 96-well plates.
22. PCR thermocycler and corresponding software (see Note 4).
23. Disposable gloves.

3. Methods

3.1. Principles of Successful Primer and Amplicon Design

As SYBR® Green I binds to dsDNA and the method is highly sensitive, it is important to design primers which are specific for the gene of interest. The following general principles can be applied to any primer/amplicon design:

1. To avoid amplification of contaminating genomic DNA, primers should, where possible, be designed to cross intron–exon boundaries. If at least one primer can anneal across the intron–exon boundary, then only cDNA will be amplified (see Note 5). Alternatively, the forward primer can be designed to hybridize to one exon and the reverse primer to the second exon. The amplicon from cDNA will be smaller than that from genomic DNA and amplified much more efficiently. The greater the difference in size of the amplification product between cDNA and genomic DNA, the more accurate the results are likely to be.
2. Primer length should be between 17–25 bp.
3. GC content should be 50–60%.
4. The primer annealing temperature (Tm) should be 50–60 °C.
5. Avoid repeats and mismatches as well as complimentary sequence stretches within and between primers.
6. Avoid sequences that will result in the formation of hairpins or primer duplexes.
7. Verify the uniqueness of the primer sequence by blasting it against the NCBI nucleotide database (www.ncbi.nlm.nih.gov/BLAST/).
8. The amplicon should be less than 200 bp (80–150 bp is optimal). There is a balance between having a short amplicon with higher PCR efficiency, and a longer one which will incorporate more SYBR Green I and give greater detection sensitivity.

Table 1
SOCS primers

Gene	Primer sequence[a]	(Tm) (°C)
CIS	F: TCCTTTCTCCTTCCATCCCG R: TGCTCCACAGCCAGCAAAG	60 60
SOCS1	F: CTCGTCCTCGTCTTCGTCCT R: GAAGGTGCGGAAGTGAGTGT	60 60
SOCS2	F: GTTGCCGGAGGAACAGTC R: AACAGTCATACTTCCCCAGTACC	60 60
SOCS3	F: ACTGAGCCGACCTCTCTCCT R: GGCAGCTGGGTCACTTTCTC	65 65
SOCS4	F: GGCGGCAGGCGCTCGGACAG R: TCTGGGCACTTTCTGGACGG	54 54
SOCS5	F: GACGGCTTAGTATCGAAGAA R: GCTTATACAATGGGTTGACC	50 50
SOCS6	F: TACGGAGGAGCTGAAGAAGC R: AGAACCATCTGGCACATTTGC	60 60
SOCS7	F: TTCGATCACAGGGTATCACC R: ACTGACAGCGATCCTCAAAC	60 60
GAPDH[b]	F: TTGTCAAGCTCATTTCCTGGT R: TTACTCCTTGGAGGCCATGTA	54 54
PBDG[b]	F: CCTGGTTGTTCACTCCCTGA R: CAACAGCATCACAAGGGTTTT	60 60

[a]Primer combinations with the corresponding annealing temperatures (Tm) for different *SOCS* and
[b]Housekeeping genes utilized in RT-qPCR

9. The GC content of the amplicon should be 40–60%.
10. Avoid amplicons with predicted secondary structure. Secondary structure can be analyzed using the following program:
    (http://molbiol-tools.ca/Repeats_secondary_structure_Tm.htm).
11. Sequences of forward and reverse primers for the SOCS genes are provided in Table 1.

3.2. Total RNA Purification and cDNA Synthesis (Two Step RT-qPCR)

RNA quality, as assessed by purity and integrity, has a significant impact on RT-qPCR performance (12). Genomic DNA contamination can be an issue with any RNA purification method, as amplification from a trace amount of DNA can lead to misinterpretation of the results, especially if there is variation in the amount of genomic DNA between the samples (see Subheading 3.1 and Note 1). We use a two-step qPCR where the mRNA sample is reverse transcribed in a separate tube to the PCR reaction. This allows the expression of multiple genes to be analyzed from one reverse transcribed sample.

1. Purify RNA using the RNAeasy or RNAeasy Plus kits, according to the manufacturer's instructions, essentially as follows. Alternate methods for RNA purification can also be used (see Note 1).
2. For mouse cells, we recommend starting with 1×10^6 cells/sample; however, it can be adjusted depending on the cell type. Lyse cells in 350 μl of RLT lysis buffer containing 143 mM of ß-mercaptoethanol (1:100 dilution from stock) for 5 min at room temperature (see Note 6). Harvest lysates into Eppendorf tubes, add 300 μl of 70% EtOH, mix by pipetting and load onto the column. Alternatively, lysates can be stored in RLT buffer at −80 °C for at least 6 months.
3. Elute RNA from the column in 10 μl of nuclease-free water (depending on the expected yield/Ct value, the volume can be increased up to 50 μl) (see Note 7).
4. Determine the concentration of RNA by measuring the absorbance at 260 nm (A_{260}) in a spectrophotometer. Take 1–5 μg of total RNA for cDNA synthesis.
5. Prepare Reaction Mix I, multiplied by the number of samples (see Note 8), mix by vortexing and spin briefly, 30 s at ≥8,000 × *g*.
6. Prepare Reaction Mix II (see Note 9), multiplied by the number of samples, mix by vortexing and centrifuge for 30 s at ≥8,000 × *g*.
7. Add 2 μl of *Reaction Mix I* to 10 μl of total RNA, mix by pipetting and incubate at 65 °C for 5 min (see Note 11).
8. Transfer the samples to ice (+4 °C), cool down for 5 min and spin briefly at ≥8,000 × *g*.
9. Add 8 μl of *Reaction Mix II*, pipette, spin briefly and incubate at 50 °C for 40 min to 1 h.
10. Terminate the reaction by heating the samples at 80 °C for 5 min, chill on ice, and centrifuge briefly at ≥8,000 × *g*.
11. Dilute the cDNA if necessary by adding 80 μl of nuclease-free water (see Note 10).

3.3. Preparation of the Standard Curve and Sample Assessment

There are several ways to obtain the amplicon for the standard curve: (a) A DNA fragment based on the cDNA sequence amplified by your primers can be ordered from a commercial company; (b) The cDNA can be amplified from a tissue or cells in which your gene of interest is highly expressed; (c) You can use a recombinant DNA plasmid, which contains the cloned gene of interest. For (b) and (c), the DNA fragment should be gel purified and its concentration determined by measuring the absorbance at 260 nm (A_{260}). The sequences of standard amplicons for the SOCS genes are provided in Table 2.

Table 2
Standard amplicons

Gene	Amplicon sequence[a]	Size of Amplicon (bp)
CIS	tcctttctccttccatcccgccgaactccgactctcgagccgccgttgtctctgggacatg-gtcctttgcgtacagggatcttgtcctttgctggctgtggagca	105
SOCS1	ctcgtcctcgtcttcgtcctcgccagcggcccccgtgcgtccccggccctgcccggcg-gtcccagccccagcccctggcgacactcacttccgcaccttc	100
SOCS2	gttgccggaggaacagtcccccgaggcggcgcgtctggcgaaagccctgcgcgagct-cagtcaaacaggatggtactggggaagtatgactgtt	94
SOCS3	tgagcgtcaagacccagtcggggaccaagaacctacgcatccagtgtgag-gggggcagcttttcgctgcagagtgacccccgaagpcacgcagccagttccccgct-tcgactgtg	114
SOCS4	ggcggcaggcgctcggacagctccgcttgagctgagctcggagagatccgtcca-gaaagtgcccaga	67
SOCS5	gacggcttagtatcgaagaaggggtggatcccctcccaacgcacaaatacacaccttt-gaagctactgcacaggtcaacccattgtataagc	93
SOCS6	tacggaggagctgaagaagcttgcaaaacaggggtggtattggggccccatcacacgct-gggaggcagaggggaagttggcaaatgtgccagatggttct	100
SOCS7	ttcgatcacagggtatcacccatcacactagaatggagcactatagagggactttcagct-tatggtgccatcccaagtttgaggatcgctgtcagt	96
GAPDH	ttgtcaagctcatttcctggtatgacaatgaatacggctacagcaacagggtggtggac-ctcatggcctacatggcctccaaggagtaa	89
PBDG	cctggtcgttcactccctgaaggatgtgcctaccatactacctcctggctttaccattg-gagccatctgcaaacgggaaaacccttgtgatgctgttg	98

[a]Sequences of standard amplicons and corresponding sizes of amplified fragments for the *SOCS* and housekeeping genes used in RT-qPCR

1. Prepare a dilution series of each standard amplicon. We recommend a tenfold dilution range (see Note 12). Add 450 μl of nuclease-free water to seven Eppendorf tubes. Transfer 50 μl from the 0.0025 μM amplicon solution to the second tube to obtain a 0.00025 μM solution. Mix by pipetting/vortexing, spin briefly, change the tip and transfer 50 μl into another tube to obtain 0.001 pmol/4 μl etc.
2. For each gene, prepare a Master Mix by adding 0.5 μl each of Forward and Reverse primers to 5 μl of SYBR® Green I Master Mix. Multiply by the number of samples. Using a multichannel pipette, transfer 6 μl of the Master mix as required, to each well of a 384-well plate.
3. Add 4 μl of Standard amplicon or cDNA template (unknown sample) to each well on the plate (generally we run our standard and samples in technical triplicate and duplicates,

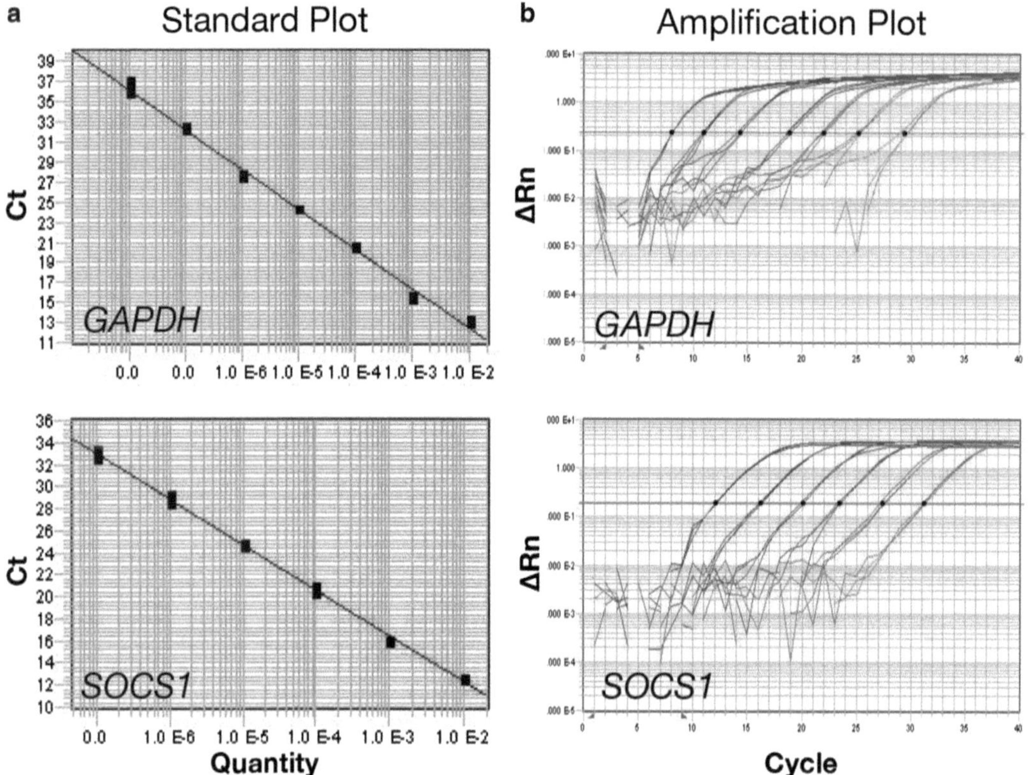

Fig. 1. Example standard curves (**a**) and amplification (**b**) plots for *GAPDH* and *SOCS1*. Plots were generated by tenfold serial dilution of the corresponding standard amplicons followed by analysis on an ABI 7900 thermocycler. The slope of the standard curve gives the efficiency of the PCR reaction (Efficiency = $10^{(-1/\text{slope})} - 1$). If the slope is −3.32 than PCR efficiency is 100% (24). In this instance the slope is −4.0 for both *GAPDH* and *SOCS1*. ΔRn is a measure of the fluorescence associated with SYBR green.

respectively; with three biological replicates for each sample) (see Note 13).

4. Seal the plate with transparent tape (optical adhesive film).
5. Spin the plate for 30 s at >8,000 × *g* at room temperature.
6. Run on a PCR thermocycler: (95°C for 10 min, 94°C for 15 s, Tm for 25 s, 72°C for 10 s) for 40 cycles.
7. Analyze data (Subheading 3.4).

3.4. Standard Curve Quantification and Normalization to a Reference Gene

The SYBR® Green I system coupled with a STD curve using a known amount of amplicon enables relative quantification (amount/cell number or total RNA). Once the PCR reaction is complete, the Ct value is converted to an amount of cDNA using the standard curve (for an example see Fig. 1 and Table 3).

Table 3
Example data analysis

Sample	Treatment	*RG*	Ct	Quantity	Average	*SOCS*	Ct	Quantity	Average	*SOCS1/ GAPDH*	Average	STDev
1a	Medium	*GAPDH*	18.15	6.17E-04		*SOCS-1*	30.27	3.51E-07				
1b	Medium	*GAPDH*	17.61	8.72E-04	7.45E-04	*SOCS-1*	29.55	5.41E-07	4.46E-07	5.99E-04		
2a	Medium	*GAPDH*	17.87	7.38E-04		*SOCS-1*	29.50	5.55E-07				
2b	Medium	*GAPDH*	17.82	7.63E-04	7.50E-04	*SOCS-1*	30.76	2.64E-07	4.09E-07	5.45E-04		
3a	Medium	*GAPDH*	17.79	7.75E-04		*SOCS-1*	29.66	5.06E-07				
3b	Medium	*GAPDH*	17.81	7.66E-04	7.71E-04	*SOCS-1*	30.63	2.85E-07	3.95E-07	5.13E-04	*5.52E-04*	*4.34E-05*
1a	IL-4	*GAPDH*	17.73	8.08E-04		*SOCS-1*	19.41	2.17E-04				
1b	IL-4	*GAPDH*	17.79	7.78E-04	7.93E-04	*SOCS-1*	19.83	1.70E-04	1.93E-04	2.44E-01		
2a	IL-4	*GAPDH*	18.28	5.65E-04		*SOCS-1*	20.05	1.49E-04				
2b	IL-4	*GAPDH*	17.89	7.27E-04	6.46E-04	*SOCS-1*	20.06	1.49E-04	1.49E-04	2.30E-01		
3a	IL-4	*GAPDH*	17.86	7.43E-04		*SOCS-1*	20.34	1.26E-04				
3b	IL-4	*GAPDH*	18.27	5.72E-04	6.57E-04	*SOCS-1*	20.49	1.15E-04	1.20E-04	1.83E-01	*2.19E-01*	*3.18E-02*

In order to correct for variation between samples, the amount of cDNA should be normalized to a housekeeping or reference gene (RG), the expression of which does not alter under the studied conditions (21–23). There is an extensive list of genes that have been used as RGs (21), but in each case the chosen gene should be tested for regulation under your experimental conditions.

We prefer glyceraldehyde 3-phosphate dehydrogenase (*GAPDH*) for the normalization of *SOCS* gene expression in BMDM. The amount of expressed *GAPDH* is easily detectable and unchanged with stimulation (Table 3 and Fig. 2.) (23). We have successfully used porphobilinogen deaminase (PBGD) in other systems, but the level of *PBGD* expression in unstimulated BMDM is very low (10^{-8} pmol). Here we use IL-4-induction of *SOCS1* (13), to provide an example of the raw data and subsequent analysis generated by RT-qPCR (Table 3).

1. Convert the Ct value to an actual amount using the standard curve.
2. Calculate the average amount obtained from the values for each technical replicate (duplicates/triplicate).
3. Normalize the average amount of the *SOCS* gene to the average amount of *GAPDH* for each sample to obtain the relative expression (see Note 14).
4. Take the average relative expression of the biological replicates (the second column from right, Table 2).
5. Calculate the Standard Deviation (STD) for each group of samples.
6. The relative expression level can be represented as in Fig. 2.

4. Notes

1. If primers have been designed based on the intron–exon boundaries, we recommend using the RNAeasy kit from QIAGEN (Valencia, CA). In the case of intronless genes such as *SOCS1*, *SOCS3*, or *SOCS5*, we would recommend using the RNAeasy Plus kit from QIAGEN. In the RNAeasy Plus system, which is designed for mammalian cells, RNA-free DNase is bound to the membrane in the shredding column, significantly decreasing contamination of genomic DNA in the RNA samples. Alternatively, DNAase treatment of the samples can be carried out.
2. We recommend using FastStart Universal SYBR Green Master mix (Rox) (Roche, Mannheim, Germany) due to high sensitivity, reproducibility and reliability (there is minimal variation between batches).

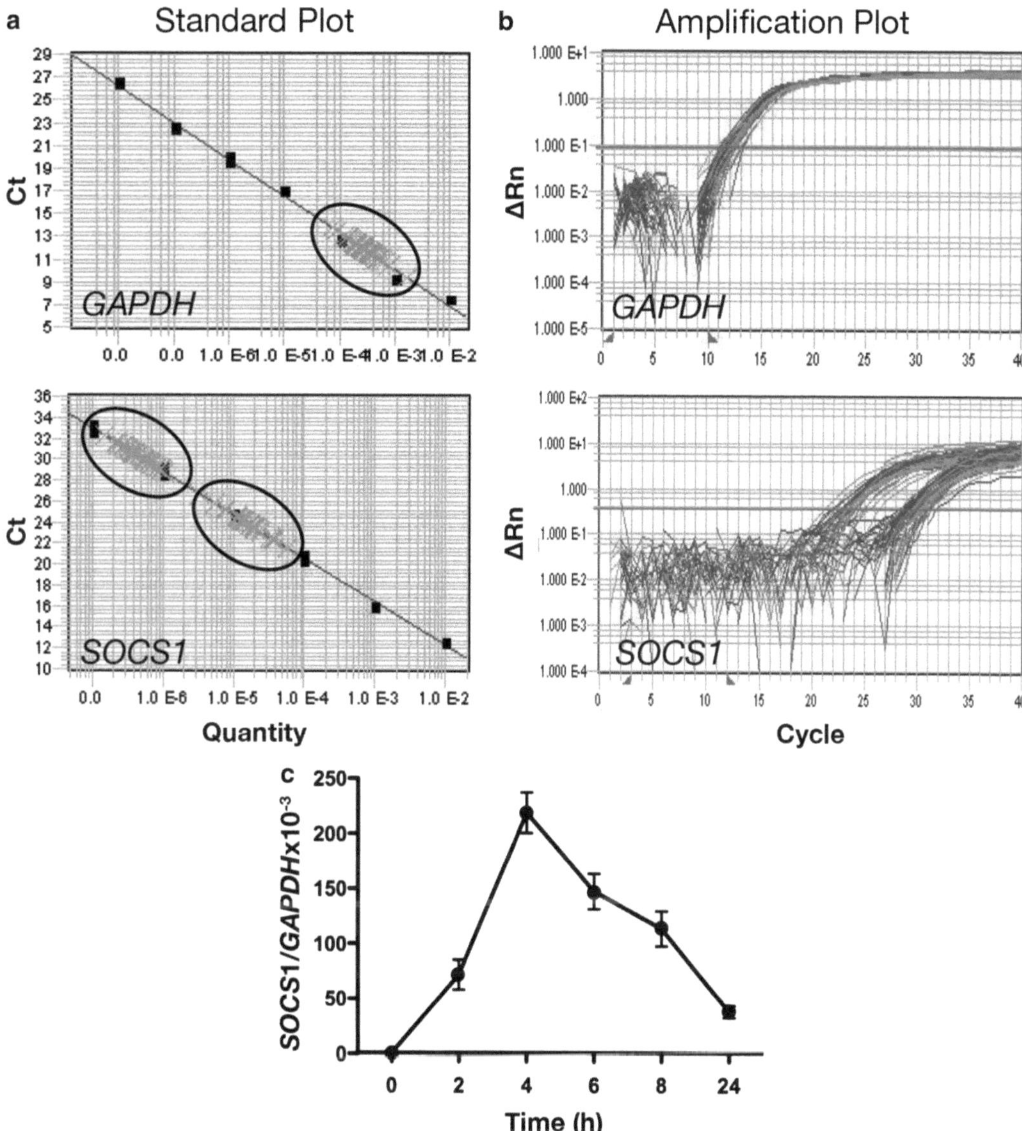

Fig. 2. Upregulation of *SOCS1* in IL-4-stimulated macrophages. 1.5×10^6 BMDM ($n = 3$) were stimulated with 10 ng/ml IL-4 for 0, 2, 4, 6, 8, and 24 h. Cells were lysed, RNA purified and reverse transcribed, and *GAPDH* and *SOCS1* expression measured by RT-qPCR. Standard curve (**a**) and Amplification (**b**) plots for *GAPDH* and *SOCS1* showing positioning of the IL-4-stimulated sample series. In (**a**), the *circles* highlight one group of GAPDH samples and two distinct groups of *SOCS1* samples (upper group with the higher Ct value reflects *SOCS1* expression in untreated cells, whilst the group with the lower Ct value reflects expression in IL-4-stimulated BMDM). In (**b**) the *horizontal line* indicates the threshold. (**c**) The relative expression level of *SOCS1* in IL-4-stimulated macrophages. The average amount of SOCS1 has been normalized to the average amount of GAPDH and is expressed as mean ± S.D.

3. We recommend the SuperScript® III First-Strand Synthesis System (Invitrogen, CA) for first strand DNA synthesis from poly A(+) RNA due to the longer life-time and increased thermostability of reverse transcriptase (up to 55 °C). As a result, it

enables high cDNA yield and full-length transcripts to be obtained from as little as 1 pg of starting material.

4. We use the ABI Prism 7900 PCR thermocycler and SDS.2 software to analyze our samples.
5. Considering that up to 20% of human genes are either single exon genes or have one or more pseudogenes (basically intronless) (12), it can be problematic to rely only on intron spanning primer design. If your gene of interest undergoes alternative splicing, this should also be taken into consideration when designing primers.
6. As qPCR is a very sensitive technique, it is important to have consistent amounts of RNA in each sample within a dataset. To obtain reproducible results, the same number of cells/volume of samples/organ weight should be taken. RLT is a lysis buffer containing guanidine thiocyanate; it is part of the RNAeasy Plus Mini kit or can be purchased separately from the same company. Add ß-mercaptoethanol to the lysis buffer before use. Dispense it in a fume hood wearing gloves.
7. We recommend converting RNA into cDNA immediately after purification to avoid RNA degradation upon freezing–thawing. Alternatively, RNA can be stored at –80°C.
8. To take into account variations in pipetting, prepare up to 10% excess volume for all master mixes.
9. We recommend including at least one extra sample without RT to determine the level (if any) of genomic DNA in the sample. A RG can be used as a positive control to confirm sample integrity.
10. The optimal dilution factor can vary and we recommend that it be optimized according to your source of RNA and Q-PCR settings.
11. The total PCR reaction volume is 10 μl, and it is therefore desirable to pipette no less than 2 μl to avoid significant variation.
12. All of the SOCS amplicons (Table 2) can be ordered from commercial sources. The standard curve should cover the complete range of expected expression. If the expression level is unknown, we recommend initially using a wide-ranging STD curve. Depending on the expression level of your gene of interest, and to increase accuracy, a twofold dilution can also be used. If the annealing temperature of the primer pairs for different amplicons is the same or similar (±2 °C), multiple standard curves can be run simultaneously.

13. Change the tip each time when collecting 4 μl, even from the same sample.
14. For simplicity, we divide one value by the other (SOCS÷RG) to obtain a relative amount. This allows for the expression of multiple genes to be directly compared in each sample and across experiments, as they are being normalized to the same housekeeping gene. However, if absolute quantification (copy number) is required, this needs to be approached slightly differently, see Ginzinger et al. (24), for more details.

Acknowledgments

This research was supported by the NHMRC Australia (Program Grant 461219 and fellowship to S.E.N.) and an NIH Grant (CA022556). It was made possible through Victorian State Government Operational Infrastructure Support and Australian Government NHMRC IRIISS.

References

1. Starr R, Willson TA, Viney EM et al (1997) A family of cytokine-inducible inhibitors of signalling. Nature 387:917–921
2. Hilton DJ, Richardson RT, Alexander WS et al (1998) Twenty proteins containing a C-terminal SOCS box form five structural classes. Proc Natl Acad Sci U S A 95:114–119
3. Wormald S, Zhang JG, Krebs DL et al (2006) The comparative roles of suppressor of cytokine signaling-1 and -3 in the inhibition and desensitization of cytokine signaling. J Biol Chem 281:11135–11143
4. Matsumoto A, Masuhara M, Mitsui K et al (1997) CIS, a cytokine inducible SH2 protein, is a target of the JAK-STAT5 pathway and modulates STAT5 activation. Blood 89:3148–3154
5. Greenhalgh CJ, Rico-Bautista E, Lorentzon M et al (2005) SOCS2 negatively regulates growth hormone action in vitro and in vivo. J Clin Invest 115:397–406
6. Yasukawa H, Misawa H, Sakamoto H et al (1999) The JAK-binding protein JAB inhibits Janus tyrosine kinase activity through binding in the activation loop. EMBO J 18: 1309–1320
7. Sasaki A, Yasukawa H, Suzuki A et al (1999) Cytokine-inducible SH2 protein-3 (CIS3/SOCS3) inhibits Janus tyrosine kinase by binding through the N-terminal kinase inhibitory region as well as SH2 domain. Genes Cells 4:339–351
8. Babon JJ, Kershaw NJ, Murphy JM, et al (2012) SOCS3 binds to a site unique to JAKs and inhibits their kinase activity via a novel, non-competitive mechanism. Immunity. In Press
9. Zhang JG, Farley A, Nicholson SE et al (1999) The conserved SOCS box motif in suppressors of cytokine signaling binds to elongins B and C and may couple bound proteins to proteasomal degradation. Proc Natl Acad Sci U S A 96:2071–2076
10. Kamura T, Sato S, Haque D et al (1998) The Elongin BC complex interacts with the conserved SOCS-box motif present in members of the SOCS, ras, WD-40 repeat, and ankyrin repeat families. Genes Dev 12:3872–3881
11. Babon JJ, Sabo JK, Zhang JG et al (2009) The SOCS box encodes a hierarchy of affinities for Cullin5: implications for ubiquitin ligase formation and cytokine signalling suppression. J Mol Biol 387:162–174
12. Yasukawa H, Ohishi M, Mori H et al (2003) IL-6 induces an anti-inflammatory response in

the absence of SOCS3 in macrophages. Nat Immunol 4:551–556

13. Dickensheets H, Vazquez N, Sheikh F et al (2007) Suppressor of cytokine signaling-1 is an IL-4-inducible gene in macrophages and feedback inhibits IL-4 signaling. Genes Immun 8:21–27
14. Yoshikawa H, Matsubara K, Qian GS et al (2001) SOCS-1, a negative regulator of the JAK/STAT pathway, is silenced by methylation in human hepatocellular carcinoma and shows growth-suppression activity. Nat Genet 28:29–35
15. Tamiya T, Kashiwagi I, Takahashi R et al (2011) Suppressors of cytokine signaling (SOCS) proteins and JAK/STAT pathways: regulation of T-cell inflammation by SOCS1 and SOCS3. Arterioscler Thromb Vasc Biol 31:980–985
16. Akhtar LN, Benveniste EN (2011) Viral exploitation of host SOCS protein functions. J Virol 85:1912–1921
17. Mullis K, Faloona F, Scharf S et al (1986) Specific enzymatic amplification of DNA in vitro: the polymerase chain reaction. Cold Spring Harb Symp Quant Biol 51(Pt 1):263–273
18. Saiki RK, Scharf S, Faloona F et al (1985) Enzymatic amplification of beta-globin genomic sequences and restriction site analysis for diagnosis of sickle cell anemia. Science 230:1350–1354
19. Higuchi R, Fockler C, Dollinger G et al (1993) Kinetic PCR analysis: real-time monitoring of DNA amplification reactions. Biotechnology (N Y) 11:1026–1030
20. Huggett J, Dheda K, Bustin S et al (2005) Real-time RT-PCR normalisation; strategies and considerations. Genes Immun 6:279–284
21. Radonic A, Thulke S, Mackay IM et al (2004) Guideline to reference gene selection for quantitative real-time PCR. Biochem Biophys Res Commun 313:856–862
22. Vandesompele J, De Preter K, Pattyn F et al (2002) Accurate normalization of real-time quantitative RT-PCR data by geometric averaging of multiple internal control genes. Genome Biol 3(7):RESEARCH0034
23. Cook NL, Vink R, Donkin JJ et al (2009) Validation of reference genes for normalization of real-time quantitative RT-PCR data in traumatic brain injury. J Neurosci Res 87:34–41
24. Ginzinger DG (2002) Gene quantification using real-time quantitative PCR: an emerging technology hits the mainstream. Exp Hematol 30(6):503–512.

Chapter 18

Detection of Endogenous SOCS1 and SOCS3 Proteins by Immunoprecipitation and Western Blot Analysis

Jian-Guo Zhang and Sandra E. Nicholson

Abstract

The suppressors of cytokine signalling (SOCS) protein family consist of eight members (SOCS 1–7, and CIS). SOCS1 and SOCS3 are the best-studied family members and have been shown to act as negative feedback inhibitors of the JAK/STAT signalling pathway. To study the physiological roles of the SOCS proteins, it is necessary to establish methods for detecting endogenous proteins often expressed at low levels in cells after cytokine induction. To facilitate the detection of endogenous SOCS1 and SOCS3 proteins, we have generated in-house antibodies specific to these proteins, which we have used together with commercially available antibodies. Here, we describe the methods for immunoprecipitating SOCS1 and SOCS3 proteins from mouse tissue extracts and their subsequent detection by Western blot analysis. These methods can also be applied to the detection of SOCS1 and SOCS3 in cell lines.

Key words: SOCS1, SOCS3, Immunoprecipitation, Western blot

1. Introduction

Cytokine signalling is initiated by binding of cytokines to their specific cognate cell surface receptors and leads to activation of the JAK/STAT pathway and subsequent gene transcription and translation, including the expression of the SOCS proteins (1–5). SOCS1 and SOCS3 are rapidly induced in response to cytokines and other stimuli such as TLR ligands, and form part of a classic negative feedback loop, acting to inhibit those cytokines that induce them. In addition, the induced SOCS proteins can inhibit other signalling pathways (cross-talk), thereby regulating opposing cytokines. Genetic deletion of either SOCS1 or SOCS3 results in catastrophic inflammation due to excessive IFNγ, IL-4 (SOCS1), G-CSF, and IL-6 signalling (SOCS3) (6–9). In general, SOCS1 and SOCS3 cannot be detected without the appropriate stimulation, with SOCS1 being primarily induced in response to type I

Sandra E. Nicholson and Nicos A. Nicola (eds.), *JAK-STAT Signalling: Methods and Protocols*, Methods in Molecular Biology, vol. 967, DOI 10.1007/978-1-62703-242-1_18, © Springer Science+Business Media New York 2013

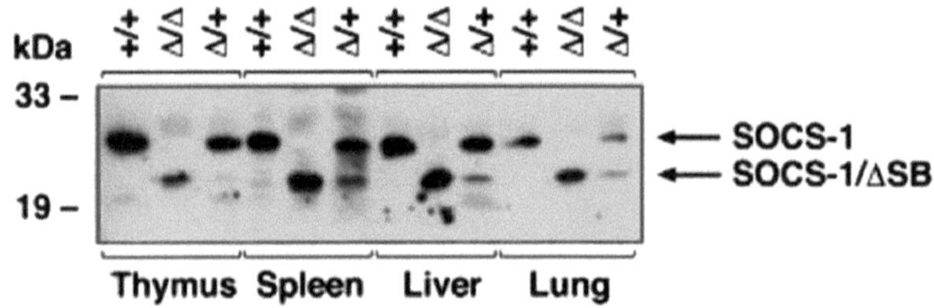

Fig. 1. Detection of SOCS1 protein expression in mouse tissue by immunoprecipitation and Western blot analysis. Cell lysates were prepared from thymus, spleen, liver, and lung tissue of mice injected with 5 μg of IFNγ for 4 h. SOCS1 proteins were immunoprecipitated with anti-SOCS1 monoclonal antibody 2E1/2D4 and then Western blotted with a biotinylated anti-SOCS1 antibody 4H1. Δ/Δ refers to mice that lack the SOCS1 SOCS box (Figure was reproduced from ref. 21).

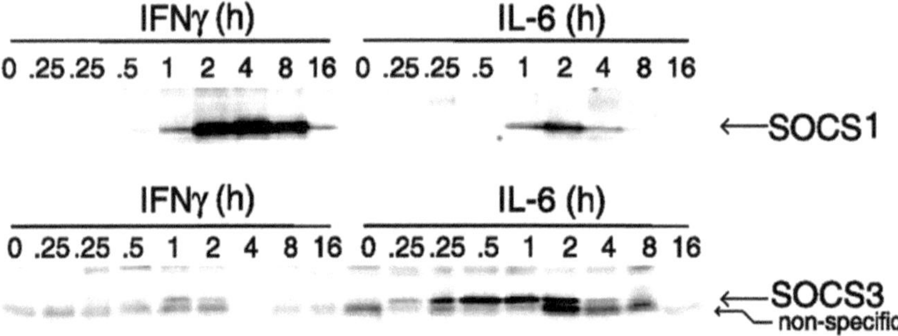

Fig. 2. Kinetics of SOCS1 and SOCS3 protein expression in response to IFNγ and IL-6 in the liver. Cell lysates were prepared from livers of mice injected with either IFNγ or IL-6 over a 16 h time course and immunoprecipitated with anti-SOCS1 and SOCS3 antibodies and Western blot analysis was then performed (Figure was reproduced from ref. 10).

and type II interferons (10–12) (Figs. 1 and 2), IL-4 (13) and cytokines that signal through the IL-2 common gamma chain (14), whilst SOCS3 is up-regulated by those cytokines that signal through gp130 (10, 15, 16) and by lipopolysaccharide (LPS) (17), among others (Fig. 2). SOCS3 in particular, is transiently expressed with a short half-life, due in part to the presence of a PEST motif inserted within the SH2 domain (11).

The detailed mechanisms of how the SOCS1 and SOCS3 proteins inhibit the JAK/STAT pathway still remain unclear. Because of recruitment of the Elongin BC complex through the conserved C-terminal SOCS box (18, 19), it has generally been accepted that SOCS1 and SOCS3 act as adapters for an E3 ubiquitin ligase complex and inhibit the JAK/STAT pathway by targeting the activated receptors and JAKs for polyubiquitination and subsequent proteasome-mediated degradation (3–5, 20). Genetic deletion of the SOCS1 or SOCS3 SOCS box results in an ameliorated phenotype, indicating that the E3 ligase activity is only one part of the inhibitory mechanism (15, 21). The SOCS-SH2 domain recruits the SOCS protein to the activated receptor complex where SOCS1 and SOCS3 interact directly with JAK to inhibit further enzymatic activity.

Inhibition requires the short region adjacent to the SH2 domain, which is known as the "kinase inhibitory region" or "KIR" (22, 23). We now know that SOCS3 interacts in a noncompetitive manner with a "GQM" motif located within the atypical insertion loop of the JAK1, JAK2 and TYK2, but not JAK3, kinase domains (24).

The ability to detect the endogenous SOCS proteins has been hampered by the relatively low level of protein expressed and the difficulty in generating antibodies of sufficient affinity to detect proteins in cell lysates. Here we describe methods that we have used for the detection of endogenous SOCS1 and SOCS3 by immunoprecipitation and Western blot.

2. Materials

2.1. Reagents

1. In-house mouse monoclonal antibodies against SOCS1: clones 2D4, 2E1, and 4H1. In-house mouse monoclonal antibody against SOCS3: clone 1B2 (see Note 1). All four monoclonal antibodies are IgG_1. Upon request, 2D4 and 2E1 can be obtained from the authors through a Materials Transfer Agreement. Due to commercial agreements, 4H1 and 1B2 and their biotinylated forms need to be purchased from MBL International or Millipore.
2. In-house rabbit anti-SOCS3 polyclonal antiserum (see Note 1).
3. Purified rabbit anti-SOCS3 polyclonal antibody (IBL; Immuno-Biological Laboratories, Cat. No. 18391 and 18395).
4. Streptavidin-HRP (Millipore).
5. Donkey anti-rabbit-immunoglobulin (Ig) conjugated to horse-radish peroxidase (HRP) (GE Healthcare).
6. Sheep anti-mouse-Ig conjugated to HRP (GE Healthcare).
7. Interferon (IFN)-γ and interleukin (IL)-6 (see Note 2).
8. Streptavidin conjugated to HRP.
9. 2× reducing SDS-PAGE sample loading buffer: 2.5% (w/v) SDS, 25% glycerol, 125 mM Tris–HCl, pH 6.8, 0.01% (w/v) bromophenol blue, 100 mM dithiothreitol (DTT).
10. Tween-20.
11. Triton X-100.
12. Skim milk.
13. Bovine serum albumin.
14. Phosphate-buffered saline (PBS): 20 mM phosphate, pH 7.4, 150 mM NaCl.
15. Complete protease inhibitor cocktail tablets (Roche Applied Bioscience) (see Note 3).

16. Lysis buffer: 1% (v/v) Triton X-100, 50 mM Tris–HCl, pH 7.4, 150 mM NaCl, 1 mM EDTA. Before use, this buffer is further supplemented with 2 mM Na_3VO_4 (see Note 4), 10 mM NaF, 1 mM phenylmethylsulphonyl fluoride (PMSF), and complete protease inhibitor (one tablet per 50 mL).
17. Protein A-Sepharose beads (GE Healthcare).
18. 1× Tris–glycine sodium dodecyl sulfate (SDS) electrophoresis running buffer: 25 mM Tris base, 192 mM glycine, 0.1% (w/v) SDS, pH 8.3.
19. 1× NuPAGE 2-(*N*-morpholino)ethanesulfonic acid (MES) SDS electrophoresis running buffer (for Novex Bis-Tris gels only): 50 mM MES, 50 mM Tris base, 0.1% (w/v) SDS, 1 mM EDTA, pH 7.3.
20. 1× Western transfer buffer: 20 mM Tris base, 150 mM glycine, 20% (v/v) methanol.
21. Precision plus protein standards (Bio-Rad).
22. 13% SDS polyacrylamide gels or 4–12% NuPAGE Novex precast gels.
23. Methanol.
24. Polyvinylidene difluoride (PVDF) membrane (Millipore or In Vitro Technologies).
25. Antibody diluent buffer: PBS, 1% (w/v) BSA, 0.1% (v/v) Tween-20.
26. Wash buffer: PBS, 0.1% (v/v) Tween-20.
27. Amersham enhanced chemiluminescent (ECL) Hyperfilm (GE Healthcare).
28. ECL substrate reagents (GE Healthcare and Millipore).
29. EZ-link Sulfo-NHS-biotin (Pierce, Cat. No. 21217).
30. NAP-5 column (GE Healthcare).
31. Column buffer: PBS, 0.02% (v/v) Tween-20, 0.02% (w/v) NaN_3.
32. Borate buffer: 1.5 M borate buffer, pH 8.6 (dissolve boric acid in deionized water and adjust to pH 8.6 with NaOH).
33. 1% (v/v) Tween-20 in deionized water.

2.2. Equipment

1. 1 and 7 mL dounce homogenizers (Wheaton).
2. Microfuge.
3. Microfuge tubes (1.5 mL).
4. Rotating wheel or other suitable device that permits end-over-end mixing.
5. Gel electrophoresis and Western transfer apparatus.
6. Electrophoresis power supply.

7. Ratek roller mixer (Ratek Instruments; BTR5-12 V).
8. Roller tubes (Sarsted) (tubes: Cat. No. 58.537, lids: Cat. No. 65.790).
9. Kodak X-OMAT film developer or suitable detection system.

3. Methods

3.1. Preparation of Mouse Liver, Lung, Spleen, and Thymus for SOCS1 and SOCS3 Immunoprecipitation

1. Adult mice weighing an average of 25 g are given a single intraperitoneal injection of either 100 μL of normal saline or 5–10 μg recombinant murine IFNγ, IL-6 or an appropriate cytokine in 100 μL of normal saline (see Note 5).
2. Mice are sacrificed by cervical dislocation or carbon dioxide gas at appropriate time intervals following injection. Tissues are collected and frozen immediately in liquid nitrogen.
3. All tissues are stored at –70°C before use (see Note 6).
4. Alternatively, ex vivo primary cells or cell lines can be incubated with the appropriate stimulus, lysed on ice for 30 min and steps 2–22 of Subheading 3.2 followed (see Note 7).

3.2. Cell Lysis/IP/ Western Blot

1. Lyse mouse spleen, thymus, liver, or lung tissue in 1 mL of ice-cold lysis buffer by homogenization using a dounce homogenizer (30 strokes) in an ice bath and leave the total lysates on ice for a further 30 min (see Note 6).
2. Spin the cell lysates at 14,000 × *g* for 15 min at 4°C.
3. Transfer the clarified supernatant (S/N) to new microfuge tubes and keep on ice.
4. Add 6 μL of anti-SOCS1 mAb 2D4 or 2E1 at 0.5 mg/mL (3 μg antibody per IP sample) or 3 μL rabbit anti-SOCS3 polyclonal antiserum into each S/N (see Subheading 2 and Notes 1 and 5) and incubate the mixtures for 30 min on ice.
5. Add 40 μL of 50% Protein A-Sepharose gel slurry to each IP sample (see Note 8) and incubate IP samples at 4°C on a rotating wheel for 2 h.
6. Spin down the beads at 10,000 × *g* for 1 min at room temperature (RT) and aspirate off the S/N.
7. Wash the beads with 3 × 1 mL of lysis buffer and repeat step 6 after each wash.
8. Add 40 μL of 2× reducing SDS sample loading buffer to the washed beads.
9. Mix the beads in SDS sample buffer and heat at 95°C for 3 min.

10. Spin down the beads as in step 6 (mix gently before spin) and recover eluates.
11. Load the eluates onto 13% SDS polyacrylamide gels or 4–12% Novex precast gels.
12. Run the SDS-PAGE initially at 80 V for 30 min and increase to 125 V until bromophenol blue dye front reaches about 1 cm from the bottom of the gel. It normally takes ~2–3 h.
13. Set up the Western transfer according to standard methods and transfer the proteins from the gel to the PVDF membrane at 100 V for 1–2 h at 4°C.
14. Roll the membrane into a Sarsted roller tube so that it is not overlapping and the protein side faces inward (see Note 9).
15. Add 5–10 mL of blocking solution (5–10% skim milk in PBS) to the membrane and incubate overnight at 4°C on a roller mixer.
16. Tip off the blocking solution and rinse the membrane three times by filling up the tube with PBS (~35 mL each rinse) to remove the blocking solution.
17. Incubate the membrane with 3 mL of 0.5 μg/mL biotinylated 4H1 for SOCS1 or 1B2 for SOCS3 (see Subheading 3.3 and Note 10) in antibody diluent buffer for 2 h at room temperature (RT).
18. Tip off the probing antibody solution and wash the membrane by filling up the tube with wash buffer for a total of ~1 h with at least six changes.
19. Incubate the membrane with 3–10 mL of streptavidin-HRP for biotinylated 4H1 or for 1B2, anti-mouse Ig-HRP both diluted at 1:10,000 in antibody diluent buffer and incubate for 1 h at RT (see Note 10).
20. Tip off the streptavidin-HRP or the secondary antibody solution and wash the membrane by filling up the tube with wash buffer for a total of ~2 h with at least six changes.
21. Set up the ECL reaction for 2 min at RT as per manufacturers' instructions.
22. Visualize the protein bands by exposing the membrane to Amersham ECL Hyperfilm for various times and developing the film using the Kodak X-OMAT film developer.

3.3. Biotinylation of Anti-SOCS1 Antibody

1. Add 8 μL of 1% (v/v) Tween-20 and 27 μL of borate buffer to 400 μL of 1 mg/mL 4H1 in PBS (the final concentration of borate buffer is ~0.1 M). Make sure the antibody is in a buffer that doesn't contain primary amine such as Tris. HEPES, PBS, and borate buffer are compatible with the biotinylation reaction. Also be aware that sometimes the biotinylation can affect the antibody activity.

2. Add 10 μL of 2.9 mg/mL EZ-link Sulfo-NHS-biotin (freshly prepared in deionized water). Mix and incubate on ice for 100 min.
3. To separate the biotinylated 4H1 from free biotinylation reagent, load ~445 μL of the biotinylation reaction mixture onto a NAP-5 column which has been previously equilibrated in column buffer.
4. Wash with 300 μL of column buffer.
5. Elute biotinylated 4H1 with 600 μL of column buffer. The biotinylated 4H1 can then be mixed 1:1 with 100% glycerol (50% final) and stored at −20°C.

4. Notes

1. In-house mouse anti-SOCS1 monoclonal antibodies, clones 2E1, 2D4, and 4H1, were raised against the N-terminal domain of murine SOCS1 (residues 2-78) and purified from culture supernatants by Protein G-Sepharose. In-house mouse anti-SOCS3 monoclonal antibody, clone 1B2, was raised against the N-terminal domain of murine SOCS3 (residues 2-44) and purified by Protein G-Sepharose. The purified antibodies can be stored in 50% glycerol in PBS at 0.5 mg/mL at −20°C. In-house rabbit anti-SOCS3 polyclonal antiserum was raised against a recombinant full-length murine SOCS3 protein (His-tagged) and stored at −20°C (avoiding freeze-thawing).
2. Human IFNγ is not active on mouse cells and vice versa. Human IL-6 is active on both human and mouse cells, whereas murine IL-6 is active on mouse cells but not on human cells.
3. Alternatively, various individual protease inhibitors can be purchased and stored as stock solutions. We would suggest final concentrations of 0.5 mM PMSF, 1 μM leupeptin, and 2 mM aprotinin.
4. To activate the sodium orthovanadate, make up a 200 mM solution and adjust the pH to 10 using either NaOH or HCl, depending on the initial pH of the solution (the solution will turn yellow). Heat in a microwave (boil) until the solution is colorless, cool to room temperature, readjust the pH, and repeat until the solution remains colorless. Aliquot and store at −20°C.
5. In our experience, it is difficult to detect endogenous SOCS1 and SOCS3 proteins by Western blot analysis alone from mouse tissue extracts without performing the immunoprecipitation step. Even in the thymus, where SOCS1 protein expression is

constitutive (highest in the absence of injected cytokines), we still have to perform an IP first, followed by Western blot analysis in order to detect the SOCS1 protein. SOCS1 protein is also readily detected by a combination of IP/Western blot from lung, spleen and liver tissue extracts prepared from mice injected with a stimulus like IFNγ (Fig. 1) (21). The highest level of SOCS1 protein expression can be seen in the liver 2 h post IFNγ injection (Fig. 2) (10, 21). For SOCS3, the highest level of SOCS3 protein expression can be seen in the liver 0.5 h post IL-6 injection (Fig. 2) (7, 10).

6. Frozen liver and lung can be cut into small pieces before dounce homogenization (keep livers frozen at all time in dry-ice during this process). After collection, thymus and spleen can be added to ice-cold PBS and pushed through a sieve to obtain suspension cells for the preparation of cell lysates. The cells prepared this way are ready to be lysed directly in lysis buffer without the need of dounce homogenization.

7. For example, in bone marrow-derived macrophages SOCS1 and SOCS3 are rapidly up-regulated following IFNγ and LPS stimulation, respectively. We would always suggest starting with a time course as the kinetics of SOCS induction varies depending on cell type and stimulus. In general, SOCS1 and SOCS3 are up-regulated at the protein level within 2–4 h of chronic stimulation (although SOCS3 can be detected as soon as 15 min with LPS). While SOCS3 can sometimes be detected on straight lysates using the IBL rabbit anti-SOCS3 antibody (for instance in macrophages which have been incubated with LPS or embryonic stem cells incubated with LIF) in our experience it is often necessary to perform an IP to enrich for SOCS proteins prior to analysis by Western blot. It is also worth noting that the IBL rabbit anti-SOCS3 antibody (C204; Code No. 18391) detects a nonspecific band of ~30 kDa in many cell lysates (16).

8. It is recommended to use cut tips when pipetting Protein A-Sepharose slurry to obtain an equal volume of the beads and also to change the tips between two pipetting events.

9. To save on expensive antibodies, we recommend the use of roller tubes from Sarsted, as 2–3 mL of antibody solution is then sufficient for probing a mini blot membrane (~7 × 9 cm) in these tubes. All the washing and ECL steps are also performed in the tubes. The diluted antibodies in the antibody diluting solution containing BSA can be kept and frozen at –20°C and reused 2–3 times.

10. The molecular masses of murine SOCS1 and SOCS3 as calculated from their amino acid sequences are 23.715 kDa and 24.776 kDa, respectively. Due to their similarity in size to the

antibody light chain (~25 kDa), it is important not to use the same antibody (or antibody from the same species) for IP and Western blotting. Several approaches can also be taken during Western blotting to minimize the secondary antibody detection of the IP antibody light chain that co-migrates with SOCS1. We prefer to use the combination of biotinylated 4H1 and streptavidin-HRP to detect SOCS1 protein. Another approach is the use of a secondary antibody which only detects the heavy chain (~50 kDa), as has been reported by Starr and colleagues, who immunoprecipitated the SOCS1 proteins with a mixture of the 2D4 and 2E1 clones, followed by Western blot detection with 4H1 and HRP-conjugated anti-mouse Fc-specific secondary antibody (25). As for SOCS3, we normally use our in-house rabbit anti-SOCS3 antiserum to IP the SOCS3 protein from tissue extracts and then Western blot with the mouse anti-SOCS3 mAb 1B2 to detect the SOCS3 protein (Fig. 2) (7, 10). In our hands, it works equally well if using 1B2 as an IP antibody (3 μg per IP sample), followed by Western blot with either in-house rabbit anti-SOCS3 polyclonal antiserum diluted at 1:1,000 or purified rabbit anti-SOCS3 antibody from IBL, used as recommended by the manufacturer (15). Where possible we would include cells or tissues from SOCS null mice as a control for antibody specificity. However if these reagents were not available, it would be important to include an isotype-matched antibody as a negative control in the IP.

Acknowledgments

This work was supported in part by the Anti-Cancer Council of Victoria, Melbourne, Australia; Australian Medical Research and Development Corp., Melbourne, Australia; The National Health and Medical Research Council (NHMRC; grant numbers 257500 and 461219), Australia; The J.D. and L. Harris Trust; The National Institutes of Health, Bethesda, MD (Grant CA-22556) and the Australia Federal Government Cooperative Research Centres Program. This work was also made possible through Victorian State Government Operational Infrastructure Support and Australian Government NHMRC IRIISS. S.E.N. was supported by an NHMRC fellowship.

References

1. Heinrich PC, Behrmann I, Muller-Newen G, Schaper F, Graeve L (1998) Interleukin-6-type cytokine signalling through the gp130/Jak/STAT pathway. Biochem J 334:297–314
2. Baker SJ, Rane SG, Reddy EP (2007) Hematopoietic cytokine receptor signaling. Oncogene 26:6724–6737
3. Alexander WS, Hilton DJ (2004) The role of suppressors of cytokine signaling (SOCS) proteins in regulation of the immune response. Annu Rev Immunol 22:503–529
4. Yoshimura A, Naka T, Kubo M (2007) SOCS proteins, cytokine signalling and immune regulation. Nat Rev Immunol 7:454–465
5. Croker BA, Kiu H, Nicholson SE (2008) SOCS regulation of the JAK/STAT signalling pathway. Semin Cell Dev Biol 19(4): 414–422
6. Croker BA, Metcalf D, Robb L, Wei W, Mifsud S, DiRago L, Cluse LA, Sutherland KD, Hartley L, Williams E, Zhang JG, Hilton DJ, Nicola NA, Alexander WS, Roberts AW (2004) SOCS3 is a critical physiological negative regulator of G-CSF signaling and emergency granulopoiesis. Immunity 20:153–165
7. Croker BA, Krebs DL, Zhang JG, Wormald S, Willson TA, Stanley EG, Robb L, Greenhalgh CJ, Forster I, Clausen BE, Nicola NA, Metcalf D, Hilton DJ, Roberts AW, Alexander WS (2003) SOCS3 negatively regulates IL-6 signaling in vivo. Nat Immunol 4:540–545
8. Kimura A, Kinjyo I, Matsumura Y, Mori H, Mashima R, Harada M, Chien KR, Yasukawa H, Yoshimura A (2004) SOCS3 is a physiological negative regulator for granulopoiesis and granulocyte colony-stimulating factor receptor signaling. J Biol Chem 279:6905–6910
9. Starr R, Metcalf D, Elefanty AG, Brysha M, Willson TA, Nicola NA, Hilton DJ, Alexander WS (1998) Liver degeneration and lymphoid deficiencies in mice lacking suppressor of cytokine signaling-1. Proc Natl Acad Sci U S A 95:14395–14399
10. Wormald S, Zhang JG, Krebs DL, Mielke LA, Silver J, Alexander WS, Speed TP, Nicola NA, Hilton DJ (2006) The comparative roles of suppressor of cytokine signaling-1 and -3 in the inhibition and desensitization of cytokine signaling. J Biol Chem 281:11135–11143
11. Babon JJ, McManus E, Yao S, DeSouza DP, Willson TA, Nicola NA, Baca M, Nicholson SE, Norton RS (2006) Structure of mouse SOCS3 reveals the structural basis of the extended SH2 domain function and identifies an unstructured insertion that regulates in vivo stability. Mol Cell 22:205–216
12. Fenner JE, Starr R, Cornish AL, Zhang JG, Metcalf D, Schreiber RD, Sheehan K, Hilton DJ, Alexander WS, Hertzog PJ (2006) Suppressor of cytokine signaling 1 regulates the immune response to infection by a unique inhibition of type I interferon activity. Nat Immunol 7:33–39
13. Dickensheets H, Vazquez N, Sheikh F, Gingras S, Murray PJ, Ryan JJ, Donnelly RP (2007) Suppressor of cytokine signaling-1 is an IL-4-inducible gene in macrophages and feedback inhibits IL-4 signaling. Genes Immun 8:21–27
14. Cornish AL, Davey GM, Metcalf D, Purton JF, Corbin JE, Greenhalgh CJ, Darwiche R, Wu L, Nicola NA, Godfrey DI, Heath WR, Hilton DJ, Alexander WS, Starr R (2003) Suppressor of cytokine signaling-1 has IFN-gamma-independent actions in T cell homeostasis. J Immunol 170:878–886
15. Boyle K, Zhang JG, Nicholson SE, Trounson E, Babon JJ, McManus EJ, Nicola NA, Robb L (2009) Deletion of the SOCS box of suppressor of cytokine signaling 3 (SOCS3) in embryonic stem cells reveals SOCS box-dependent regulation of JAK but not STAT phosphorylation. Cell Signal 21:394–404
16. Yasukawa H, Ohishi M, Mori H, Murakami M, Chinen T, Aki D, Hanada T, Takeda K, Akira S, Hoshijima M, Hirano T, Chien KR, Yoshimura A (2003) IL-6 induces an anti-inflammatory response in the absence of SOCS3 in macrophages. Nat Immunol 4:551–556
17. Stoiber D, Kovarik P, Cohney S, Johnston JA, Steinlein P, Decker T (1999) Lipopolysaccharide induces in macrophages the synthesis of the suppressor of cytokine signaling 3 and suppresses signal transduction in response to the activating factor IFN-gamma. J Immunol 163:2640–2647
18. Zhang JG, Farley A, Nicholson SE, Willson TA, Zugaro LM, Simpson RJ, Moritz RL, Cary D, Richardson R, Hausmann G, Kile BJ, Kent SB, Alexander WS, Metcalf D, Hilton DJ, Nicola NA, Baca M (1999) The conserved SOCS box motif in suppressors of cytokine signaling binds to elongins B and C and may couple bound proteins to proteasomal degradation. Proc Natl Acad Sci U S A 96:2071–2076
19. Kamura T, Sato S, Haque D, Liu L, Kaelin WG Jr, Conaway RC, Conaway JW (1998) The Elongin BC complex interacts with the conserved SOCS-box motif present in members of the SOCS, ras, WD-40 repeat, and

ankyrin repeat families. Genes Dev 12: 3872–3881

20. Kile BT, Schulman BA, Alexander WS, Nicola NA, Martin HM, Hilton DJ (2002) The SOCS box: a tale of destruction and degradation. Trends Biochem Sci 27:235–241
21. Zhang JG, Metcalf D, Rakar S, Asimakis M, Greenhalgh CJ, Willson TA, Starr R, Nicholson SE, Carter W, Alexander WS, Hilton DJ, Nicola NA (2001) The SOCS box of suppressor of cytokine signaling-1 is important for inhibition of cytokine action in vivo. Proc Natl Acad Sci U S A 98:13261–13265
22. Yasukawa H, Misawa H, Sakamoto H, Masuhara M, Sasaki A, Wakioka T, Ohtsuka S, Imaizumi T, Matsuda T, Ihle JN, Yoshimura A (1999) The JAK-binding protein JAB inhibits Janus tyrosine kinase activity through binding in the activation loop. EMBO J 18: 1309–1320
23. Sasaki A, Yasukawa H, Suzuki A, Kamizono S, Syoda T, Kinjyo I, Sasaki M, Johnston JA, Yoshimura A (1999) Cytokine-inducible SH2 protein-3 (CIS3/SOCS3) inhibits Janus tyrosine kinase by binding through the N-terminal kinase inhibitory region as well as SH2 domain. Genes Cells 4:339–351
24. Babon JJ, Kershaw NJ, Murphy JM, Varghese LN, Laktyushin A, Young SN, Lucet SL, Norton RS, Nicola NA (2012) SOCS3 binds to a site unique to JAKs and inhibits their kinase activity via a novel, non-competitive mechanism. Immunity 36(2):239–250
25. Starr R, Fuchsberger M, Lau LS, Uldrich AP, Goradia A, Willson TA, Verhagen AM, Alexander WS, Smyth MJ (2009) SOCS-1 binding to tyrosine 441 of IFN-gamma receptor subunit 1 contributes to the attenuation of IFN-gamma signaling in vivo. J Immunol 183:4537–4544

Chapter 19

In Vitro Ubiquitination of Cytokine Signaling Components

Jeffrey J. Babon, Artem Laktyushin, and Nadia J. Kershaw

Abstract

The eight SOCS (Suppressor of Cytokine Signaling) proteins encoded in the human genome all contain a C-terminal domain, the SOCS box, that allows them to function as E3 ubiquitin ligases and thereby catalyze the ubiquitination of components of the JAK/STAT signaling pathway. This activity is key to their function as cytokine signaling inhibitors as, once ubiquitinated, signaling molecules are degraded by the proteasome which allows the cell to return to its basal (unstimulated) state. SOCS based E3s are a subset of the CRL (Cullin-Ring-Ligase) family of ubiquitin ligases with the SOCS protein acting as the substrate recruitment module and interacting specifically with Cullin5, the E3 scaffold. Included here are protocols for the expression and purification of SOCS-based E3 complexes and their use in in vitro ubiquitination assays to characterize potential substrates. We have currently verified two components of the JAK/STAT pathway as substrates for ubiquitination using this method.

Key words: JAK, SOCS, Ubiquitin, Ubiquitination, E3 ligase, Cullin, In vitro

1. Introduction

In humans, the SOCS family consists of eight members: SOCS1-7 and CIS (Cytokine Inducible SH2 domain containing protein) (1). The SOCS proteins were first identified on the basis of their ability to down-regulate JAK/STAT signaling (2–4). Although there are differences in the mechanism of signaling repression across the SOCS family, one method employed by all SOCS proteins is to promote the proteasome mediated degradation of signaling molecules by catalyzing their poly-ubiquitination (5). Determining which signaling molecules are true substrates for ubiquitination is not trivial and the use of in vitro ubiquitination systems, as described here, is an important method of verification.

Ubiquitination, an extremely common protein posttranslational modification, is a process whereby ubiquitin, a small 9 kDa protein, is covalently attached to a target protein (6). This attachment occurs

Sandra E. Nicholson and Nicos A. Nicola (eds.), *JAK-STAT Signalling: Methods and Protocols*, Methods in Molecular Biology, vol. 967, DOI 10.1007/978-1-62703-242-1_19, © Springer Science+Business Media New York 2013

by the formation of a peptide bond between the C-terminal carboxylate of ubiquitin and the ε-amine of a target lysine. The process is catalyzed by an enzymatic cascade consisting of an activating (E1), conjugating (E2) and ligating (E3) enzyme (7). In many cases, E3 ligases will catalyze the attachment of a second ubiquitin molecule to a lysine residue on the first ubiquitin. This process can continue and give rise to polyubiquitin chains covalently bound to substrates. The fate of a protein once ubiquitinated depends upon whether it is monoubiquitinated or polyubiquitinated and upon the topology of the polyubiquitin chain (8).

The Cullin-Ring-Ligase (CRL) family of ubiquitin ligases (9) is a class of E3 ligases that consist of multiple subunits all built around a Cullin scaffold (10). The C-terminal domain of this scaffold binds activated ubiquitin whilst the N-terminal domain binds substrate (9). Both of these binding events are mediated by further subunits. For example, the interaction between Cullin and the activated ubiquitin is mediated by a RING domain protein, either Rbx1 or Rbx2, whilst the interaction with substrate is mediated by an extremely diverse array of substrate recruiting proteins. The Cullin scaffold utilized by the SOCS family of proteins is Cullin5, and this is unique in its preference for Rbx2 as its associated RING domain protein (11, 12). Whilst other Cullin proteins associate with the RING protein Rbx1, it has been shown in organisms as diverse as humans and *Drosophila* that Rbx2 is the preferred subunit of Cullin5-based E3 ubiquitin ligases (11, 12). The interaction between the SOCS protein and Cullin5 is also dependent upon the adapter complex ElonginBC (13, 14). Being tethered to the scaffold of the E3 ligase by its C-terminal SOCS box, the SOCS protein is then able to present anything bound to its upstream domains, usually an SH2 domain, as a substrate for ubiquitination. Taken together, a SOCS based E3 ligase then consists of five subunits: SOCS-Cullin5-Rbx2-ElonginB-ElonginC.

Here we demonstrate how to reconstruct a full E1-E2-E3 cascade in vitro using purified, recombinant components. In particular we wished to construct a system that would allow direct visualization of results (i.e., via Coomassie staining of SDS-PAGE gels) and avoid the need for western blotting so that the ubiquitination state and stoichiometry of each component could be readily tracked. All components are expressed in *Escherichia coli* using relatively standard protocols but they mostly require co-expression with other subunits in order to be properly folded. For example, Cullin5 is co-expressed as two separate domains, in conjunction with Rbx2, which all associate inside the *E. coli* cell to form a functional protein, mirroring the procedure to produce Cullin1 in bacteria as first demonstrated by Zheng, Schulman, and colleagues (9). Likewise SOCS proteins are co-expressed with elonginBC which again associate inside the *E. coli* cell to form a ternary complex (13).

2. Materials

1. Molecular biology reagents require for cloning such as restriction enzymes, thermostable polymerase, T4 DNA ligase, equipment for agarose gel electrophoresis and bacterial growth, antibiotics (ampicillin, kanamycin, and chloramphenicol) and *E. coli* strain BL21(DE3).
2. Equipment and reagents for SDS-PAGE (sodium dodecyl-sulfate polyacrylamide gel electrophoresis).
3. Plasmids pGEX-4T and pACYC-DUET (Novagen).
4. Media for *E. coli* growth, we prefer superbroth: 3.2% tryptone, 2% yeast extract, 0.5% NaCl, 5 mM NaOH.
5. Ni-NTA resin (Qiagen) and glutathione-sepharose resin (GE Healthcare).
6. Lysis Buffer 1: Phosphate-buffered saline (PBS), 5 mM dithiothreitol (DTT).
7. Lysis Buffer 2: 20 mM Tris–HCl, pH 7.5, 150 mM NaCl, 2 mM DTT.
8. Buffer A: Lysis buffer 2 plus 10 mM imidazole–HCl, pH 7.5.
9. Buffer B: Lysis buffer 2 plus 30 mM imidazole–HCl, pH 7.5.
10. Buffer C: Lysis buffer 2 plus 500 mM imidazole–HCl, pH 7.5.
11. Thrombin protease (Roche).
12. ATP (Adenosine Triphosphate).
13. Phenylmethylsulfonyl fluoride (PMSF).
14. DNAse 1.
15. Hen egg-white lysozyme.
16. Bovine ubiquitin.
17. Ubiquitination buffer: 20 mM Tris–HCl, pH8.0 150 mM NaCl, 2 mM DTT, 2 mM ATP, 4 mM $MgCl_2$.
18. Human El enzyme (Biomol International).
19. Thrombin cleavage buffer: Tris-buffered saline pH 7.5, 5 mM DTT, 2 mM $CaCl_2$.
20. Superdex 200 26/60 Gel filtration column (GE Healthcare).
21. Superdex 75 26/60 Gel filtration column (GE Healthcare).
22. Fast-Performance Liquid Chromatography (FPLC) system.
23. 30 kDa MWCO centrifugal filters
24. SDS-PAGE loading buffer: 100 mM Tris–HCl, pH 6.8, 200 mM DTT, 4% (w/v) SDS, 0.2% (w/v) bromophenyl blue, 20% (v/v) glycerol.

25. 10× Ubiquitination assay buffer: 200 mM Tris–HCl, pH 8.0, 1 M NaCl, 1 mM DTT, 40 mM $MgCl_2$, 20 mM ATP) (see Note 15).

3. Methods

The methods described below outline: (1) Expression and purification of the five-component E3 ligase in *E. coli* cells, (2) Expression and purification of the E2 ubiquitin conjugating enzyme in *E. coli* cells and (3) Assay for substrate ubiquitination.

3.1. Expression and Purification of the SOCS-Cullin5-Rbx2-ElonginBC E3 Ubiquitin Ligase

This procedure expresses the E3 ligase in two halves. The first half consists of co-expressed SOCS with elonginBC whilst the second half consists of co-expressed Cullin5 (as two separate domains) and Rbx2 (see Note 1). Here we present methodology for the expression and purification of a SOCS3 based E3 ligase; however, the same procedure can be used to produce other SOCS based E3 ligases.

3.1.1. Expression and Purification of SOCS-ElonginBC

1. Clone SOCS3 into pGEX-4T using appropriate restriction enzyme sites and standard molecular biology protocols. Likewise clone elonginB and elonginC into the first and second multiple cloning sites (mcs) of pACYC-DUET respectively (see Note 2). The domain boundaries used are as follows:

SOCS3	22-225	(Genpept NP_031733, see Note 3)
ElonginB	1-118	(Genpept AAH56983)
ElonginC	17-112	(Genpept NP_080732)

   This will generate GST-SOCS3 and untagged elongins B and C.

2. Co-transform the two plasmids into *E. coli* strain BL21(DE3) using standard protocols (see Note 4). Desired transformants are selected by plating on an LB agar plate containing 100 μg/mL ampicillin and 35 μg/mL chloramphenicol.
3. Innoculate 10 mL superbroth containing 100 μg/mL ampicillin and 35 μg/mL chloramphenicol with a single colony overnight at 37°C.
4. Inoculate 2 L of superbroth containing 100 μg/mL ampicillin and 35 μg/mL chloramphenicol with the overnight culture. Grow cells to an OD_{600} of 0.9 at 37°C with 170 rpm shaking. Reduce temperature to 18°C and then induce expression with the addition of 1 mM isopropyl-*β*-D-thiogalactoside (IPTG) and continue induction overnight (see Note 5).

5. Harvest cells by centrifugation at 6,000 ×*g* and resuspend cell pellet in 50 mL Lysis buffer. Add 1 mM PMSF, 500 units DNAse 1, and 20 mg hen egg-white lysozyme. Allow cells to lyse cells for 30 min at 4°C with gentle rotation.
6. Sonicate cells using 6 × 10 s pulses with a microtip at a medium power level (40–60 W). Perform sonication in an ice–water bath to ensure pellet does not overheat.
7. Centrifuge cell lysate at 13,000 ×*g* for 30 min. Prepare a column containing 2 mL of packed glutathione-sepharose resin.
8. Pass the supernatant over the glutathione-sepharose column and allow to drain by gravity flow.
9. Wash column with 50 mL lysis buffer.
10. Resuspend glutathione-sepharose resin in 15 mL thrombin cleavage buffer. Add 0.5 mg of thrombin and allow protease digestion to occur overnight at 4°C with continuous rotation (see Note 6).
11. Collect supernatant, which contains SOCS/elonginBC, add 1 mM PMSF and store at 4°C (see Note 7).

3.1.2. Expression and Purification of Cullin5/Rbx2

1. Clone the N-terminal domain (NTD) of Cullin5 into pGEX-4 T using appropriate restriction enzyme sites and standard molecular biology protocols. Likewise clone Rbx2 and the C-terminal domain (CTD) of Cullin5 into the first and second multiple cloning sites of pACYC-DUET respectively (see Note 8). The domain boundaries used are as follows:

Cullin5 NTD	Residues 1-384	(Swissprot Q9D5V5)
Cullin5 CTD	Residues 385-780	(Swissprot Q9D5V5)
Rbx2	Residues 1-113	(Genpept NP_035409)

This will generate GST-Cullin5 NTD, untagged Cullin5 CTD and His6-tagged Rbx2

2. Repeat steps 2–6, Subheading 3.1.1 with the exception that Lysis buffer 2 is used (see Note 9).
3. Centrifuge cell lysate at 13,000 ×*g* for 30 min. Prepare a column containing 2 mL of packed Ni-NTA resin.
4. Pass the supernatant over the Ni-NTA column and allow to drain by gravity flow.
5. Wash column with 50 mL Buffer A.
6. Wash column with 10 mL Buffer B.

At this stage the Cullin5/Rbx2 complex is bound to the Ni-Sepharose, the next section will pass the SOCS/elonginBC purified in Subheading 3.1.1 over this resin to form the five-protein complex (see Note 10).

3.1.3. Purification of the Full E3 Ligase (Cullin5/Rbx2/SOCS/ElonginBC)

1. Pass the 15 mL purified SOCS3/elonginBC from Subheading 3.1.1, step 11 over the Ni-NTA resin (containing bound Cullin5/Rbx2) from Subheading 3.1.2, step 6 (see Note 11).
2. Wash resin with 10 mL Buffer B.
3. Elute complex with 10 mL Buffer C.
4. Pass the eluate over a 2 mL glutathione-sepharose column and allow to drain by gravity flow.
5. Wash column with 50 mL lysis buffer 2 (see Note 12).
6. Resuspend glutathione-sepharose resin in 15 mL thrombin cleavage buffer. Add 0.5 mg of thrombin and allow protease digestion to occur overnight at 4°C with continuous (but not vigorous) rotation.
7. Collect supernatant, which contains the full E3 ligase, concentrate to 2 mL, and load onto a Superdex 200 26/60 Gel filtration column equilibrated in Lysis buffer 2.
8. Analyze fractions by SDS-PAGE, pool desired fractions, concentrate to 5 mg/mL using 30 kDa MWCO centrifugal filters and snap-freeze in liquid nitrogen (see Note 13).

3.2. Expression and Purification of E2 Ubiquitin Conjugating Enzyme (UbcH5a) in E. coli

1. Clone the UbcH5a into pGEX-4T using appropriate restriction enzyme sites and standard molecular biology protocols. Use the full-length sequence:

UbcH5a	Residues 1-147	(Genpept AAH15997)

2. Repeat steps 2–10, Subheading 3.1.1 with the exception that chloramphenicol is not used.
3. Collect supernatant, which contains UbcH5a, concentrate to 2 mL, and load onto a Superdex 75 26/60 Gel filtration column equilibrated in Lysis buffer.
4. Analyze fractions by SDS-PAGE, pool desired fractions, concentrate to 5 mg/mL, and snap-freeze in liquid nitrogen (see Note 14).

3.3. In Vitro Ubiquitination Assay

The final composition of each ubiquitination reaction will be as follows:

20 mM Tris–HCl, pH = 8.0.
100 mM NaCl.
2 mM ATP.
4 mM $MgCl_2$.
0.1 mM DTT.
0.1 μM E1.
2.5 μM E2.
2.5 μM E3.

50 μM Ubiquitin.
1 mg/mL Substrate protein.

Reactions are usually performed in 20 μL for 2–30 min.

1. Prepare 10× ubiquitination assay buffer. See Note 15.
2. Prepare a 20 μL reaction containing 2 μL 10× ubiquitination buffer, 2.5 μM E2, 2.5 μM E3, 50 μM ubiquitin, 1 mg/mL substrate protein, make up the reaction volume to 20 μL using MilliQ water (see Note 16).
3. Pre-warm reaction mixture at 37°C.
4. Take a 10 μL aliquot and add 10 μL of SDS-PAGE loading buffer. This is the zero timepoint.
5. The reaction is initiated by adding 0.1 μL of 0.5 mg/mL E1 enzyme (see Note 17).
6. Allow reaction to proceed for the desired time (2–30 min). The reaction can be scaled up and aliquots taken at specific times to observe the temporal course of ubiquitination.
7. Stop reaction with the addition of SDS-PAGE loading buffer and boil for 5 min.
8. Analyze reaction by loading onto a 4–12% Bis-Tris SDS-polyacrylamide gel (see Note 18). Separate by electrophoresis and stain gel with Coomassie Brilliant Blue using standard protocols. Ubiquitination of substrate is visible as a ladder of higher molecular weight bands above the substrate that are approximately 10 kDa in separation (Fig. 1b). If the substrate is large this may be visible as a smear of stained protein. There should be a concomitant reduction in the amount of stained substrate, and ubiquitin. Note that Cullin5 itself is ubiquitinated in this procedure (see Note 19), which can complicate analysis, therefore a zero substrate reaction should be performed as a control to allow for this.

4. Notes

1. This procedure is necessary as most of these components are unstable when expressed in isolation.
2. We suggest using the NcoI site in the first mcs of pACYC-DUET to clone elonginB, as this avoids the His6 tag on that vector which is unnecessary when using this protocol.
3. Note that SOCS3 will only express in the soluble fraction when a large unstructured loop (the PEST motif) is removed (see ref. 13 for details and full sequence), otherwise it enters inclusion bodies. Other SOCS proteins, which do not contain a

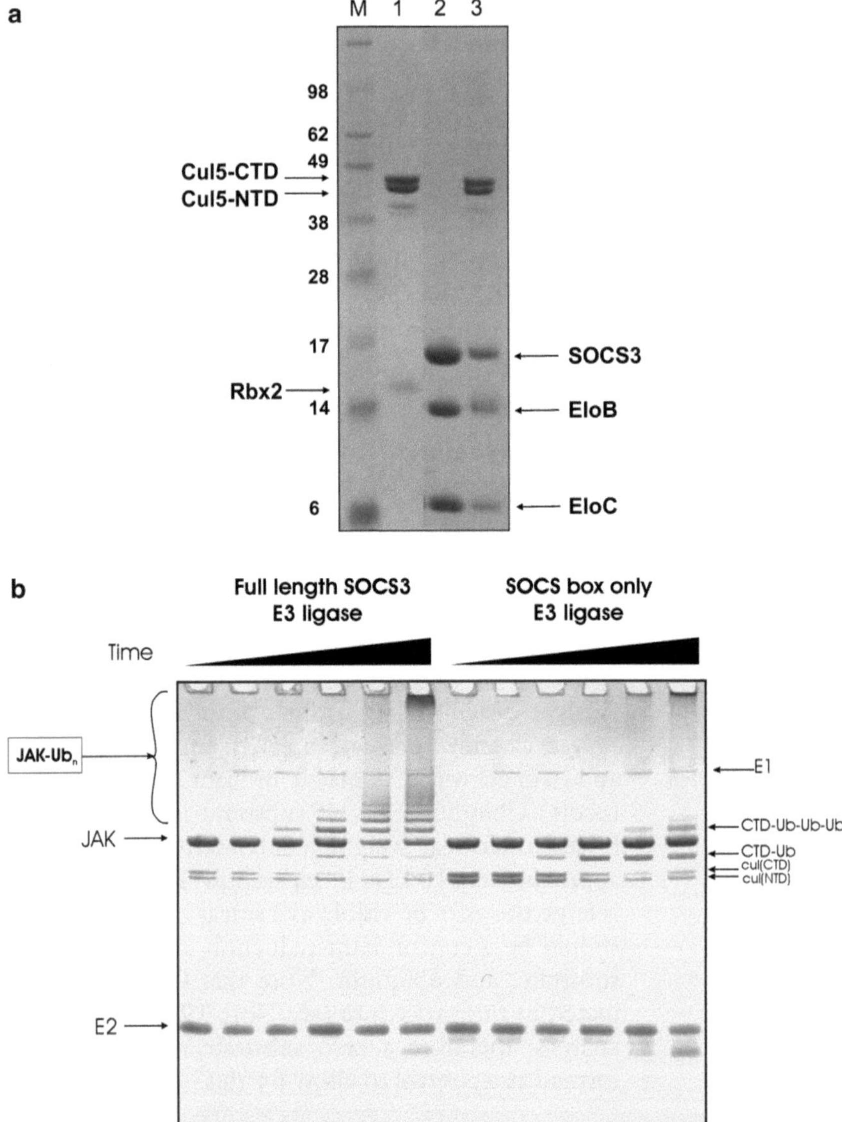

Fig. 1. (**a**) SDS-PAGE analysis of the purified E3 ligase components. *Lanes 1–2* are purified Cul5/Rbx2 and SOCS3/elonginBC respectively. The full E3 ligase is shown in *lane 3* after gel filtration purification. (**b**) A fragment of SOCS3 containing the KIR, SH2 domain, and SOCS box (residues 22-225) was able to induce the polyubiquitination of JAK2 *(lanes 1–6)* whereas the SOCS3 SOCS box alone *(lanes 7–12)* was not. The reactions were incubated for 0, 5, 10, 15, 60, and 120 min (left to right in each case). All results are visualized by Coomassie staining following SDS-PAGE. Protein components are indicated. Note that Cul5 itself is efficiently ubiquitinated on Lys724. *SOCS3, elonginB, and elonginC are present at concentrations too low to be observed by Coomassie staining in this experiment.

PEST motif in their SH2 domain, can be substituted for SOCS3 at this point. We have successful incorporated SOCS2, SOCS4 and SOCS5 into this system, using the same domain boundaries as SOCS3. Other SOCS box containing proteins, such as members of the ASB, WSB and SPSB families (1, 15–17) may also be used.

4. If using a heat-shock protocol, doubling the DNA/competent cell incubation time and the recovery time can be advantageous when co-transforming more than one plasmid.
5. SOCS3, even in complex with elonginBC, is insoluble when expressed at 37°C, therefore 18°C is necessary.
6. Do not allow rotation to occur too vigorously or else SOCS will begin to precipitate
7. SOCS3/BC begins to slowly precipitate over a few days. A high concentration of DTT can delay but not prevent this aggregation. At this point it can be concentrated to 5 mg/mL and snap-frozen in liquid nitrogen after the addition of glycerol to 10%. Stored at −80°C in this way the complex is stable indefinitely.
8. Don't use the Nco1 site in the first mcs of pACYC-DUET to clone Rbx2 as the His6 is necessary when using this protocol.
9. The use of high concentrations of reducing agent, such as DTT, is necessary to ensure that Cullin5 stays monomeric; however, the Ni-NTA resin can only cope with concentrations of DTT <2 mM. Therefore, it is suggested that all steps post imidazole elution are performed in 5 mM DTT.
10. Ideally, the further purification steps will be performed immediately as Cul5/Rbx2 is prone to disulphide-induced aggregation over time. However, it can be eluted from the beads and snap-frozen at this point. This is especially useful if multiple SOCS box containing E3 ligases are to be tested, as the purified Cul5/Rbx2 will associate with any purified SOCS/elonginBC complex in vitro to form a functional ligase.
11. It is important for the SOCS3/BC to be in excess at this stage to obtain the highest yield of full E3 ligase.
12. This step is required to remove excess Rbx2. This order of steps in the purification protocol has been developed to ensure that the final stoichiometry of the five-protein complex is correct (i.e., 1:1:1:1:1), even in the absence of gel filtration.
13. See Fig. 1a for what the full E3 ligase should look like on a Coomassie stained SDS-PAGE gel. It is important to only use those fractions that can be seen to have the correct stoichiometry.
14. UbcH5a expresses in very high yields and is extremely stable.
15. The ATP should be prepared fresh and not stored for more than 2 days unless frozen.
16. We have verified two substrates using this system: JAK2 and the gp130 receptor. Note the SOCS family mostly binds

substrates via their SH2 domain and hence many substrates, such as gp130, require phosphorylation before they can be ubiquitinated. A generic tyrosine kinase can be used to perform this. Whether other SOCS box containing protein families, i.e., ASBs and WSB, require posttranslational modification of their substrates in order to bind them is unclear.

17. El can be expressed in insect cells (18); however, as such small amounts are used per reaction we find it more cost-efficient to purchase it.
18. SDS-PAGE analysis should be performed under reducing conditions as lysine ubiquitination is insensitive to the presence of reducing agent.
19. Cullin5 is very efficiently ubiquitinated at Lysine724 (see Fig. 1b). Lysine724 is the site of Neddylation in vivo. In our hands, ubiquitination of this lysine has no effect, neither stimulatory nor inhibitory, on the ability of the E3 ligase to ubiquitinate substrate.

Acknowledgments

This work was made possible through Victorian State Government Operational Infrastructure Support and Australian Government NHMRC IRIISS. This research was supported by an NHMRC Program Grant (461219), NIH Grant (CA022556), NHMRC CDA (JJB, 516777), and NHMRC Project Grant (1011804).

References

1. Hilton DJ, Richardson RT, Alexander WS, Viney EM, Willson TA, Sprigg NS, Starr R, Nicholson SE, Metcalf D, Nicola NA (1998) Twenty proteins containing a C-terminal SOCS box form five structural classes. Proc Natl Acad Sci U S A 95:114–119
2. Starr R, Willson TA, Viney EM, Murray LJ, Rayner JR, Jenkins BJ, Gonda TJ, Alexander WS, Metcalf D, Nicola NA, Hilton DJ (1997) A family of cytokine-inducible inhibitors of signalling. Nature 387:917–921
3. Endo TA, Masuhara M, Yokouchi M, Suzuki R, Sakamoto H, Mitsui K, Matsumoto A, Tanimura S, Ohtsubo M, Misawa H, Miyazaki T, Leonor N, Taniguchi T, Fujita T, Kanakura Y, Komiya S, Yoshimura A (1997) A new protein containing an SH2 domain that inhibits JAK kinases. Nature 387:921–924
4. Naka T, Narazaki M, Hirata M, Matsumoto T, Minamoto S, Aono A, Nishimoto N, Kajita T, Taga T, Yoshizaki K, Akira S, Kishimoto T (1997) Structure and function of a new STAT-induced STAT inhibitor. Nature 387:924–929
5. Zhang JG, Farley A, Nicholson SE, Willson TA, Zugaro LM, Simpson RJ, Moritz RL, Cary D, Richardson R, Hausmann G, Kile BJ, Kent SB, Alexander WS, Metcalf D, Hilton DJ, Nicola NA, Baca M (1999) The conserved SOCS box motif in Suppressor of cytokine signaling binds to elongins B and C and may couple bound proteins to proteasomal degradation. Proc Natl Acad Sci U S A 96:2071–2076
6. Dye BT, Schulman BA (2007) Structural mechanisms underlying posttranslational modification by ubiquitin-like proteins. Annu Rev Biophys Biomol Struct 36:131–150

7. Hochstrasser M (2006) Lingering mysteries of ubiquitin-chain assembly. Cell 124:27–34
8. Pickart CM (2001) Mechanisms underlying ubiquitination. Annu Rev Biochem 70:503–533
9. Zheng N, Schulman BA, Song L, Miller JJ, Jeffrey PD, Wang P, Chu C, Koepp DM, Elledge SJ, Pagano M, Conaway RC, Conaway JW, Harper JW, Pavletich NP (2002) Structure of the Cul1-Rbx1-Skp1-F boxSkp2 SCF ubiquitin ligase complex. Nature 416:703–709
10. Petroski MD, Deshaies RJ (2005) Function and regulation of cullin-RING ubiquitin ligases. Nat Rev Mol Cell Biol 6:9–20
11. Kamura T, Maenaka K, Kotoshiba S, Matsumoto M, Kohda D, Conaway RC, Conaway JW, Nakayama KI (2004) VHL-box and SOCS-box domains determine binding specificity for Cul2-Rbx1 and Cul5-Rbx2 modules of ubiquitin ligases. Genes Dev 18:3055–3065
12. Reynolds PJ, Simms JR, Duronio RJ (2008) Identifying determinants of cullin binding specificity among the three functionally different *Drosophila melanogaster* Roc proteins via domain swapping. PLoS One 3:e2918
13. Babon JJ, Sabo JK, Soetopo A, Yao S, Bailey MF, Zhang JG, Nicola NA, Norton RS (2008) The SOCS box domain of SOCS3: structure and interaction with the elonginBC-cullin5 ubiquitin ligase. J Mol Biol 381:928–940
14. Babon JJ, Sabo JK, Zhang JG, Nicola NA, Norton RS (2009) The SOCS box encodes a hierarchy of affinities for Cullin5: implications for ubiquitin ligase formation and cytokine signalling suppression. J Mol Biol 387:162–174
15. Kile BT, Schulman BA, Alexander WS, Nicola NA, Martin HM, Hilton DJ (2002) The SOCS box: a tale of destruction and degradation. Trends Biochem Sci 27:235–241
16. Krebs DL, Hilton DJ (2001) SOCS proteins: negative regulators of cytokine signaling. Stem Cells 19:378–387
17. Hilton DJ (1999) Negative regulators of cytokine signal transduction. Cell Mol Life Sci 55:1568–1577
18. Li T, Pavletich NP, Schulman BA, Zheng N (2005) High-level expression and purification of recombinant SCF ubiquitin ligases. Methods Enzymol 398:125–142

Part IV

Structural Analysis of the JAK/STAT Pathway

Chapter 20

Production and Crystallization of Recombinant JAK Proteins

Isabelle S. Lucet and Rebecca Bamert

Abstract

JAK kinases are critical mediators in development, differentiation, and homeostasis and accordingly, have become well-validated targets for drug discovery efforts. In recent years, the integration of X-ray crystallography in kinase-focused drug discovery programs has provided a powerful rationale for chemical modification by allowing a unique glimpse of a bound inhibitor to its target. Such structural information has not only led to an improved understanding of the key drivers of potency and specificity of several JAK-specific compounds but has greatly facilitated and accelerated the design of compounds with improved pharmacokinetic properties.

JAK kinases are traditionally difficult candidates to express in significant quantities, generally requiring eukaryotic expression systems, protein engineering, mutations to yield soluble, homogeneous samples suitable for crystallization studies. Here we review the key methods utilized to express, purify, and crystallize the JAK kinases and provide a detail description of the methods that we have developed to express, purify, and crystallize recombinant JAK1 and JAK2 proteins in the presence of small molecule inhibitors.

Key words: JAK kinases, JAK kinase domain, Baculovirus expression, Protein purification, Phosphorylation, Crystallization

1. Introduction

The Janus Kinases (JAK) are a pivotal family of protein tyrosine kinases that play prominent roles in numerous cytokine signaling pathways with aberrant JAK activity associated with a variety of hematopoietic malignancies, cardiovascular diseases and immune-related disorders. Accordingly, the JAKs represent key cellular and molecular targets for drug discovery opportunities (1).

Members of the JAK family each share a unique domain structure, with a C-terminal protein tyrosine kinase (PTK) domain (known as the JAK homology-1 (JH1) domain) immediately adjacent to a

Sandra E. Nicholson and Nicos A. Nicola (eds.), *JAK-STAT Signalling: Methods and Protocols*, Methods in Molecular Biology, vol. 967, DOI 10.1007/978-1-62703-242-1_20, © Springer Science+Business Media New York 2013

regulatory domain (known as kinase-like domain or JH2), and five additional JAK homology domains (JH3-JH7) comprising the SH2-like domain and the FERM domain (2). The JH1 domain that contains the conserved ATP-binding site is critically important for JAK physiological functions and has represented the most obvious opportunity for JAK drug discovery and development.

To date, the crystal structures of the JH1 domain of all four family members (JAK1, JAK2, JAK3 and TYK2) have been elucidated and a total of 23 structures of the JH1 domain in complex with small molecule inhibitors are publically available (Table 1). The JAK JH1 kinase domain exhibits a typical kinase bilobal fold with a small N-terminal lobe incorporating a curled β-sheet of five strands and one α-helix (αC), and a larger C-terminal predominantly helical lobe with eight α-helices. The ATP-binding cleft resides between the two lobes, with a deep hydrophobic pocket able to accommodate the adenine base, behind a polar pocket that binds the ribose and triphosphate groups and coordinates magnesium. It is in this cleft that the ATP-competitive inhibitors bind.

JAK kinases, like other protein tyrosine kinases (PTKs), exist in either a catalytically inactive state or catalytically active state. These conformational states are largely governed by the phosphorylation of tandem conserved tyrosine residues within the activation loop that results in the expulsion of the activation loop from the active site. In addition, this conformational switch repositions the highly conserved Asp-Phe-Gly motif (DFG motif) in the proximity of the active site (DFG in), allowing a shift in the position of the αC helix. All 23 reported structures of JAK kinase domains depict the active conformation of the kinases. It is important to note that while a few of the reported TYK2/JAK3 structures are of the non-phosphorylated state, these structures still adopt an active conformation, with the activation loop and DFG motif configuration consistent with phospho-transfer activity (3, 4). As expected from the high level of sequence conservation between the four members of the JAK family, the structural similarity between the JAK kinase domains is very high, with only five points of difference in and around the ATP binding site, highlighting the difficulties faced in designing specific competitive inhibitors that discriminate between the JAK kinases. While there is still a compelling need to expand the JAK chemical space available for drug design through the determination of a structure of any of the JAK kinase domains in an inactive conformation (5), these publically available structures of JAK kinase domains have provided a crucial starting point for structure-based kinase inhibitor design and have enabled the discovery of small molecule inhibitors that discriminate sufficiently between members of the JAK family. Indeed, a number of JAK2 and JAK3-targeted inhibitors are currently in clinical trials for the treatment of a range of diseases caused by V617F-related myeloproliferative disorders (JAK2 inhibitors: e.g., TG101348, CYT387,

Table 1
Crystallization of JAK kinases with drug-like inhibitors

JAK kinases	Inhibitor	Construct used for expression and purification method	Phosphorylation status	Crystallization condition	Space group	Resolution (Å)	PDB ID	References
JAK2	CMP-6	Residues 835-1132/ GST Baculovirus GST resin/Thrombin cleavage Size exclusion S75	pYpY	28 % PEG 8000, 0.2 M Ammonium acetate, 0.1 M Sodium citrate, pH 6.0	$P4_1$	2	2B7A	(7)
JAK2	CP-690,550	Residues 835-1132/ His Baculovirus Nickel affinity/TEV cleavage Size exclusion S75	pYpY	18 % PEG 8000, 100 mM Cacodylate, pH 6.5/6.7, 100 mM Magnesium acetate, 100 mM KCl-4C	$P4_1$	2.4	3FUP	(13)
JAK2	5-bromo-1*H*-indazole-3-amine	Not disclosed	pYpY	28–32 % PEG 4000, 100 mM, 200 mM Ammonium acetate, Sodium citrate pH 6.5–6.8	$C222_1$	1.9	3E62	(23)
JAK2	5-phenyl-1*H*-indazole-3-amine	Not disclosed	pYpY	28–32 % PEG 4000, 100 mM, 200 mM Ammonium acetate, Sodium citrate pH 6.5–6.8	$C222_1$	1.9	3E63	(23)

(continued)

Table 1 (Continued)

JAK kinases	Inhibitor	Construct used for expression and purification method	Phosphorylation status	Crystallization condition	Space group	Resolution (Å)	PDB ID	References
JAK2	4-(3-amino-1*H*-indazol-5-yl)-*N*-*tert*-butylbenzenesulfonamide	Not disclosed	pYpY	28–32 % PEG 4000, 100 mM, 200 mM Ammonium acetate, Sodium citrate pH 6.5–6.8	$C222_1$	1.8	3E64	(23)
JAK2	(3*S*)-3-(4-hydroxyphenyl)-1,5-dihydro-1,5,12-triazabenzo[4,5] cycloocta[1,2,3-cd] inden-4(3*H*)-one	T842-G1132/His Baculovirus Nickel affinity γ-Phenyl ATP-Sepharose–ADP elution Thrombin cleavage Size exclusion Co-crystallization	pYpY	2.1–1.5 D-L malic acid, pH 7.0	$C222_1$	2.1	3JY9	(14)
JAK2	(3*S*)-1-[6-(2-aminopyrazolo[1,5-a] pyrimidin-3-yl)pyrimidin-4-yl]-*N*,*N*diethylpiperidine-3-carboxamide	As per ref. 14	pYpY	2.–1.5 D-L malic acid	$C222_1$	2.6	3I07	(20)
JAK2	3-(6-[[(1*S*)-1-(4-fluorophenyl)ethyl] amino]pyrimidin-4-yl) pyrazolo[1,5-A] pyrimidin-2-amine	As per ref. 14	pYpY	2.–1.5 D-L malic acid	$C222_1$	2.1	3I0K	(14)

JAK2	3-chloro-4-(4*H*-3,4,7-triazadibenzo[cd,f]azulen-6-yl)phenol	As per ref. 14	pYpY	2.1–1.5 D-L malic acid, pH 7.0	$C222_1$	2.2	3KCK	(14)
JAK2	NVP-BSK805	Not disclosed	pYpY	1.2 M Sodium citrate, 0.1 M Hepes, pH 7.5	$C222_1$	1.8	3KKR	(24)
JAK2	*N*-methyl-4-[3-(3,4,5-trimethoxyphenyl)quinoxalin-5-yl]benzenesulfonamide	Not disclosed	pYpY	30 % PEG4000, 0.1 M Ammonium sulfate, 0.1 M Sodium citrate, pH 6.2	$P4_1$	2.0	3LPB	(25)
JAK2	5-Chloro-*N2*-[(1*S*)-1-(5-Fluoropyrimidin-2-yl)ethyl]-*N4*-(5-methyl-1*H*-pyrazol-3-yl)pyrimidine-2,4-diamine	Residues 835-1132 with mutations K943A and K945A/His *Escherichia coli* BL21 cells Staurosporine added before Lysis Ni-NTA Lambda phosphatase/TEV protease Resource Q ion exchange Size exclusion S75 Crystallization with staurosporine Soaking with compounds of interest	pYpY	28 % PEG3350, 200 mM Ammonium acetate, 100 mM Sodium citrate, pH 6.0	$C222_1$	1.8	2XA4	

(continued)

Table 1 (Continued)

JAK kinases	Inhibitor	Construct used for expression and purification method	Phosphorylation status	Crystallization condition	Space group	Resolution (Å)	PDB ID	References
JAK2	2-(2,6-difluoro-4-methoxyphenyl)-1-(4-{4-[(3-methyl-1*H*-pyrazol-5-yl)amino] pyrrolo[2,1-f][1,2,4] triazin-2-yl}piperazin-1-yl)ethanone	Not disclosed	pYpY	35 % PEG3350, 0.1 M sodium chloride, 0.1 M MES, pH 6.5	$P4_1$	2.5	3Q32	(26)
JAK1	CP-690,550	Residues 841-1154	pYpY	0.1 M Hepes, pH 7.5, 0.1 M KSCN, 10 mM DTT, 25–27 % PEG 3350	$C222_1$	1.9	3EYG	(13)
JAK1	CMP-6	Residues 841-1154	pYpY	0.1 M Hepes, pH 7.5, 0.1 M KSCN, 10 mM DTT, 25–27 % PEG 3350	$C222_1$	2.0	3EYH	(13)
JAK3	CP-690,550	Residues 812–1124/ His non cleavable Baculovirus S21 cells Ni-NTA Desalt with HiPrep 26/10 column HiTrapTM Q-Sepharose HP column Inhibitor added Size exclusion 16/60 Superdex 200	YY	20 % PEG-3350, 0.2 M Ammonium sulfate, 0.02 % Phenylurea, pH 7.5	$P2_12_12_1$	2.0	3LXK	(3)

JAK3	CMP-6	Residues 812–1124/ His non cleavable Baculovirus S21 cells Ni-NTA Desalt with HiPrep 26/10 column HiTrapTM Q-Sepharose HP column Inhibitor added Size exclusion 16/60 Superdex 200	YY	20 % PEG-3350, 0.2 M Ammonium sulfate, 0.02 % phenylurea, pH 7.5	$P2_12_12_1$	1.75	3LXL	(3)
JAK3	AFN941	Residues 811-1124/ GST Baculovirus Hi5 cells Glutathione sepharose affinity resin Mono Q column Co-crystallization	pYpY	Malonate, glycerol, pipes, pH 6.0	$P2_12_12_1$	2.55	1YVJ	(12)
JAK3	3-(1*H*-indol-3-yl)-4-[2-(4-oxopiperidin-1-yl)-5-(trifluoromethyl) pyrimidin-4-yl]-1*H*-pyrrole-2,5-dione	Not disclosed		8 % PEG 4000, 0.05 M MES pH 6.5, 0.01 M $MgCl_2$	$P2_12_12_1$	2.2	3PJC	(27)
TYK2	CMP-6	Residues 888–1182/ His C936A, C1142A, Q969A, E971A, K972A Baculovirus S21 cells Ni-NTA, pH maintained to 8.2	YpY	10 % Ethanol, 0.1 M Tris, and 0.3 M Magnesium chloride, pH 8.0	$P2_12_12_1$	1.6	3LXP	(3)

(continued)

Table 1 (Continued)

JAK kinases	Inhibitor	Construct used for expression and purification method	Phosphorylation status	Crystallization condition	Space group	Resolution (Å)	PDB ID	References
		Desalt with HiPrep 26/10 column Size exclusion 16/60 Superdex 200 in presence of compound Co crystallization						
TYK2	CP-690,550	Residues 888–1182/ N-ter His C936A, C1142A, Q969A, E971A, K972A Baculovirus S21 cells Ni-NTA—pH of eluate adjusted to 5.5 Desalt with HiPrep 26/10 column Size exclusion 16/60 Superdex 200 in presence of compound Co-crystallization	YpY	10 % Ethanol, 0.1 M Tris, 0.3 M Magnesium chloride, pH 8.0	$P2_12_12_1$	2.5	3LXN	(3)

TYK2	CMP-6	Residues 885–1176/ C-ter His Asp1023Asn Baculovirus sf9 cells Ni-NTA Size exclusion 16/60 Superdex 75 Compound added at 1 mM Sepharose FF S column Flow through collected and concentrated	YY	16 % PEG 3350 0.1 M KSCN 4 % Hexafluoro-2-propanol	$P2_12_12_1$	2.2	3NZ0	(4)
TYK2	CMP-1	Same as above	YY	15 % PEG 3350 0.2 M Sodium acetate	$P2_12_12_1$	2.4	3NYX	(4)

AZD1480) and as immunosuppressants for organ transplants and rheumatoid arthritis (JAK3 inhibitors, e.g., CP690,550).

In this chapter we provide the reader with a summary of the current methods utilized to express, purify, and crystallize the JAK kinases and provide a detailed description of the methods that we have developed to express, purify, and crystallize JAK1 and JAK2 kinase domains in active conformations and in complex with small molecule inhibitors.

2. Overview of the Current Method Utilized to Express, Purify, and Crystallize the JAK Kinase Domains

2.1. Overall Structure of the JAK Kinase Domain

The JAK kinase domain (JH1 domain) is about 37 kDa and displays a typical kinase bilobal fold containing N-and C-terminal lobes connected by a hinge region that dictates the opening of the inter-lobe angle. An overview of the crystal structures of the JAKs is shown in Fig. 1. The N-terminal lobe comprises a curled β sheet of five antiparallel β-strands and one a-helix (αC). The C-terminal lobe is mainly α-helical with eight α-helices and three 3/10 helices, and three pairs of antiparallel β strands. While the JAK kinase domains appear to conform in most respects to the structures of other PTKs, a loop structure located in the C-term lobe, termed the JAK-specific insertion loop, JSI (6) or "Lip" (7), does not resemble a feature observed in any other kinase. This loop is a conserved feature of the JAK family and most likely plays an important role in the regulation of the kinase (6).

It is well established that kinases are conformationally flexible molecules. Most significant conformational differences between members of the JAK family are centered around the hinge region that dictates the inter-lobe opening angle, the glycine loop that participates in a network of interactions with ATP/ATP-competitive small molecule inhibitors, the activation loop that is regulated by phosphorylation events on tandem tyrosines and the JAK-specific insertion loop known to be important for regulation. A detailed comparison of the different structures can be found in the following references (3, 7, 8).

To limit the inherent flexibility of a kinase, the presence of small molecule inhibitors within the ATP binding cleft located between the two-lobes of the kinase has often proven necessary to stabilize the kinase domain and facilitate its crystallization. As such, all JAK kinase domain structures reported have been solved in the presence of small molecule inhibitors.

2.2. Protein Constructs and Expression System

The requirement of high protein concentrations for crystallization studies makes it essential to obtain pure homogeneous protein in sufficient amounts. In general, the JAK kinase domains (see Table 1

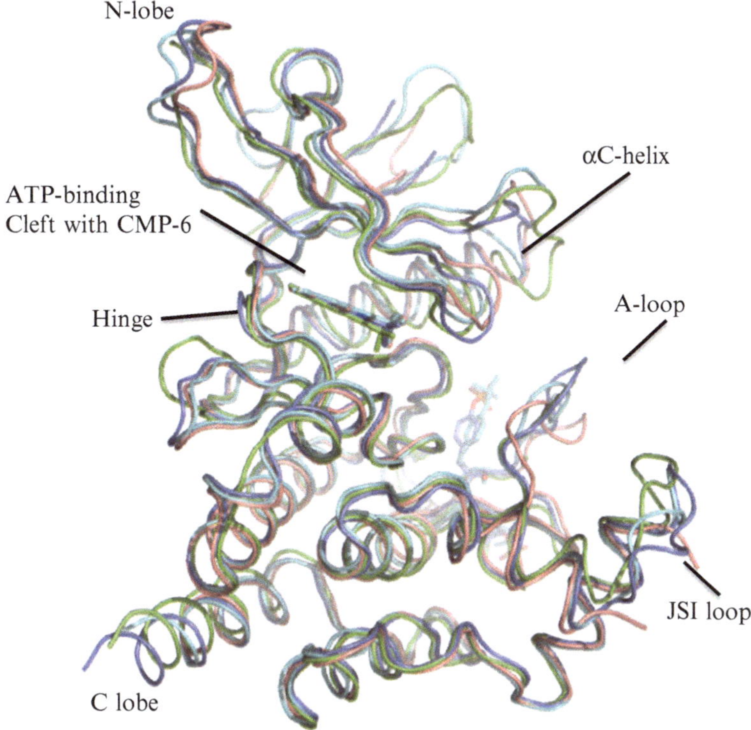

Fig. 1. Overview of the crystal structures of JAKs: structural alignment of JAK1 (3EYG), JAK2 (2B7A), JAK3 (3LXL), and TYK2 (3LXP) crystal structures solved in an active conformation and in the presence of CMP-6. The N-terminal lobe comprises a five-stranded antiparallel β-sheet and one α-helix (αC). The larger C-terminal lobe is predominantly helical with eight α-helices and three 3/10 helices, with one main pair of antiparallel β-strands (β7–β8). Locations of the hinge region, glycine loop, activation loop (A-loop) and JAK-specific insertion loop (JSI) are indicated. JAK1 is colored in *pink*, JAK2 in *blue*, JAK3 in *cyan*, and TYK2 in *green*.

for domain boundaries) have been expressed with an N-terminal TEV/thrombin cleavable His-tag to facilitate protein purification. The recombinant JAK proteins have been expressed in the baculovirus-insect cell expression system using either sf9 or sf21 cells. However, it is worth noting that Ioannidis et al. have recently reported the expression of JAK2 kinase domain in *Escherichia coli* cells (BL21) after introducing two surface mutations (K943A and K945A) (9). Hall et al. have also reported the introduction of two surface mutations on the JAK2 kinase (M1073S and F1076T) to improve solubility when expressed in insect cells, although during the course of our work we have not noticed any solubility issues. In the case of the TYK2 kinase domain (residues 888-1182), three surface mutations were introduced (Q969A, E971A, K972A) and two cysteines were mutated to alanines C936A and C1142A to facilitate crystallization (3, 10). In another study, an active-site mutation was generated (D1023N) (4). No surface mutations have been reported to facilitate the purification and crystallization of JAK1 and JAK3 kinase domains.

Two strategies have been employed to obtain sufficient amounts of purified JAK protein for crystallization studies: (1) Expression using large-scale production (>10 L up to 30 L) in BioWave reactor bags (3, 4, 11); (2) Expression in the presence of small molecule inhibitors. Indeed, we have shown in our laboratory that addition of an inhibitor at micromolar concentrations to the expression media can in some cases not only greatly increase the level of expression of a JAK kinase, but also influence its phosphorylation state, allowing the production of mostly one phosphorylation state rather than a mixture of phosphorylation states (*see* Subheading 2.3 below). This method will be further explored later in this chapter.

2.3. Purification Methods for the Different Phosphorylation States of the JAK Kinases

Obtaining a pure homogeneous protein in terms of phosphorylation state is critical for crystallization of the JAK kinases (as per most of the kinases). It is well known that phosphorylation can result in multiple conformational states of the kinases and that conformational heterogeneity largely hinders crystal growth. The separation of the different phosphorylation states is usually achieved with the use of ion-exchange chromatography. Thus far, all reported JAK1 and JAK2 structures are of the doubly phosphorylated states. The JAK3 structures are of a doubly phosphorylated state (12) and a non-phosphorylated state (3), whilst TYK2 structures are of either a mono-phosphorylated state (3) or a non-phosphorylated state (4).

Most protocols for purifying JAK kinases initially employ Ni^{2+}-NTA agarose affinity chromatography, making use of the 6-His N-terminal tag. In most studies, lysis buffers contained high NaCl concentrations (up to 500 mM), glycerol (5–20 %), 0.1 % detergent (Tween/thesit) to reduce non specific binding of protein contaminants onto the Ni^{2+}-NTA agarose resin, EDTA-free protease inhibitors and β-mercaptoethanol or TCEP (see Note 1). After extensive washes, the JAKs were eluted with imidazole (either by step or gradient elution) in the presence of NaCl. Purification protocols then often differ depending upon the JAK protein being studied.

2.3.1. JAK2 Kinase Domain

When expressed in the presence of small molecule inhibitors, JAK2 kinase domain eluted from Ni^{2+}-NTA agarose was simply buffer exchanged to remove the imidazole, concentrated and loaded onto a size exclusion column (Superdex 75 for chromatographic separation of the monomeric JAK2). JAK2 kinase domain recovered was active and mainly doubly phosphorylated (13) and was readily concentrated for crystallization studies.

In contrast, when no inhibitor was added during the cell culture, extra purification steps have been required to obtain pure homogeneous sample. In one study, JAK2 kinase domain has been further purified using a γ-phenyl ATP-Sepharose column, eluted in the presence of ADP and $MgCl_2$ and pooled JAK2 fractions loaded again onto a size exclusion column as a final polishing step. The purest JAK2 fractions were concentrated in the presence of ADP and $MgCl_2$ and the inhibitor of interest was added just prior to crystallization studies (14). In another study, addition of an ion exchange column (Resource Q/ Mono-Q) has been found necessary to resolve the different phosphorylation states (11).

2.3.2. TYK2 Kinase Domain

One method reported for the purification of the TYK2 kinase domain has involved the use of a desalting column (HiPrep 26/10) after the Ni^{2+} NTA-agarose affinity step and size exclusion chromatography (16/60 Superdex 200) in the presence of 1 μM of the inhibitor of interest. The final TYK2 inhibitor complex eluting as a monomer was then concentrated for crystallization studies (3).

2.3.3. JAK1 and JAK3 Kinase Domains

For both JAK1 and JAK3 kinase domains, the inclusion of an ion exchange column such as Mono Q or resource Q column after Ni^{2+} NTA-agarose step is a necessary step to isolate the doubly phosphorylated forms and guarantee crystallization. Details of JAK1 expression and purification are provided later in this chapter. In the case of JAK3, inhibitors of interest were added after the ion exchange (3, 12).

2.4. Crystallization in Complex with Small Molecule Inhibitors

The crystal structures of JAK kinases in complex with small molecule inhibitors have been crucial to elucidate the structural basis of inhibitor specificity and potency. All JAK structures have now been elucidated with two pan-JAK small molecule inhibitors, CMP-6 a JAK inhibitor developed by Merck Research Laboratories (15) but not used in the clinic and CP-690,550, a potent and selective JAK family inhibitor developed by Pfizer that was first designated as a JAK3-specific inhibitor for the treatment of transplant rejection (16). Collectively, these structural data have not only provided insight into how these inhibitors exert their specificity but have also revealed important structural differences surrounding the ATP-binding site, that have been exploited for the development of more JAK member-specific compounds see refs. 3, 13 and 8 for review.

The majority of the JAK-inhibitor complex crystals that have been reported are from crystals formed by co-crystallization (Table 1) where the inhibitor is added prior to setting up the crystallization experiment, either during co-expression (e.g., JAK2, JAK1 (7, 13)) or during purification (e.g., TYK2, JAK3 (3, 4)) or just prior to crystallization trays (e.g., JAK2 and JAK3 (11, 12)).

Another strategy is to soak preexisting crystals with the inhibitor of interest. This method has been used to generate JAK3/ CP-690550 crystals by directly adding CP-690550 at a final concentration of 5 mM to JAK3-CMP-6 co-crystals over a period of

5 days (3). A similar method has been employed recently to generate JAK2 crystals where the JAK2 kinase domain was purified and crystallized in the presence of staurosporine and the JAK2-staurosporine crystals were then soaked in the presence of an inhibitor of interest (9).

The range of crystallization conditions used to obtain crystals of JAK-inhibitor complexes can be found in Table 1. Most of the crystallization conditions for the JAK proteins were sought using the vapor-diffusion method, either using the hanging-drop method in a 24-well Linbro plate or using the sitting-drop method in a 96-well crystallization plate. The vapor diffusion method relies on the slow formation of an equilibrium between the drop of protein solution mixed with crystallizing agents and the larger reservoir solution containing the crystallizing agents at higher concentrations than in the drop. As equilibration is established between the drop and the reservoir solution, the precipitant concentration increases in the drop to a level optimal for crystallization (for a general description of the principles that govern crystallization, please refer to refs. 17, 18).

The JAK1 and JAK2 original crystallization conditions were refined by varying the pH and concentration of precipitant (7, 13). In other studies, JAK2 crystals were optimized using repeated microseeding techniques (14, 19, 20), as per JAK3 (3), whereas TYK2 crystal growth was optimized by cross-seeding with a JAK3 seed stock (3). For a more general description of the different seeding techniques, please refer to ref. 21.

3. Examples: Expression, Purification and Crystallization of JAK1 and JAK2 Kinase Domains in the Presence of Small Molecule Inhibitors

Prepare all solutions using ultrapure water.

3.1. Materials

3.1.1. Protein Expression

1. Sf9 cells (clonal isolate, derived from *Spodoptera frugiperda*) were used for expression as serum-free suspended cultures.
2. Sf900II media (Invitrogen).
3. Gentamicin sulfate: stock solution should be made at 10 mg/ml stock, filtered sterilized and stored in aliquots at −20 °C.
4. 1 and 3 L shaker flasks
5. 6-Well tissue culture plates: 75, 175, and 300 cm^2 sterile tissue culture flasks for virus stock storage.
6. Sterile serological pipettes: 5, 10, and 25 ml.
7. Hemocytometer and glass coverslips for cell counting.
8. Trappsol.
9. Dimethyl sulfoxide (DMSO), tissue culture grade.

10. Inhibitory kinase compounds (at 20 mM in 100 % DMSO).
11. Refrigerated shaker incubator, flat base platform.
12. Microscope.
13. Benchtop centrifuge.
14. Large volume refrigerated centrifuge for harvesting Insect cells for protein expression.
15. –80 °C Freezer for storage of expressed Sf9 cell pellets.

3.1.2. Purification of Recombinant JAK Proteins

Reagent and Equipment Lists:

1. All size exclusion and anion exchange chromatography have been carried out using an Akta Basic, fitted with P920 pumps, UPC900 processor and F900 fraction collector.
2. All SDS-PAGE have been run on 12 % acrylamide gels, using the BioRad Mini-Protean 3 system.
3. P1 pump (GE Science).
4. Sonicator.
5. Centrifuge, Sorvall RC5.
6. Refrigerated benchtop centrifuge.
7. Refrigerated microcentrifuge.
8. Superdex 75 16/60 size exclusion chromatography sepharose column (GE Science).
9. 10/50 Mono-Q anion exchange chromatography column (GE Science).
10. Nickel-chelating resin (Probond).
11. 50 ml Glass columns for packing Ni-IDA resin.
12. SS-34 Sorvall plastic centrifuge tubes.
13. 50 ml Falcon tubes, 15 ml Falcon tubes, 1.8 ml Eppendorf tubes.
14. Amicon Ultra-15 and Ultra-4, 30 kDa and 10 kDa MW cut-off centrifugal concentrators.
15. Microcon YM30 and YM10 Ultracell microcentrifugal concentrators (Millipore).
16. Steritop vacuum driven disposable bottle top 0.22 μM filters (Millipore).
17. Nanosep 0.22 μM centrifugal filters (Pall Life Sciences).
18. Complete EDTA-free protease inhibitor tablets (Roche).
19. 1× Bradford Reagent (BioRad).
20. V8 protease (Sigma).

3.1.3. Solutions Required for the Purification of JAK1 and JAK2 Kinase Domains

1. The following stock solutions can be prepared in advance: 5 M NaCl, 1 M Tris–HCl, pH 8.5, 1 M Tris–HCl, pH 9.0, 10 % Thesit, 2 M imidazole, pH 8.0, and 1 M dithiothreitol (DTT).

All subsequent buffers should be prepared fresh.

2. Lysis Buffer: 20 mM Tris–HCl, pH 8.5, 500 mM NaCl, 5 % Glycerol, 0.1 % Thesit, 3 mM β-mercaptoethanol, one Complete EDTA-free protease inhibitor tablet per 50 ml of lysis buffer.

Ni-NTA Buffers:

3. Wash Buffer: 20 mM Tris–HCl, pH 8.5, 500 mM NaCl, 5 % glycerol, 3 mM β-mercaptoethanol, 20 mM imidazole.
4. Elution Buffer A: 20 mM Tris–HCl, pH 8.5, 500 mM NaCl, 5 % glycerol, 3 mM β -mercaptoethanol, 100 mM imidazole.
5. Elution Buffer B: 20 mM Tris–HCl, pH 8.5, 500 mM NaCl, 5 % glycerol, 3 mM β -mercaptoethanol, 200 mM imidazole.
6. Elution Buffer C: 20 mM Tris–HCl, pH 8.5, 500 mM NaCl, 5 % glycerol, 3 mM β -mercaptoethanol, 400 mM imidazole.

   All buffers used for size exclusion and anion exchange chromatography should be filtered (0.22 μm) just before use.

Gel Filtration Buffers:

7. JAK2 gel filtration buffer: 20 mM Tris–HCl, pH 8.5, 250 mM NaCl, 1 mM DTT, 0.5 mM EDTA.
8. JAK1 gel filtration buffer: 20 mM Tris–HCl, pH 8.9, 200 mM NaCl, 3 mM DTT, 0.5 mM EDTA.

Mono Q Buffers for JAK1 Purification:

9. Buffer A: 20 mM Tris–HCl, pH 8.9, 0.5 mM EDTA, 3 mM DTT.
10. Buffer B: 20 mM Tris–HCl, pH 8.9, 500 mM NaCl, 0.5 mM EDTA, 3 mM DTT.

3.1.4. Crystallization

1. Crystallization plates (Hampton Research).
2. Commercial crystallization kits are available from the following companies (Hampton Research, Jenna Bioscience, Sigma-Aldrich, Emerald BioStructures, Qiagen, Molecular Dimensions).
3. All chemicals to make buffer and salt solutions were obtained from Sigma-Aldrich.
4. Stock solutions for all buffers should be standardized (all pH buffering solutions at 1 M, all salt solutions at 2 M). Filter buffer and salt stock solutions through a 0.22 μM filter and store in sterile containers (50 ml Falcon tube). These stock solutions are then used to create the individual crystallization conditions.
5. Polyethylene glycol (PEG) (Hampton Research). All PEG solutions were made up at 50 % (w/v) in MilliQ water with

0.02 % sodium azide. PEG solutions should be stored in the dark or in dark containers.

6. Reservoir solution A: 15–18 % (w/v) PEG 8000, 100 mM cacodylate, pH 6.5/6.7, 100 mM magnesium acetate, 100 mM KCl.
7. Reservoir solution B: 25–27 % PEG (polyethylene glycol) 3350, 0.1 M Hepes, pH 7.5, 0.1 M KSCN, 10 mM DTT.
8. Siliconized coverslips (Hampton Research).
9. Silicon vacuum grease.
10. Controlled environment in regards to temperature and stability (absence of vibration).

3.2. Methods

3.2.1. Cloning into pFASTBac and Generation of Bacmid DNA

The kinase domains of human JAK1 (residues 841–1154 (PubMed AAI32730)] and JAK2 [residues 835–1132] (PubMed NM_004972)] were cloned into pFastBac HTC, which allows the protein to be expressed fused to a 6-His cleavable tag.

The generation of recombinant Bacmid DNA should be performed carefully following all instructions from the Invitrogen Bac-to-Bac® Baculovirus Expression System handbook.

3.2.2. Transfection into Sf9 Cells and Amplification of Recombinant Virus (P1, P2, and P3 Virus Stocks)

Isolate recombinant bacmid DNA containing JAK1 and JAK2 inserts and use them to transfect Spodoptera frugiperda (Sf9) insect cells. Isolation of P1 and generation of P2 and P3 virus stocks should again be performed carefully, following all instructions from the Bac-to-Bac® Baculovirus Expression System handbook. The average titer of the P2 stocks generated is ~10^7 pfu/ml. Using this, a consistent supply of P3 stocks can be generated by adding no more than 5 ml of P2 stocks into 500 ml culture of Sf9 cells at a cell density of ~2×10^6 cells/ml, with ~100 % viability, to reach a very low multiplicity of infection (MOI) (<0.1 pfu/cell). For more general detail on how to generate high titer virus stocks, refer to the Bac-to-Bac® Baculovirus Expression System handbook and (22). Cells are incubated at 27 °C with shaking until they are well infected (normally 4–5 days). All virus stocks are stored in the dark (wrapped in foil) at 4°C.

3.2.3. Protein Expression

Preparation of Inhibitors CMP-6 and CP-690,550:

1. Dissolve inhibitor in 100% DMSO at a stock concentration of 20 mM. Store frozen at 4°C.
2. Inhibitor aliquots should be removed from refrigeration and allowed to thaw at room temperature, prior to adding to the Sf9 cells.
3. Dilute each inhibitor in 0.1 M Trappsol (see Note 2). For JAK2 expression, 50 μl of inhibitor (CMP-6 or CP-690,550) is added to 1 ml 0.1 M Trappsol at room temperature (1/20 dilution). For JAK1 expression, 20 μl of CMP-6 or 25 μl of

CP-690,550 is added to 1 ml 0.1 M Trappsol at room temperature (1/50 and 1/40 dilution respectively). Mix and allow to stand for 5–10 min whilst preparing for infection of Sf9 cells with the P3 viral stock.

Once cells are infected with P3, the Trappsol-inhibitor complex is added to the cells to create a final concentration of 1 μM inhibitor in the case of JAK2, or 400 nM of CMP-6 and 500 nM of CP-690,550 in the case of Jak1 (see Notes 3 and 4).

Infection of Sf9 Cells

1. One day prior to infection with P3 virus stock, split Sf9 cells to a density of ~1×10^6 cells/ml in 1 L Sf900II medium supplemented with gentamycin (10 μg/ml final) per 3 L flask. Incubate overnight at 27°C with shaking.
2. The following day check cell density. Cells should be at ~$1.5–2 \times 10^6$ cells/ml, with ~100 % viability.
3. Prepare inhibitors as described above.
4. Whilst waiting for the Trappsol-inhibitor complex to form, infect cells with 70 ml P3 virus stock/L of cells to achieve a relatively high MOI (>10). Add the Trappsol/inhibitor complex and return cells to the 27°C shaker for 48 h.
5. Check cell density and viability upon harvest. Cell density should be at ~$2–4 \times 10^6$ cells/ml, with 50–70% of cells viable.
6. In the case of JAK1, the cells must be at 2.5×10^6 for infection (hence split to ~1.5×10^6 the day prior). Add only 25–40 ml of P3 viral stock, and incubate for 48 h post-infection, prior to harvesting.

3.2.4. Purification

Carry out all purification steps at 4°C unless otherwise specified.

Di-phosphorylated JAK2–Inhibitor Purification

1. Harvest the cells from 1 L of Sf9 cells and resuspend in 40 ml of Lysis buffer.
2. Lyse cells by sonication (three times 30 s with interval of 1 min on ice). Transfer lysate into SS34 tubes.
3. Centrifuge lysate at 32,000 × *g* for 60 min.
4. Filter the supernatant (critical) using a Steritop 0.22 μm filter unit.

   *Take a sample of the filtered supernatant for SDS-PAGE gel analysis.
5. Load the filtered supernatant onto a nickel resin column pre-equilibrated in wash buffer.

   * Take a sample of the unbound material for SDS-PAGE gel analysis.

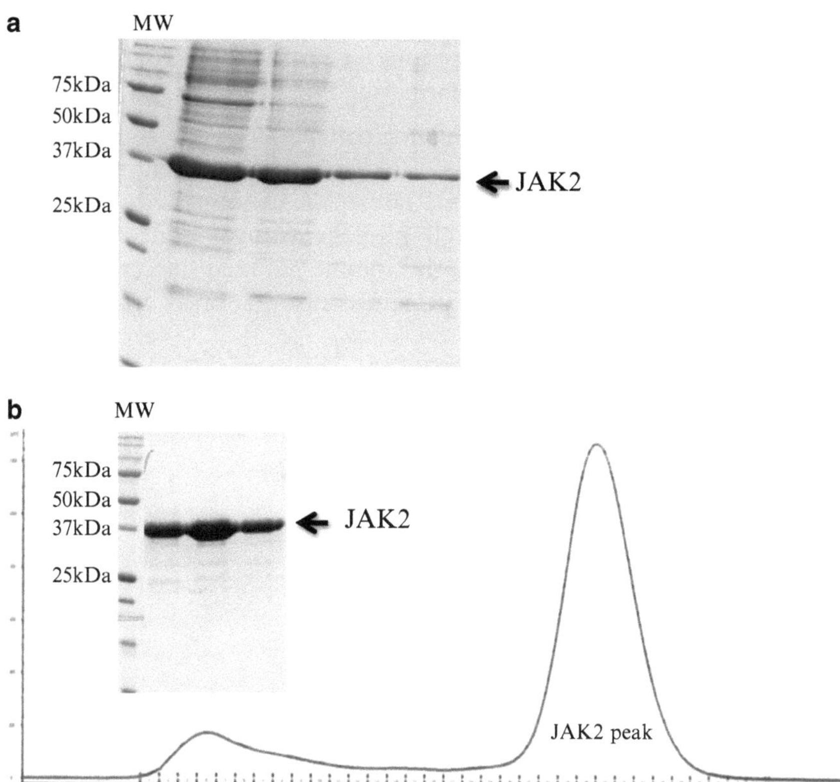

Fig. 2. (**a**) Coomassie-stained SDS-PAGE gel showing JAK2 fractions eluted from Ni^{2+}-NTA resin. (**b**) Size exclusion chromatography trace of the purification of JAK2 following elution from Ni^{2+}-NTA resin. The JAK2 protein containing peak, (as evidenced by the band of ~37 kDa seen in the accompanying Coomassie-blue stained SDS-PAGE), elutes at ~65ml post-injection from the S75 16/60 size exclusion column.

6. Wash beads with 50 ml of wash buffer. * Retain a sample of the first 5 ml of wash flow-through for SDS-PAGE gel analysis.
7. Elute proteins with stepwise imidazole concentration in 3 ml elution volumes as follows: 3 × 3 ml fractions with Elution buffer A, followed by 3 × 3 ml fractions with Elution Buffer B and 3 × 3 ml fractions Elution buffer C.

   * Take a sample of each elution fraction for SDS-PAGE gel analysis.
8. Run samples of the eluted fractions on an SDS-PAGE gel to determine which fractions to pool for the next step purification (see Fig. 2a).
9. Pool fractions containing His-JAK2 and concentrate to 5 ml using Amicon Ultra-15 30 kDa centrifugal concentrators.
10. Filter and load the concentrated fractions onto a Superdex 75 gel-filtration column (HiLoad 16/60) pre-equilibrated in JAK2 gel filtration buffer. Run at 1 ml/min and commence fraction collection at 40 ml post-injection. The JAK2-inhibitor

complex typically elutes as a monomer at ~65 ml. Run the corresponding fractions on an SDS-PAGE gel to determine the purity and quantity (see Fig. 2b).

11. Pool JAK2-inhibitor complex fractions and concentrate using Amicon Ultra-4, 10 kDa MW cut-off centrifugal concentrators to around 1 ml. Transfer the concentrated sample to Microcon YM10 Ultracell microcentrifugal concentrators and then concentrate to the volume required for crystallization trials so that the final protein concentration is around 8–10 mg/ml

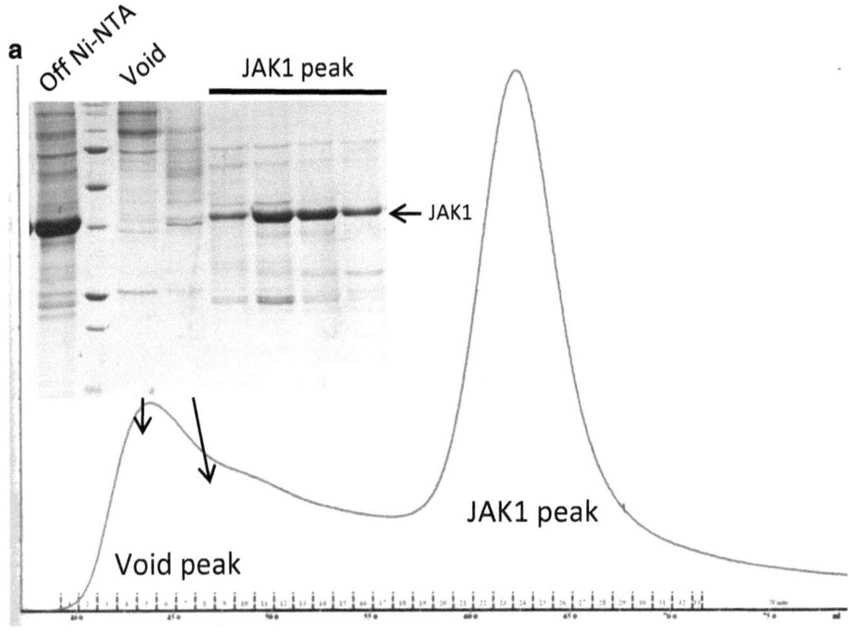

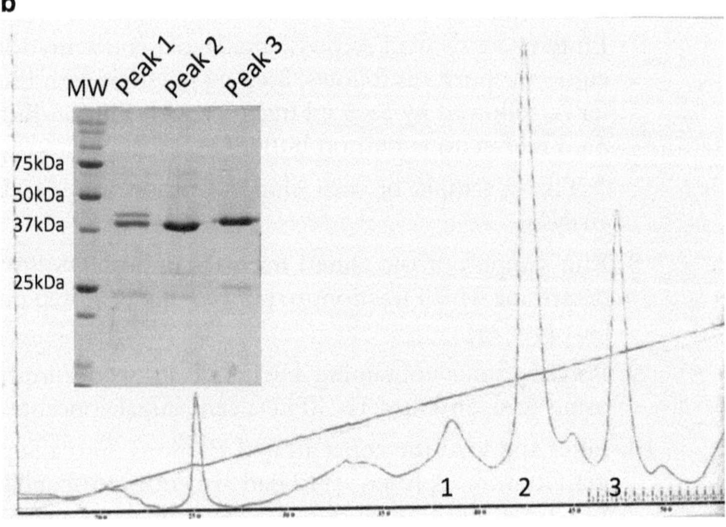

Fig. 3. (continued)

(see Note 5). Filter the concentrated sample using Nanosep 0.22 μm centrifugal filters. Determine the protein concentration using a Bradford assay (see Note 6) and comparison to bovine serum albumin (BSA) standards. After purification, the protein is stable for a couple of weeks when stored as a filtered, concentrated sample at 4°C, prior to setting up crystallization trays.

12. The typical yield of pure JAK2 protein recovered is around 8–10 mg/L of cell culture when expressed in the presence of CMP-6 and around 5 mg/L when expressed in the presence of CP-690,550 (see Note 4).

Di-phosphorylated JAK1–Inhibitor Purification

Carry out steps 1–9 as described in the JAK2-inhibitor purification protocol above.

10. Load the pooled, concentrated fractions (~2–5 ml) onto a Superdex 75 gel-filtration column (HiLoad 16/60) that has been pre-equilibrated in JAK1 gel filtration buffer. Run at 1 ml/min and commence fraction collection at 40 ml post-injection. The JAK1-inhibitor complex typically elutes at 61 ml post-injection (see Fig. 3a). Run the corresponding fractions on an SDS-PAGE gel to determine purity and quantity.

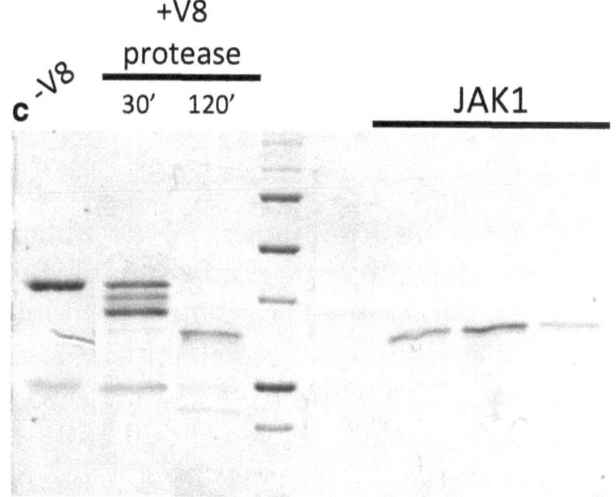

Fig. 3. (**a**) Size exclusion chromatography trace of the purification of JAK1 following elution from Ni^{2+}-NTA resin. The JAK1 protein containing peak, (as evidenced by the band of ~37 kDa seen in the accompanying Coomassie-blue stained SDS-PAGE), elutes at ~61 ml post-injection from the S75 16/60 size exclusion column. (**b**) Anion exchange chromatography trace of the separation of different phospho-forms of the kinase domain of JAK1 that result from recombinant baculovirus expression. Peaks 1, 2, and 3 all contain JAK1 protein, (as evidenced by the band of ~37 kDa seen in the accompanying Coomassie-blue stained SDS-PAGE), however they differ in the level of phosphorylation present on the activation loop tyrosines. Peak 1 is completely unphosphorylated, peak 2 is mono-phosphorylated, and peak 3 is the di-phosphorylated form used for further purification and crystallization studies. Di-phosphorylated JAK1 (peak 3) elutes at ~150 mM NaCl from the 10/50 MonoQ column. (**c**) Cleaved, purified di-phosphorylated recombinant JAK1. This Coomassie-stained SDS-PAGE gel shows the final purification step undertaken with JAK1. Following isolation of the di-phosphorylated His-tagged species (using anion exchange, *lane 1* above) the protein is incubated with V8 protease for 2 h at 30 °C (*lane 2*—partially digested, *lane 3*—fully digested) and then loaded onto a final, polishing size exclusion chromatography column (S75 16/60). The protein elutes as a single peak at ~65–70 ml post-injection (*lanes 6–8* above) and is then pooled and concentrated for crystallization.

11. Pool the JAK1 fractions containing ~37 kDa monomeric JAK1 protein and concentrate to 0.4 ml using an Amicon Ultra-4, 10 kDa MW cut-off centrifugal concentrator. Dilute 10-fold with Buffer A before loading onto a Mono Q 5/50 GL column equilibrated with 4 % Buffer B.
12. Wash the MonoQ column with 10 column volumes of 4 % Buffer B at 1 ml/min. Elute protein with a linear gradient of 20–200 mM NaCl (4–40% Buffer B), over 40 min at a 1 ml/min flow rate (see Fig. 3b). Pool the JAK1 protein (eluting at ~150 mM NaCl) and estimate the protein concentration using a Bradford Assay against a BSA standard curve. V8 protease is added at a concentration of 1 U protease per 20 μg of JAK1 protein and the mixture incubated at 30 °C for 2 h before loading directly onto a Superdex 75 gel-filtration column (volume ~1–2 ml), as described above (steps 10–11) (see Note 7).
13. Fractions are collected from 40–100 ml post-injection, with the cleaved, di-phosphorylated JAK1 eluting at ~65 ml post-injection.
14. Di-phosphorylated JAK1-inhibitor complex-containing fractions should be assessed using SDS-PAGE gel electrophoresis (see Fig. 3c), pooled and concentrated using Amicon Ultra-4, 10 kDa MW cut-off and Microcon YM10 Ultracell microcentrifugal concentrators to a final concentration of 8 mg/ml for crystallization trials.
15. The typical yield was lower than that for JAK2 as the protein undergoes more purification steps, and less than 50% of the original monomeric JAK1 obtained from the first size exclusion chromatography step is of the di-phosphorylated species. Often ~0.2 mg purified, monomeric, di-phosphorylated JAK1 is recovered from an initial 3 L expression culture.

3.2.5. Crystallization

JAK2-CMP6/JAK2–CP-690,550 (and the majority of JAK2-Inhibitor Complexes)

The initial JAK2 crystals obtained in our laboratory were crystallized with the hanging drop method using PEG 8000 as a precipitant and at 4°C (see Note 8) (7).

1. Filter the protein sample just prior to use, using a Nanosep 0.22 μM spin filter (see Note 9).
2. Prepare 1000 μl of reservoir solution A (see Note 10) and place in the reservoir of a Linbro plate with the rim of each well greased with silicon grease.
3. Mix 0.5 μl of protein solution with 0.5 μl of reservoir solution on the coverslip and invert the coverslip over the reservoir solution (see Note 11).

4. Crystals are grown from precipitate at 4°C and are formed after 3–6 days. Crystals are flash-frozen prior to data collection using 5% glycerol as a cryoprotectant (see Note 12).

JAK1–CMP6 and JAK1–CP-690,550

1. Concentrate the purified JAK1/CMP6 complex to 10 mg/ml.
2. Filter protein sample just prior use using Nanosep 0.22 μm spin filter.
3. Prepare 1000 μl of reservoir solution B and place in the reservoir of a Linbro plate with the rim of each well greased with silicon grease.
4. Mix 1 μl of protein solution with 1 μl of reservoir solution on the coverslip and invert the coverslip over the reservoir solution.
5. Crystals were visible from precipitate at 20°C after 1 day, and were flash-frozen prior to data collection using 30% PEG 3350 as a cryoprotectant, within one further day of formation (i.e., 24–48 h after tray setup). The crystals formed rapidly, but also deteriorated quite rapidly, and hence needed to be frozen soon after formation.

3.3. Notes

1. To avoid reduction of the Ni^{2+} on the column, do *not* use strong reducing agents, such as DTT, in the lysis buffer or during the Ni^{2+}-NTA agarose affinity chromatography step. TCEP is to be used at a final concentration of 1 mM or less. β-mercaptoethanol can be used at a final concentration of 3 mM or less (check Ni^{2+}-NTA agarose manufacturer's recommendations).
2. Preparation of 0.1 M Trappsol (MW ~1,400 Da). 1.4 g of Trappsol is weighed into a 10 ml sterile disposable tube and milliQ water added to bring the total volume to 10 ml. The solution is gently agitated until cleared and then filter sterilized (0.22 μM) into 1 ml aliquots for storage at 4 °C. Allow to come to room temperature before addition of inhibitor.
3. Importance of phosphorylation state for JAK1 crystallization. Extensive trials were conducted to determine the ideal cell density, P3 virus volume and inhibitor concentration required for maximal expression of di-phosphorylated JAK1. This contrasts to JAK2, where these parameters were more flexible and yields of consistently di-phosphorylated JAK2 as the sole species present were reliable with an initial infection cell density from $1.5–2 \times 10^6$ cells/ml. For JAK1, a relatively high cell density at infection (2.5×10^6 cells/ml) was required, and addition of only 25–40 ml of P3 was sufficient to maximize expression of JAK1. The addition of inhibitor at 400 nM for CMP-6 and 500 nM for CP-690,550, optimized the proportion of

the di-phosphorylated species present. Higher inhibitor concentrations (>500 nM) prevent phosphorylation (resulting in the unphosphorylated species predominating), whereas lower concentrations (<300 nM) resulted in overall low expression (akin to the yield with no inhibitor added).

4. The level of JAK2 expression and the amount of pure di-phosphorylated JAK2 is directly proportional to the affinity of the inhibitor used during expression. High affinity inhibitors (e.g., CMP-6, CP-690,550) with an affinity below 20 nM typically yield 5–10 mg of di-phosphorylated JAK2 per L of cell culture. If the inhibitor used has an affinity between 20 and 50 nM, it is recommended to increase the final concentration of inhibitor during cell expression to 1.5 μM. If the affinity of the inhibitor used is between 50 to 100 nM, it is recommended to increase the final concentration of inhibitor during cell expression to 2 μM and to increase the volume of expression to 4 L so that sufficient amount of protein can be purified for crystallization studies.
5. When the inhibitor used has a low affinity (>50 nM and above), it is strongly recommended to add an additional 2 μM of inhibitor just prior the final concentration step of the protein. Perform a 1/10 dilution of the inhibitor from the stock solution into the gel filtration buffer and add 1 μl of the 1/10 dilution into 1 ml of pre-concentrated protein solution.
6. Bradford is recommended for the determination of protein concentration. Do not determine the protein concentration by measuring the absorbance at 280 nm as most of the inhibitors have an absorbance at 280 nm and this can lead to false over-estimation of protein concentration.
7. In contrast to JAK2 kinase domain, which crystallized readily in the presence of the N-term His-tag, it was necessary to treat JAK1 kinase domain with V8 protease, which removes the N-term His-tag as well as the first 23 residues (residues 841-863) of JAK1 kinase domain, to improve the diffraction quality of JAK1 crystals.
8. It was also possible to crystallize JAK2 at room temperature in 28 % (w/v) PEG 8000, 0.2 M ammonium acetate and 0.1 M sodium citrate, which is the original published crystallization condition (7). However, crystals grown at 4 °C in 18 % PEG 8000, 100 mM Cacodylate, pH 6.5/6.7, 100 mM magnesium acetate, and 100 mM KCl were found to be more stable and exhibited better diffraction.
9. Use freshly purified protein when possible.
10. Ensure that the reservoir solution is well mixed when working with highly viscous solution such as PEGs.

11. Most crystallization experiments with JAK2 have been successful at pH values between 6.3 and 6.9 and PEG 8000 concentrations from 12 to 22 % (w/v). Therefore it is recommended to perform the initial screen by varying the pH from 6.3 to 6.9 with increments of 0.2 units, and the PEG 8000 concentration from 12 to 22 % with increments of 2 %. Once crystals have formed, refine the conditions and increase the size of the drops to at least 2 μl to grow larger crystals.
12. Larger crystals have been found to have better diffraction quality.

Acknowledgments

This work was supported by the Australian Research Council and the National Health and Medical Research Council Industry Fellowship (I.L.). We thank Dr Onisha Patel and Dr Neal Williams for conducting part of the experiments described in this chapter. We thank Prof Jamie Rossjohn and Prof Andrew Wilks for supporting the research.

References

1. Wilks AF (2008) The JAK kinases: not just another kinase drug discovery target. Semin Cell Dev Biol 19:319–328
2. Harpur AG, Andres AC, Ziemiecki A et al (1992) JAK2, a third member of the JAK family of protein tyrosine kinases. Oncogene 7:1347–1353
3. Chrencik JE, Patny A, Leung IK et al (2010) Structural and thermodynamic characterization of the TYK2 and JAK3 kinase domains in complex with CP-690550 and CMP-6. J Mol Biol 400:413–433
4. Tsui V, Gibbons P, Ultsch M et al (2011) A new regulatory switch in a JAK protein kinase. Proteins 79:393–401
5. Liu Y, Gray NS (2006) Rational design of inhibitors that bind to inactive kinase conformations. Nat Chem Biol 2:358–364
6. Haan C, Kroy DC, Wuller S et al (2009) An unusual insertion in Jak2 is crucial for kinase activity and differentially affects cytokine responses. J Immunol 182:2969–2977
7. Lucet IS, Fantino E, Styles M et al (2006) The structural basis of Janus kinase 2 inhibition by a potent and specific pan-Janus kinase inhibitor. Blood 107:176–183
8. Alicea-Velazquez NL, Boggon TJ (2011) The use of structural biology in Janus kinase targeted drug discovery. Curr Drug Targets 12:546–555
9. Ioannidis S, Lamb ML, Wang T et al (2011) Discovery of 5-Chloro-N(2)-[(1S)-1-(5-fluoropyrimidin-2-yl)ethyl]-N(4)-(5-methyl-1H-pyr azol-3-yl)pyrimidine-2,4-diamine (AZD1480) as a Novel Inhibitor of the Jak/Stat Pathway. J Med Chem 54:262–276
10. Korniski B, Wittwer AJ, Emmons TL et al (2010) Expression, purification, and characterization of TYK-2 kinase domain, a member of the Janus kinase family. Biochem Biophys Res Commun 396:543–548
11. Hall T, Emmons TL, Chrencik JE et al (2010) Expression, purification, characterization and crystallization of non- and phosphorylated states of JAK2 and JAK3 kinase domain. Protein Expr Purif 69:54–63
12. Boggon TJ, Li Y, Manley PW et al (2005) Crystal structure of the Jak3 kinase domain in complex with a staurosporine analog. Blood 106:996–1002
13. Williams NK, Bamert RS, Patel O et al (2009) Dissecting specificity in the Janus kinases: the structures of JAK-specific inhibitors complexed to the JAK1 and JAK2 protein tyrosine kinase domains. J Mol Biol 387:219–232

14. Wang T, Duffy JP, Wang J et al (2009) Janus kinase 2 inhibitors. Synthesis and characterization of a novel polycyclic azaindole. J Med Chem 52:7938–7941
15. Thompson JE, Cubbon RM, Cummings RT et al (2002) Photochemical preparation of a pyridone containing tetracycle: a Jak protein kinase inhibitor. Bioorg Med Chem Lett 12:1219–1223
16. Changelian PS, Flanagan ME, Ball DJ et al (2003) Prevention of organ allograft rejection by a specific Janus kinase 3 inhibitor. Science 302:875–878
17. Chayen NE, Saridakis E (2008) Protein crystallization: from purified protein to diffraction-quality crystal. Nat Methods 5:147–153
18. McPherson A (1999) Crystallization of biological macromolecules. Cold Spring Harbor Laboratory, Cold Spring Harbor, NY
19. Wang T, Ioannidis S, Almeida L et al (2011) In vitro and in vivo evaluation of 6-aminopyrazolyl-pyridine-3-carbonitriles as JAK2 kinase inhibitors. Bioorg Med Chem Lett 21: 2958–2961
20. Wang T, Ledeboer MW, Duffy JP et al (2010) A novel chemotype of kinase inhibitors: discovery of 3,4-ring fused 7-azaindoles and deazapurines as potent JAK2 inhibitors. Bioorg Med Chem Lett 20:153–156
21. Bergfors T (2003) Seeds to crystals. J Struct Biol 142:66–76
22. King LA, Hitchman R, Possee RD (2007) Recombinant baculovirus isolation. Methods Mol Biol 388:77–94
23. Antonysamy S, Hirst G, Park F et al (2009) Fragment-based discovery of JAK-2 inhibitors. Bioorg Med Chem Lett 19:279–282
24. Baffert F, Regnier CH, De Pover A et al (2010) Potent and selective inhibition of polycythemia by the quinoxaline JAK2 inhibitor NVP-BSK805. Mol Cancer Ther 9: 1945–1955
25. Pissot-Soldermann C, Gerspacher M, Furet P et al (2010) Discovery and SAR of potent, orally available 2,8-diaryl-quinoxalines as a new class of JAK2 inhibitors. Bioorg Med Chem Lett 20:2609–2613
26. Harikrishnan LS, Kamau MG, Wan H et al (2011) Pyrrolo[1,2-f]triazines as JAK2 inhibitors: achieving potency and selectivity for JAK2 over JAK3. Bioorg Med Chem Lett 21: 1425–1428
27. Thoma G, Nuninger F, Falchetto R et al (2011) Identification of a potent Janus Kinase 3 inhibitor with high selectivity within the Janus Kinase family. J Med Chem 54: 284–288

Chapter 21

Bacterial Expression, Purification, and Crystallization of Tyrosine Phosphorylated STAT Proteins

Florence Baudin and Christoph W. Müller

Abstract

Signal Transducer and Activator of Transcription (STAT) proteins are latent cytoplasmic transcription factors that become activated by phosphorylation at a C-terminal tyrosine residue. Upon activation STAT proteins translocate to the nucleus and bind to their specific target sites. Here, we describe the recombinant expression of tyrosine phosphorylated STAT proteins in bacteria. This method allows the production of large amounts of activated STAT proteins for structural and biochemical studies including the high-throughput screening of chemical libraries.

Keywords: STAT proteins, STAT purification, tyrosine phosphorylation, DNA oligonucleotide purification, Protein–DNA complex, EMSA, Protein–DNA co-crystallization.

1. Introduction

Signal Transducers and Activators of Transcription (STATs) constitute a family of eukaryotic transcription factors involved in cytokine signaling (1). So far, seven mammalian STATs have been identified: STAT1, STAT2, STAT3, STAT4, STAT5a, STAT5b, and STAT6. Upon activation, their latent, cytoplasmic form is activated by phosphorylation of a single tyrosine residue at their C-terminal end. This activation step is triggered by the binding of growth factors, hormones or cytokines to specific cell-surface receptors. After tyrosine phosphorylation STAT proteins homo- or heterodimerize, translocate to the nucleus, and regulate gene expression by binding to specific response elements (2, 3).

Genetic, biochemical, and structural information revealed a common organization of STAT proteins (Fig. 1). The N-terminal

Sandra E. Nicholson and Nicos A. Nicola (eds.), *JAK-STAT Signalling: Methods and Protocols*, Methods in Molecular Biology, vol. 967, DOI 10.1007/978-1-62703-242-1_21, © Springer Science+Business Media New York 2013

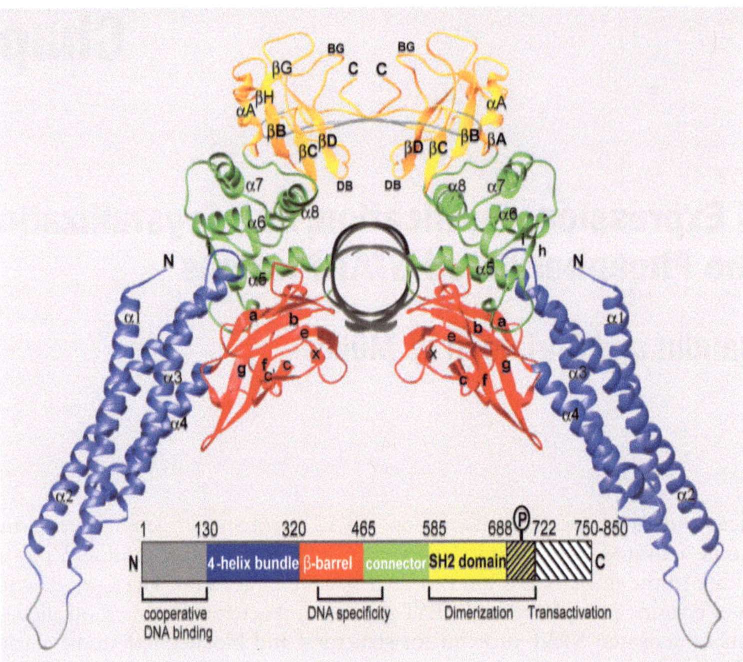

Fig. 1. Core structure of the STAT3 dimer bound to it DNA target site (8). The DNA duplex (*black*) is shown in the center with the view along the DNA axis. The STAT3 dimer contacts DNA using several loops protruding from the central β-barrel and the linker domain. The inset shows the domain structure of STAT3β.

part of the STATs, named N-domain, roughly 130 amino acids in length, is involved in dimerization, tetramerization, and protein–protein interaction (4, 5). The crystal structure of the STAT4 N-domain revealed a unique hook-shaped architecture comprising eight helices and forming stable dimers (6, 7). The crystal structures of the central core domain of human STAT1 and murine STAT3β dimers in complex with DNA have also been reported (8, 9). The central core domain is approximately 600–700 residues long, is highly conserved and contains a four-helix bundle domain important for protein–protein interaction (10–15), a DNA binding domain adopting an immunoglobulin fold (16), an all-helical linker domain (17), and an Src homology 2 (SH2) domain (18). The SH2 domain mediates the interaction with the cytoplasmic domain of the receptor through phosphotyrosine binding as well as dimerization of the STATs through reciprocal binding of the phosphotyrosine residue following the SH2 domain. The C-terminal extension of the STATs contains long stretches of acidic residues which constitute the trans-activation domain (19, 20). This region is poorly ordered and only becomes structured upon interaction with binding partners. So far, complex structures of a short peptide of the STAT6 trans-activation domain bound to the PAS-B domain of NCoA-1 (21) and small regions of the trans-activating domains of STAT1 and STAT2 bound to the CBP/p300 TAZ2 and TAZ1 domain, respectively (22) have been determined.

Both, STAT1 (9) and STAT3 (8) crystal structures lack the N-domain and the C-terminal trans-activation domain, and their core domains were co-crystallized with duplex DNA corresponding to an M67 variant of a cis-inducible element of the c-fos promoter that binds to both STAT1 and STAT3 dimers with high affinity (23). The overall crystal structures of both STATs are very similar and resemble a "nutcracker" or a "pair of pliers" that grip the DNA from both sides. The crystal structure of a phosphorylated homodimer of *Dictyostelium* STAT (Dd-STATa) in its unbound form was solved at 2.7 Å (24). The domain organization of Dd-STATa resembles that of STAT1 and STAT3, except that the coiled-coil domain only comprises three helices and adopts an inverted orientation compared to the mammalian STATs. The overall structure of Dd-STATa adopts a fully extended "open" conformation rather than the "closed" conformation observed in STAT1 and STAT3 bound to DNA. This suggests that the Dd-STATa dimer undergoes a large conformational change from an "open" to a "closed" conformation upon DNA recognition. In addition, two crystal structures of unphosphorylated human STAT1 (1–683) and mouse STAT5a (residues 128–712) have been determined (25, 26). Both structures show an antiparallel, unphosphorylated dimer structurally different from the DNA-bound, phosphorylated dimer. This conformation has been suggested to be the predominant cytosolic conformation prior to cytokine stimulation. Accordingly, phosphorylation of the C-terminal tyrosine residue would cause a major conformational change of STAT dimers from an "antiparallel" to a "parallel" conformation.

Different STAT transcription factors including STAT3 but also STAT1 and STAT5 are constitutively active in various types of human cancers and have therefore become attractive targets for anticancer drugs (27). Random screening of chemical libraries or structure-based design has been used as alternative approaches to identify initial small molecules that bind with high affinity to STAT proteins and inhibit their function (28). The expression of recombinant STAT proteins in large amounts is required for detailed structural and functional analysis, but also in the search for potent, small-molecule inhibitors. Here, we present methods used in our laboratory for the recombinant expression of the core domain of STAT proteins lacking the N-domain and the C-terminal tail. We co-expressed the catalytic domain of the Elk receptor tyrosine kinase and STAT core domains in bacteria to obtain specifically tyrosine phosphorylated, dimeric STAT proteins (24, 29). In the absence of Elk receptor tyrosine kinase, the bacterial expression of the STAT core domains yields unphosphorylated, monomeric STAT protein. In addition, we briefly summarize the strategies for co-crystallization of STAT proteins with double-stranded DNA.

2. Materials

2.1. Protein Expression

The following protocol describes the expression and purification of STAT proteins specifically tyrosine-phosphorylated by co-expression with Elk receptor tyrosine kinase. Phosphorylation occurs almost quantitatively at a C-terminal tyrosine residue (Y705 in Stat3β and Y702 in Dd-STATa), presumably because they are mobile and therefore easily accessible for phosphorylation. Expression without Elk receptor tyrosine kinase leads to the expression of unphosphorylated, monomeric STAT proteins that can be purified using similar procedures. We have applied the described approach to STAT3β, STAT5 and Dd-STATa, but similar approaches can presumably also be applied for the expression and purification of other STAT proteins. In general, all chemicals should be of ACS reagent grade or 99.0 % purity (see Note 1).

1. Expression vector for protein expression (for example pET32a (Novagen) (see Note 2).
2. Competent cells *Escherichia coli* strain BL21(DE3)TKB1 (Agilent technologies, Catalog # 200134).
3. LB medium: dissolve 10 g tryptone, 5 g yeast extract, and 10 g NaCl in water to a final volume of 1 L, adjust pH to 7.4–7.5. Autoclave.
4. 2× YT medium: dissolve 16 g tryptone, 10 g yeast extract, and 5 g NaCl in water to a final volume of 1 L, adjust pH to 7.4–7.5. Autoclave.
5. Selective antibiotics ampicillin 100 mg/mL stock in double distilled water (ddH_2O), tetracycline 12 mg/mL stock, sterile filtered (0.2 μm), stored at −20 °C.
6. 1 M Isopropyl-β-d-1-thiogalactoside (IPTG), sterile filtered: dissolve 2.38 g of IPTG in deionized water, make the final volume up to 10 mL, sterilize the IPTG stock solution by filtering.

2.2. Protein Kinasing

1. Kinasing buffer: 1× modified M9 medium supplemented with 0.1 % (w/v) casamino acids (Difco), 1.5 μM thiamine–HCl and 53 μM 3β-indoleacrylic acid (see Note 3).
2. Modified M9 medium: dissolve 6 g Na_2HPO, 3 g KH_2PO, 0.5 g NaCl, 1 g NH_4Cl in 1 L, autoclave. Then add 10 mL filter sterilized 100 mM $MgSO_4$, 20 % glucose, 10 mM $CaCl_2$ before use.
3. 3β-Indoleacrylic acid from Sigma. Prepare stock solution of 2.5 mg/mL in 95 % ethanol. Store the stock solution at −20 °C.

2.3. Protein Purification

1. Extraction buffer: 20 mM HEPES–HCl, pH 7.6, 0.1 M KCl, 10 % glycerol, 1 mM EDTA, 10 mM $MnCl_2$, 20 mM DTT, 0.5 mM PMSF.
2. PMSF stock: dissolve 0.261 g PMSF in 10 mL isopropanol, store at room temperature.
3. Polyethylenimine from Sigma. Stock preparation: dissolve polyethylenimine in ddH_2O. Stock 10 % (w/v). Adjust to pH 7 with HCl. Store protected from light at 4 °C (see Note 4). Use at 0.1 % final concentration.
4. Ammonium sulfate powder.
5. Dialysis buffer: 20 mM HEPES–HCl, pH 7.0, 200 mM NaCl, 10 mM $MgCl_2$, 5 mM DTT, 0.5 mM PMSF.
6. Gel filtration Superose 12 HR 10/30 column (GE Healthcare).
7. 10 % SDS-polyacrylamide gel (see Note 5).
8. Centricon-50 concentration tubes (Amicon).

Alternatively, the use of a His-tag at the N-terminal end of the protein offers the possibility to purify the protein on a cobalt affinity column. In this case the following reagents are required:

1. Co-TALON affinity resin (Clontech).
2. Plastic columns 10 mL capacity (Pierce).
3. Complete EDTA-free protease inhibitor cocktail tablets (Roche).
4. Imidazole stock 1 M. Dissolve 68 g in 1 L of water. Filtration using 22 μm filter.
5. His-tagged TEV protease for His-Tag removal.

2.4. Electrophoretic Mobility Shift Assay

1. Binding Buffer: 10 mM Tris–HCl, pH 7.5, 1 mM DTT, 0.2 mM PMSF, 0.1 mM EDTA, 5 % glycerol, 50 mM NaCl, 0.1 % NP-40, 5 mM $MgCl_2$.
2. 30 % (w/v) Acrylamide/bis-acrylamide stock solution (37.5:1) (BioRad) (see Note 5).
3. TBE 0.5×: Tris 6.05 g, boric acid 2.75 g, EDTA 0.465 g, water up to 1 L.
4. 3MM Whatman paper.
5. Gel dryer.
6. X-ray film (BioMax MR-film, Kodak) and cassettes for autoradiography.

3. Methods

3.1. Plasmid Construction

1. A C-terminal fragment of murine STAT3β cDNA, starting at amino acid G127 and ending at amino acid K722, is amplified by polymerase chain reaction (PCR) using primers 5′-dGGGATCTACTTCCATATGGGCCAGGCCAACCACC and 5′-dGGAATTCATCATTTCCAAACTGCATCAATGAA TGGTGTCACACAGATGAACTTGGT (restriction sites NdeI and EcoR1 underlined). The second primer is designed to start 20 bases upstream of the splice site between STAT3 and STAT3β (30, 31), because the template for the PCR reaction was a cDNA encoding STAT3. Therefore, the primer had to encode the differing seven amino acids after the splice site at P713.
2. After cleavage with *Nde*I and *Eco*RI of the resulting PCR fragment, the DNA fragment is introduced into the expression pET32a (Novagen), cut with the same restriction sites.
3. The resultant expression plasmid pET32aStat3β is transformed into the *E. coli* strain BL21(DE3)TKB1 (Agilent technologies), and plated on agarose containing 100 μg/mL ampicillin and 12 μg/mL tetracycline. Ampicillin maintains the selective pressure for the STAT3 containing plasmid; tetracycline maintains the selective pressure for the Elk kinase domain containing plasmid.

3.2. Expression of Recombinant STAT Protein

1. Pick up a single colony on an agarose plate and inoculate 10 mL of 2 × YT broth, supplemented with 100 μg/mL ampicillin and 12 μg/mL tetracycline.
2. Grow overnight at 37 °C under shaking.
3. Dilute 1/50 into 0.5 L of LB broth supplemented with the same antibiotics.
4. The culture is grown in a 2 L flask at 37 °C shaking with 180 rpm.
5. When OD_{600nm} reaches 0.3 reduce the temperature to 21 °C (see Note 6).
6. At mid-log phase at OD_{600nm} = 0.6 to 0.7, induce the protein expression by adding 0.5 mL IPTG 1 M stock solution (1 mM final concentration) (IPTG).
7. Incubate for 5 h at 21 °C. The low temperature is important to obtain soluble protein.
8. Harvest the bacterial cells by centrifugation 30 min at 2,800 × *g* with a swinging bucket rotor.
9. Redissolve the cells in 1 L of kinasing buffer (1× modified M9 containing 0.1 % (w/v) casamino acids (Difco), 1.5 μM

thiamine–HCl, 53 μM 3β-indoleacrylic acid), and supplemented with 50 μg/mL ampicillin and 12.5 μg/mL tetracycline. The 3β-indoleacrylic acid induces the expression of the Elk receptor protein-tyrosine kinase domain under the control of the *trp* promoter.

10. Incubate the culture for another 2.5 h at 37 °C.
11. Harvest the cells by centrifugation for 30 min at 2,800 × *g* using a swinging bucket rotor.
12. Store the cells at −80 °C.

3.3. Purification of Recombinant STAT Protein

1. Resuspend the cells in ice-cold extraction buffer, use 10 g cells per 10 mL buffer.
2. Break the cells by sonication. Keep the plastic beaker containing the bacterial suspension in an ice bucket during sonication.
3. The lysate is centrifuged at 27,000 × *g* for 45 min at 4 °C.
4. Keep the clear supernatant.
5. To remove nucleic acids, polyethyleneimine (PEI) was added to the ice-cooled, stirred supernatant to a final concentration of 0.1 % (w/v).
6. Stir the resulting suspension on ice for another 15 min.
7. Centrifuge for 20 min at 27,000 × *g*.
8. While the solution is stirred on ice, slowly add ammonium sulfate powder until the supernatant reaches 35 % saturation (Fig. 2, see Note 7).
9. The precipitated protein is collected by centrifugation for 20 min at 27,000 × *g*.
10. Redissolve the protein pellet in 0.3 mL of buffer 20 mM HEPES–HCl, pH 7.0, 200 mM NaCl, 10 mM $MgCl_2$, 5 mM DTT, 0.5 mM PMSF and dialyze overnight at 4 °C against 0.5 L of the same buffer.
11. The dialyzed protein is loaded on a Superose 12 HR 10/30 column (GE Healthcare) that has been equilibrated with the same buffer. The protein is eluted at a flow rate of 0.5 mL/min (Fig. 3).
12. Collect the peak fractions.
13. Analyze the fractions on a 10 % SDS-polyacrylamide gel.
14. Fractions of highest purity are pooled. Typical yields were 5 mg of >95 % pure STAT3β protein (as judged by Coomassie staining) from 1 L of starting culture.
15. Concentrate the sample using a Centricon-100 (Amicon) to a concentration of 6–7 mg/mL (see Note 8).

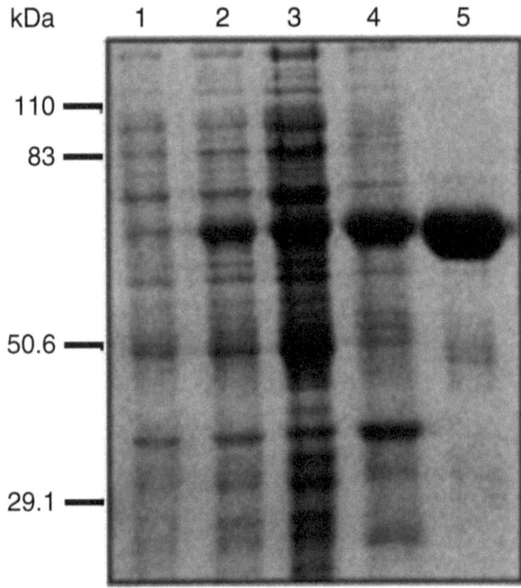

Fig. 2. Purification of recombinant, phosphorylated STAT3β. A 10 % SDS-polyacrylamide gel is depicted. *Lane 1*, uninduced STAT3β transformed cells; *lane 2*, induced STAT3β transformed cells; *lane 3*, supernatant of induced cells after sonication; *lane 4*, pellet of induced cells after sonication; *lane 5*, protein precipitated with 35 % saturated ammonium sulfate at 0 °C. Figs. 2, 3 and 4 are reproduced from ref. 29.

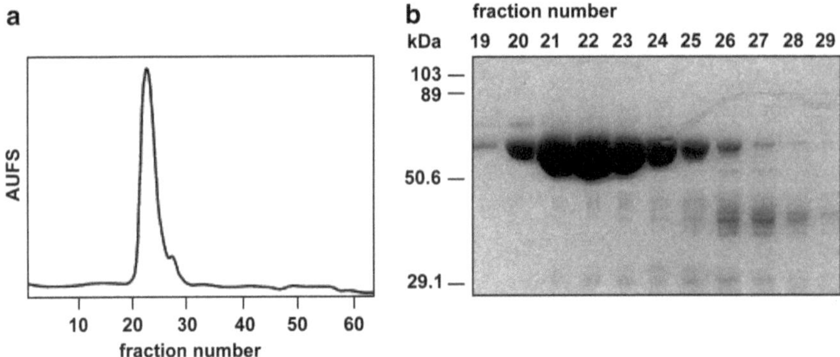

Fig. 3. Purification of STAT3β by gel filtration chromatography. (**a**) After precipitation with 35 % saturated ammonium sulfate and dialysis overnight against dialysis buffer, STAT3β was loaded on a Pharmacia Superose 12 HR 10/30 column. Flow rate, 0.5 mL/min; fraction size, 0.5 mL. (**b**) 10 % SDS-polyacrylamide gel shows fractions 19 to 29 of the chromatography run.

Alternatively, the use of a His-tag at the N-terminal end of the protein offers the possibility to purify the protein on Co-TALON affinity resin (Clontech) columns (see Note 9). In that case, after sonication and centrifugation (step 4):

5a. Add 100–500 μL of TALON resin to the clear lysate (0.5 M NaCl, 5 mM imidazole, 20 mM HEPES pH 7.5, 1 mM PMSF, 5 mM β-mercaptoethanol)

6a. Rotate the tube at 4 °C for 1–2 h.

7a. Pour the mixture in a small column (gravity flow), and extensively wash with the same buffer (in general 10× column volume).

8a. The protein is eluted with the same buffer but supplemented with 250 mM imidazole. Subsequently, the protein is dialyzed against 0.5 M NaCl, 5 mM imidazole, 20 mM Tris pH 8, 1 mM PMSF, 5 mM β-mercaptoethanol.

9a. The His-tag is removed by adding His-tagged TEV protease (0.02 % (w/w) for 20 h at 20 °C).

10a. TEV protease and uncleaved protein are removed by incubation with Ni-NTA affinity resin (Qiagen).

11a. The protein is subsequently purified over a Superose12 HR10/30 column (GE Healthcare) in 20 mM HEPES–HCl pH 7.0, 200 mM NaCl, 10 mM $MgCl_2$, 5 mM DTT, 0.5 mM PMSF. Now proceed with step 12 (above).

3.4. Determination of Protein Concentration

The purified protein was quantified by its absorbance at 280 nm. The extinction coefficient ε_{280nm} was calculated according to ref. 32: $\varepsilon_{280nm} = (5700 \times W + 1300 \times Y)/\text{molecular weight}$ with W = number of tryptophan residues, Y = number of tyrosine residues.

For STAT3, the value of ε_{280nm} was 0.74 assuming a molecular weight of 136 kDa for the dimeric protein.

3.5. Electrophoretic Mobility Shift Assay

To confirm that the STAT protein is phosphorylated and forms dimers that specifically bind their DNA target sites, binding of the STAT proteins to a DNA duplex containing the specific target site should be tested. Electrophoretic mobility shift assay (EMSA), where the reconstituted DNA–protein complex migrates on a non-denaturing gel, is commonly used to test DNA-binding. In our case, we chose two oligonucleotides containing two target sites previously characterized to bind STAT proteins: M67: 5′-TGCATTTCCCGTAAATCT-3′ (23) and a wild type acute phase response element (APRE): 5′-AGCTTCC TTCTGGGATTCCT-3′ (33). The core binding sites are underlined. The first oligonucleotide corresponds to the so-called M67 variant of a region of the c-*fos* promoter (23) and has been widely used in DNA binding studies on STAT protein (16). The second oligonucleotide contains the TT(N4)AA motif, reported to specifically bind STAT3 (34, 35).

3.5.1. Oligonucleotide Duplex Formation

1. Mix the two complementary oligonucleotides in equimolar amounts in 10 mM HEPES–HCl pH 7.5, 100 mM NaCl, 10 mM $MgCl_2$ to a final concentration of 10 mg/mL.
2. Incubate in a water bath of 10 L 90 °C that is cooled down from 90 °C to room temperature over several hours.

Alternatively, a thermocycler can be used to allow the controlled decrease in temperature.

3.5.2. Radioactive Labeling

1. Mix 5 pmol of duplex DNA with 2 μL of [γ-^{32}P] ATP (specific activity 3,000 Ci/mmol, Hartman), 1 unit of T4 polynucleotide kinase in the labeling buffer (70 mM Tris–HCl pH 7.6, 10 mM $MgCl_2$, 5 mM DTT), in a final volume of 20 μL.
2. Incubate the tube at 37 °C for 1 h.
3. Unincorporated nucleotides are removed by gel chromatography (G25 spin columns; Pharmacia, pre-equilibrated in the labeling buffer), add 30 μL labeling buffer to the 20 μL labeling mix and apply the mix to the column gel and centrifuge.
4. Apply again 50 μL buffer, centrifuge and store the radiolabeled duplex at 4 °C.

3.5.3. DNA–STAT Protein Complex Formation and EMSA

1. Prepare seven Eppendorf tubes, each tube containing 2 fmol purified STAT3βtc protein, 20 ng bovine serum albumin (BSA), and ~5,000 cpm radiolabeled oligonucleotide (corresponding to ~0.5 fmol).
2. Tube 1 is the control and contains no STAT protein.
3. To tubes 2 to 7, add respectively unlabeled DNA 5-fold (2.5 fmol), 10-fold (5 fmol), 20-fold (10 fmol) 50-fold (25 fmol) and 100-fold (50 fmol) excess (see Note 10).
4. Adjust the volume of each tube to 20 μL of reaction volume with binding buffer.
5. The reaction mix is incubated for 30 min at room temperature.
6. The 20 μL reaction products are then loaded on a 4 % polyacrylamide gel (1.5 mm thick) containing 0.25 × Tris-borate/EDTA (see Note 11).
7. Before loading the gel should be pre-run at 15 V/cm for 1 h at 4 °C.
8. After loading the gels were run for another 3.5 h at 4 °C.
9. Open the plates, transfer the gel onto Whatman paper 3MM, cover with a plastic sheet and dry for 45 min under vacuum at 70 °C.
10. Expose to X-ray film (BioMax MR-film, Kodak) for several hours.

STAT3βtc shows the strongest interaction with M67, while APRE binds weaker. In both cases the binding is specific, as the unlabeled DNA fragments could competitively bind against the

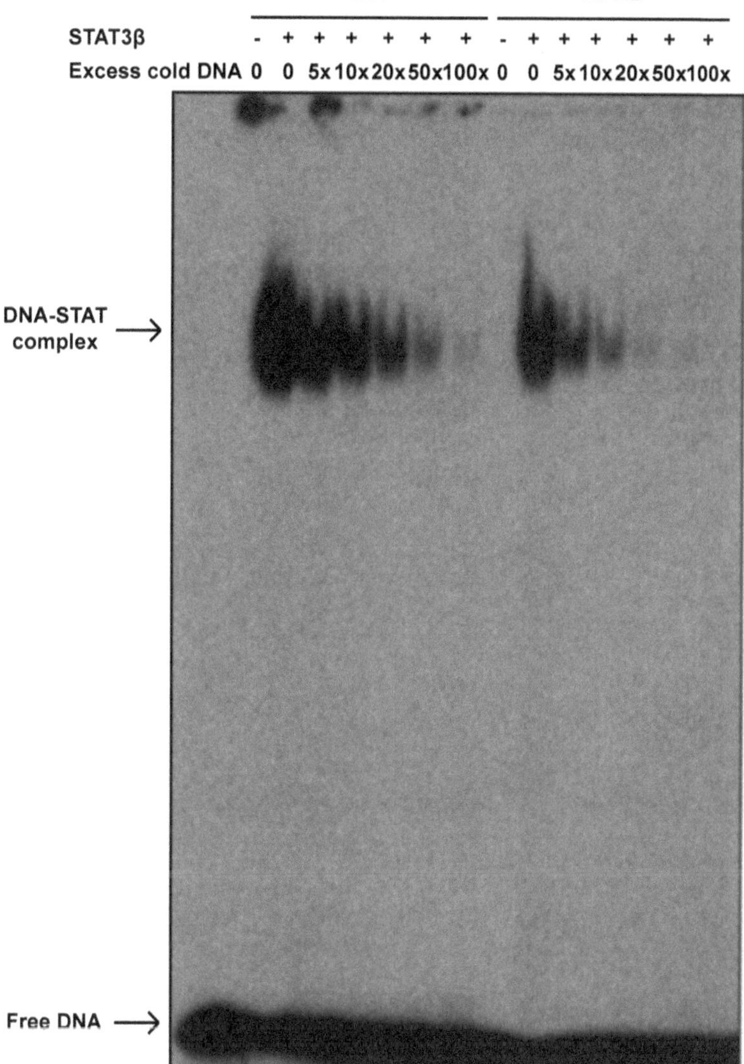

Fig. 4. STAT3β binds to mutant and wild type STAT DNA target sites. The labeled probes only (*lanes 1, 8*) or reaction mixes of STAT3β with the probes (*lanes 2–7, 9–14*) were analyzed by EMSA. Unlabeled probe was added to the reaction mixes in excess as mentioned probes only (*lanes 1, 8*) or reaction mixes of STAT3β with the probes (*lanes 2–7, 9–14*) were analyzed by EMSA. Unlabeled probe was added to the reaction mixes in excess as mentioned.

respective ^{32}P-labeled fragments to STAT3β (Fig. 4). M67 binding was still observed when the unlabeled probe was used in a 100-fold excess, while no APRE binding could be observed any more under these conditions (Fig. 4).

4. Co-crystallization of Tyrosine Phosphorylated STAT Proteins with Their DNA Target Site

The possibility to express large amounts of tyrosine phosphorylated STAT proteins in bacteria opens to the possibility to broadly screen for conditions that allow crystallization of STAT proteins in its unbound form (24) or bound to its DNA target sites. The choice of a high-affinity and biological relevant DNA target site is one important parameter for the crystallization of protein–DNA complexes. For example STAT3β was initially co-crystallized with the M67 DNA site (8) that is bound with higher affinity than the related APRE site. An often critical parameter is the choice of the correct DNA-length for co-crystallization. Lengths of the DNA oligonucleotides, composition of the ends (blunt- or overlapping), position of the target site within the oligonucleotides often have to systematically varied until well diffracting crystals can be obtained (compare also (36)). Furthermore, the stoichiometry of the protein and DNA components is yet another important parameter for the successful crystallization of protein–DNA complexes. Generally, we noticed that a slight excess (~10 %) of DNA with respect to the protein is favorable to crystallization. In the case of STAT3β, we therefore assayed STAT3β dimer–DNA ratios from 1:1 to 1:1.5. Another important consideration is the concentration of the protein–DNA complex, where we generally use 5–15 mg/mL of protein–DNA complex.

4.1. General Considerations for the Crystallization of Protein–DNA Complexes

Highly purified DNA oligonucleotides are generally a further prerequisite to obtain well-ordered crystals of a protein–DNA complex. Typically, 400–600 nmol of purified DNA are required to start crystallization trials. DNA synthesis reactions can be usually purchased at a 1 μmol scale. Further purification to separate full-length oligonucleotides from side products resulting from incomplete synthesis is generally required. Many companies also offer highly purified oligonucleotides for co-crystallization experiments. In addition, a purification protocol using anion-exchange chromatography is provided below. To screen for suitable conditions that allow the formation of protein–DNA co-crystals, we generally use the "hanging drop" or "sitting drop" method, where vapor diffusion allows the equilibration between reservoir and drop. Different concentrations of protein–DNA mixtures are placed over reservoirs containing different precipitants and are allowed to equilibrate over a period of time. A large number of variables can be tested by this method, including temperature, pH, and precipitation concentration.

4.2. Purification of DNA Oligonucleotides by Anion-Exchange Chromatography

All buffers and solutions for anion exchange should be filtered through a 0.2 μm filter before use.

Chemically synthesized oligonucleotides are purified by anion-exchange chromatography (Mono-Q HR 10/10 column, FPLC system, GE Healthcare)

1. Prepare two buffers; buffer A: 10 mM NaOH and buffer B: 10 mM NaOH, 1 M NaCl. These buffers should be filtered on 0.22 μM and degassed.
2. Add ddH_2O to the lyophilized oligonucleotide, keep the volume at 300–500 μL maximum.
3. Spin 5 min at 16,000 *g* before injecting, keep supernatant.
4. Inject the diluted oligonucleotide into the column.
5. Wash column with buffer A.
6. The entire purification procedure is carried out at a flow rate of 4.0 mL/min. Set UV detection at 260 nm. During the entire preparative purification run, all fractions (2 mL size) should be collected.
7. Sample will be eluted from the column with the following profile:

   40 mL of 10 to 40 % buffer B,

   100 mL of 40 to 75 % of buffer B,

   20 mL of 75 to 100 % of buffer B,

   Oligonucleotides generally elute between 0.4 and 0.7 M NaCl depending on the lengths of the oligonucleotides. The length of the oligonucleotide is generally proportional to the concentration of NaCl required for elution.
8. Add 100 μL 1 M Hepes pH 7.5 in each 2 mL fraction to neutralize the NaOH solution as soon as it leaves the column (see Note 12).
9. Peak fractions absorbing at 260 nm are pooled.
10. Fill up a dialysis tubing of 3.5 kDa molecular weight cut-off (Spectrapore).
11. Dialyse the pooled fractions against 4 L demineralized water in the cold room for 3–4 h.
12. Change the water and dialyze again for 5–8 h.
13. Freeze the sample and lyophilize.
14. Resuspend the oligonucleotide in 40–80 μL ddH_2O.
15. Determine the concentration of single-stranded oligonucleotide (25 OD at 260 nm corresponds to ~1 mg/mL of single-stranded oligonucleotide).
16. Adjust concentration to 20 mg/mL (see Note 13).

4.3. Annealing the DNA Strands

1. Mix the two complementary oligonucleotides in equimolar amounts (1:1 stoichiometry) at a final stock concentration of 10 mg/mL. Example: 20 μL oligo1 + 20 μL oligo2 + 40 μL annealing buffer (10 mM HEPES pH 7.5, 100 mM NaCl, 10 mM $MgCl_2$, 1 mM DTT).
2. Incubate in a 10 L water bath preheated 90 °C that is allowed to cool down from 90 °C to room temperature over several hours.

4.4. Co-crystallization of STAT Proteins

Crystallization conditions can widely vary between different protein–DNA complexes and even between closely related systems. For this reason using broad screens varying crystallization conditions and lengths of oligonucleotides are generally recommended. An example the conditions used for the crystallization of the STAT3β-DNA complex are given:

1. Mix 1 μL STAT3 10 mg/mL (=0.074 mM) protein dimer and μL of previously annealed DNA duplex diluted to ~1 mg/ml (= 0.085 mM). The sequences of M67 oligonucleotides are: 5′-TGCATTTCC CGTAAATCT-3′ and 5′-AGATTTACGGG AAATGCA-3′.
2. To this 2 μL complex is added 2 μL of reservoir containing 0.1–0.4 M NaCl, 5 mM $MgSO_4$, 50 mM MES, pH 5.6–6.0, 0.1 M ammonium acetate, 10 % glycerol (v/v).
3. Invert the cover slip and seal it above a well containing the same precipitant solution.
4. Place the plate at room temperature.
5. Bipyramidal crystals grew within one week to an average size of 400 μm × 400 μm × 150 μm.
6. To verify that the crystal contains both protein and DNA, remove several crystals from the drop, wash them in reservoir solution, and then dissolve them in water to run on denaturing SDS-PAGE gel.
7. For visualization of crystal components, use silver stain which allows DNA and protein detection in very small amounts (see Note 14).

5. Notes

1. All solutions should be prepared in deionized water.
2. We used pET32a (Novagen) as the protein expression vector for Stat3β. However, the choice of the protein expression vector is not critical as long as it is compatible with the TKB1 strain.
3. Modified M9 buffer and kinasing buffer are also described in the manual of the TKB1 cells (Agilent).
4. Ideally, prepare fresh polyethylenimine for every STAT preparation. Store protected from light at 4 °C.
5. Use appropriate safety precautions when handling acrylamide.
6. For the soluble expression of STAT proteins in E. *coli*, growing the bacteria at low temperature is very important. Therefore, the temperature of the bacterial culture should be lowered before induction.

7. Ammonium sulfate powder should be added stepwise and very slowly during STAT preparation, while the solution is stirred on ice.
8. For the storage of recombinant STAT proteins, the protein should be snap-frozen in liquid nitrogen and kept at −80 °C.
9. In case Co^{2+} resin is used during the preparation of His-tagged STAT protein, using fresh beads will always give the highest protein yield.
10. During EMSA analyses, a DNA duplex with unrelated sequence of the same length may be used to visualize the level of nonspecific protein binding.
11. For EMSA analyses, pouring the gel the day before the experiment will increase the quality of the DNA–protein migration.
12. During DNA oligonucleotide purification it is important to neutralize the eluted oligonucleotide as soon as possible. Therefore, 100 μL 1 M HEPES–HCl pH 7.5 in each 2 mL fraction are already present in the tube, when the oligonucleotide elutes from the Mono-Q HR 10/10 column.
13. The purified DNA oligonucleotide should be kept as concentrated as possible to avoid large volumes when forming the DNA–protein complex. DNA concentration will also decrease by two when annealing buffer is added.
14. Silver staining is a very sensitive detection method (detection 1–10 ng per band). Gloves must be worn during the entire procedure (even for loading the gel) and only extensively and freshly cleaned glassware, and Milli-Q water, should be used to avoid contamination.

Acknowledgments

We thank former lab members Stefan Becker, Montserrat Soler-Lopez, and Carlo Petosa for setting up the initial protocols for STAT3β and Dd-STATa.

References

1. Levy DE, Darnell JE Jr (2002) Stats: transcriptional control and biological impact. Nat Rev Mol Cell Biol 3:651–662
2. Schindler C, Darnell JE Jr (1995) Transcriptional responses to polypeptide ligands: the JAK-STAT pathway. Annu Rev Biochem 64:621–651
3. Shuai K, Horvath CM, Huang LH, Qureshi SA, Cowburn D, Darnell JE Jr (1994) Interferon activation of the transcription factor Stat91 involves dimerization through SH2-phosphotyrosyl peptide interactions. Cell 76:821–828
4. Liao J, Fu Y, Shuai K (2000) Distinct roles of the NH2- and COOH-terminal domains of the protein inhibitor of activated signal transducer and activator of transcription (STAT) 1 (PIAS1) in cytokine-induced PIAS1-Stat1

interaction. Proc Natl Acad Sci U S A 97:5267–5272

5. Zhang JJ, Vinkemeier U, Gu W, Chakravarti D, Horvath CM, Darnell JE Jr (1996) Two contact regions between Stat1 and CBP/p300 in interferon gamma signaling. Proc Natl Acad Sci U S A 93:15092–15096
6. Vinkemeier U, Moarefi I, Darnell JEJ, Kuriyan J (1998) Structure of the amino-terminal protein interaction domain of STAT-4. Science 279:1048–1052
7. Chen X, Bhandari R, Vinkemeier U, Van Den Akker F, Darnell JE Jr, Kuriyan J (2003) A reinterpretation of the dimerization interface of the N-terminal domains of STATs. Protein Sci 12:361–365
8. Becker S, Groner B, Muller CW (1998) Three-dimensional structure of the Stat3b homodimer bound to DNA. Nature 394:145–151
9. Chen X, Vinkemeier U, Zhao Y, Jeruzalmi D, Darnell JEJ, Kuriyan J (1998) Crystal structure of a tyrosine phosphorylated STAT-1 dimer bound to DNA. Cell 93:827–839
10. Dumoutier L, de Meester C, Tavernier J, Renauld JC (2009) New activation modus of STAT3: a tyrosine-less region of the interleukin-22 receptor recruits STAT3 by interacting with its coiled-coil domain. J Biol Chem 284:26377–26384
11. Horvath CM, Stark GR, Kerr IM, Darnell JE Jr (1996) Interactions between STAT and non-STAT proteins in the interferon-stimulated gene factor 3 transcription complex. Mol Cell Biol 16:6957–6964
12. Lufei C, Ma J, Huang G, Zhang T, Novotny-Diermayr V, Ong CT, Cao X (2003) GRIM-19, a death-regulatory gene product, suppresses Stat3 activity via functional interaction. EMBO J 22:1325–1335
13. Nakajima H, Brindle PK, Handa M, Ihle JN (2001) Functional interaction of STAT5 and nuclear receptor co-repressor SMRT: implications in negative regulation of STAT5-dependent transcription. EMBO J 20:6836–6844
14. Zhang X, Wrzeszczynska MH, Horvath CM, Darnell JE Jr (1999) Interacting regions in Stat3 and c-Jun that participate in cooperative transcriptional activation. Mol Cell Biol 19:7138–7146
15. Zhu M, John S, Berg M, Leonard WJ (1999) Functional association of Nmi with Stat5 and Stat1 in IL-2- and IFNgamma-mediated signaling. Cell 96:121–130
16. Horvath CM, Wen Z, Darnell JE Jr (1995) A STAT protein domain that determines DNA sequence recognition suggests a novel DNA-binding domain. Genes Dev 9:984–994
17. Yang E, Wen Z, Haspel RL, Zhang JJ, Darnell JE Jr (1999) The linker domain of Stat1 is required for gamma interferon-driven transcription. Mol Cell Biol 19:5106–5112
18. Fu XY (1992) A transcription factor with SH2 and SH3 domains is directly activated by an interferon alpha-induced cytoplasmic protein tyrosine kinase(s). Cell 70:323–335
19. Moriggl R, Gouilleux-Gruart V, Jahne R, Berchtold S, Gartmann C, Liu X, Hennighausen L, Sotiropoulos A, Groner B, Gouilleux F (1996) Deletion of the carboxyl-terminal transactivation domain of MGF-Stat5 results in sustained DNA binding and a dominant negative phenotype. Mol Cell Biol 16:5691–5700
20. Qureshi SA, Leung S, Kerr IM, Stark GR, Darnell JE Jr (1996) Function of Stat2 protein in transcriptional activation by alpha interferon. Mol Cell Biol 16:288–293
21. Razeto A, Ramakrishnan V, Litterst CM, Giller K, Griesinger C, Carlomagno T, Lakomek N, Heimburg T, Lodrini M, Pfitzner E, Becker S (2004) Structure of the NCoA-1/SRC-1 PAS-B domain bound to the LXXLL motif of the STAT6 transactivation domain. J Mol Biol 336:319–329
22. Wojciak JM, Martinez-Yamout MA, Dyson HJ, Wright PE (2009) Structural basis for recruitment of CBP/p300 coactivators by STAT1 and STAT2 transactivation domains. EMBO J 28:948–958
23. Wagner BJ, Hayes TE, Hoban CJ, Cochran BH (1990) The SIF binding element confers *sis*/PDGF inducibility onto the c-*fos* promoter. EMBO J 9:4477–4484
24. Soler-Lopez M, Petosa C, Fukuzawa M, Ravelli R, Williams JG, Muller CW (2004) Structure of an activated dictyostelium STAT in its DNA-unbound form. Mol Cell 13:791–804
25. Mao X, Ren Z, Parker GN, Sondermann H, Pastorello MA, Wang W, McMurray JS, Demeler B, Darnell JE Jr, Chen X (2005) Structural bases of unphosphorylated STAT1 association and receptor binding. Mol Cell 17:761–771
26. Neculai D, Neculai AM, Verrier S, Straub K, Klumpp K, Pfitzner E, Becker S (2005) Structure of the unphosphorylated STAT5a dimer. J Biol Chem 280:40782–40787
27. Yu H, Pardoll D, Jove R (2009) STATs in cancer inflammation and immunity: a leading role for STAT3. Nat Rev Cancer 9:798–809
28. Berg T (2008) Inhibition of transcription factors with small organic molecules. Curr Opin Chem Biol 12:464–471
29. Becker S, Corthals GL, Aebersold R, Groner B, Muller CW (1998) Expression of a tyrosine

phosphorylated, DNA binding Stat3beta dimer in bacteria. FEBS Lett 441:141–147

30. Schaefer TS, Sanders LK, Nathans D (1995) Cooperative transcriptional activity of Jun and Stat3 beta, a short form of Stat3. Proc Natl Acad Sci U S A 92:9097–9101
31. Caldenhoven E, van Dijk TB, Solari R, Armstrong J, Raaijmakers JA, Lammers JW, Koenderman L, de Groot RP (1996) STAT3beta, a splice variant of transcription factor STAT3, is a dominant negative regulator of transcription. J Biol Chem 271: 13221–13227
32. Cantor CR, Schimmel PR (1980) Biophysical chemistry, part II, techniques for the study of biological structure and function. W.H. Freeman and Company, San Francisco, CA, pp 380–381
33. Ito T, Tanahashi H, Misumi Y, Sakaki Y (1989) Nuclear factors interacting with an interleukin-6 responsive element of rat alpha 2-macroglobulin gene. Nucleic Acids Res 17:9425–9435
34. Fujitani Y, Nakajima K, Kojima H, Nakae K, Takeda T, Hirano T (1994) Transcriptional activation of the IL-6 response element in the junB promoter is mediated by multiple Stat family proteins. Biochem Biophys Res Commun 202:1181–1187
35. Seidel HM, Milocco LH, Lamb P, Darnell JE Jr, Stein RB, Rosen J (1995) Spacing of palindromic half sites as a determinant of selective STAT (signal transducers and activators of transcription) DNA binding and transcriptional activity. Proc Natl Acad Sci U S A 92:3041–3045
36. Cramer P, Muller CW (1997) Engineering of diffraction-quality crystals of the NF-kappaB P52 homodimer:DNA complex. FEBS Lett 405:373–377

Index

Sandra E. Nicholson and Nicos A. Nicola (eds.), *JAK-STAT Signalling: Methods and Protocols*, Methods in Molecular Biology, vol. 967, DOI 10.1007/978-1-62703-242-1, © Springer Science+Business Media New York 2013

MIX
Papier aus verantwortungsvollen Quellen
Paper from responsible sources
FSC® C105338

If you have any concerns about our products,
you can contact us on
ProductSafety@springernature.com

In case Publisher is established outside the EU,
the EU authorized representative is:
Springer Nature Customer Service Center GmbH
Europaplatz 3, 69115 Heidelberg, Germany

Printed by Libri Plureos GmbH
in Hamburg, Germany